T0188165

Microoptics Technology
Second Edition

OPTICAL ENGINEERING

Founding Editor
Brian J. Thompson
University of Rochester
Rochester, New York

Microoptics Technology

Second Edition

Nicholas F. Borrelli

Corning Incorporated
Corning, New York, U.S.A.

CRC Press
Taylor & Francis Group
Boca Raton London New York

CRC Press is an imprint of the
Taylor & Francis Group, an **informa** business

CRC Press
Taylor & Francis Group
6000 Broken Sound Parkway NW, Suite 300
Boca Raton, FL 33487-2742

First issued in paperback 2019

© 2009 by Taylor & Francis Group, LLC
CRC Press is an imprint of Taylor & Francis Group, an Informa business

No claim to original U.S. Government works

ISBN-13: 978-0-8247-5921-6 (hbk)
ISBN-13: 978-0-367-39353-3 (pbk)

Library of Congress Cataloging-in-Publication Data
A catalog record for this book is available from the Library of Congress.

Visit the Informa Web site at
www.informa.com

and the Informa Healthcare Web site at
www.informahealthcare.com

To the future and those in
whose hands it lies,
Barbie, Steve, Matthew, and Noah

Preface to Second Edition

It has been five years since the first publication of the book on Microoptics Technology, but what a five years it has been. It is arguable that in these last five years never has optical technology experienced such a burst of activity and optimism. Although the financial aspect of the optimistic projection, the so-called bubble, burst, nonetheless the technical community produced a body of significant real results. In other words, although Wall Street may have come out empty, the optical technology community advanced with new materials, devices, and systems. Although bubbles burst, hardware sustains.

The second edition of this book speaks to some of the new things that emerged out of the flurry of activity that was fueled by the bubble. I have added two new chapters and substantially enlarged another that hopefully supports this assertion. I am happy to find out that much of the content of the first edition served as a good basis for the kind of optical devices that were to be involved in the telecommunication portion of the effort. I have tried to add new things that emerged over the last five years to further add to this conclusion. I fully expect that in the next five years we will begin to see some of these innovative ideas begin to creep into our ways of doing things.

I gratefully thank the many people who have helped me in this writing. In particular, I thank David Morse, the Director of Research of the Laboratory at Corning Incorporated for given me the opportunity to spend the time on the book. I thank Karl Koch and the whole photonic crystal fiber group for their advice and a great deal of material to draw from, to Charlene Smith for critically reading the new chapters, and to Shiela Hughey for her diligence, patience, and good humor in helping with the drawings.

Contents

1

Introduction

The emergence and rapid growth of the microelectronic industry, and its concomitant drive for the miniaturization of electronic devices, together with the optical fiber telecommunication industry and its need to couple light in and out of single mode waveguides, has created in their wake, a new area of optics termed "micro-optics". What is meant by this term is optical elements of dimension of a millimeter or smaller. Primarily, this has meant lenses, or elements that act as lenses, as well as structures that redirect, polarize, and otherwise alter some state or direction of light. These would include elements like, mirrors, gratings, polarizers, and the like, made in some "micro" form. If one includes optical waveguide structures as well, this broader classification can be considered under the name, "integrated optics". Another interesting aspect of the evolution of micro-optics has been, not only has the microelectronic industry supplied the need for tiny optical devices, but has been able, in some cases, to supply the technology by which it can be fabricated. We will see in a later chapter that the very microfabrication technology used to fabricate microcircuits can be used to produce patterns that provide optical function through diffraction of light. This is but one way in which small optical elements, primarily lenses and lens arrays can be fabricated.

One can find discussions of devices that contain micro-optic elements, often done in a comprehensive, but narrow way [1]. However, here the device and its performance are stressed, and the elements themselves are not discussed in any real detail. Moreover, the possible alternative methods are not compared and discussed. There are monographs dealing with particular technologies, like gradient index, and consequently they do not deal with the alternatives [2]. There are a few encyclopedic-type publications that discuss optical elements, but never deal with the optical elements and the methods of fabrication in any detail [3,4]. In this book we try to stress the optical elements and how they are fabricated as well as try to give sufficient applications where the reader can appreciate the interplay and implications of a particular approach. We will attempt to address some of these fabrication and application methods in some detail. In addition, we will introduce some of the new micro-optic phenomena that may find its way into applications in the near future.

The exposition will be broken down into *four* parts. In the first part, we will deal with the important optical element called the "microlens". These microlenses are found in numerous optical devices and therefore will be given significant attention. Chapters 2–4 will deal with the different ways that microlenses can be fabricated along with a brief review of the underlying optic principles, Here, we will emphasize the properties and fabrication methods.

The structure of the first part of the book will be to separate the lens fabrication techniques into the type of lens. For example, the categories refractive lenses, gradient index lenses, and diffractive lenses will constitute individual chapters. For each of these lens types, we will give a brief optics background, sufficient for the reader to appreciate the advantages, limitations, and problems that subsequently arise as a consequence of the optical imaging principles. This is particularly important when one is trying to compare, say, the efficiency of a gradient index lens to a diffractive lens.

Although the function is the same, to gather light from an object and direct it to an image, the formulation of the imaging is done in a different way, thus the reader must be familiar with at least, the terminology, if not the optical principles. Under each lens category, we will further discuss the various fabrication methods. For example, under the refractive lens heading, we will discuss methods such as molded, photosensitive glass, etching, etc. The advantages and disadvantages of the fabrication method will be discussed. Because many of these fabrication methods are the property of commercial companies, we will give the general approach to the fabrication rather than a recipe. Where we can, we will refer specifically to the commercial vendors.

In the next part, Chapters 5 and 6, we will deal with the variety of application of microlens arrays. We will cover the one-to-one imaging application of lens arrays in Chapter 5. The major applications will be as lens bars for scanning and reading documents. In Chapter 6, we will cover the wide range of two-dimensional lens arrays that find application in many diverse areas. In these two chapters, it will become clear the distinction in applications where the arrays are used to image collectively, that is act like a conventional larger lens, and those where each microlens acts independently. In the fabrication of a microlens array, in addition to the attention that must be paid to develop the properties of individual lens itself, equal attention must be paid to the manner in which exact positioning of the lenses relative to some fixed point can be accomplished. In a single lens, one can imagine a mounting fixture that permits the accurate positioning of the lens relative to the light source, or fiber. However, if a lens array is to be aligned to, for example, a laser diode array with a well-specified center-to-center distance as a consequence of its fabrication, alignment may never be possible, if the spacing between the lenses was not maintained during its fabrication. One is dealing with maintaining alignment of microns over centimeters. This adherence to dimensional stability over dimensions of many centimeters can make the defining difference in the choice of what method to use for any given application. This adds another aspect to the manufacture of lens arrays which brings in the temperature of the process and how it affects the geometric stability of the substrate. Since some common substrate materials, in particular glass, undergo some degree of irreversible volume change upon heating, the extent of which depends on the temperature relative to the fictive temperature, the maximum temperature achieved during the fabrication can be critical. Because of this more critical requirement, we will be mainly dealing in this book with the methods by which lens arrays are made.

Because the impetus for micro-optic elements has come simultaneously from two different directions, microelectronics and optical telecommunications, the optical lens design and performance as well as the size and layout are different enough to influence the optimum fabrication method. This will be made clear in the subsequent discussions of the individual fabrication techniques as they relate to applications. Consider the following areas that have been of interest over the past few years.

Compact optical devices requiring imaging optics to be confined to a small space. Examples of this are document readers, bar code readers, and scanners. These particular applications require erect one-to-one imaging. The advantage of using a lens bar-shaped lens array for essentially a-line-at-a-time imaging operation, over a conventional lens is the shortness of the working distance that can be achieved, an yet cover an 8.5″ document. Total conjugate distances (distance from object to image) as short as 10 mm are achievable over paper-size distances See Fig. 1.1 for a schematic representation of this function.

Optical device interface with microelectronic structures, like CCD detector arrays—the dimension of such structures requires small closely spaced lenses, registered precisely to the electronic elements. An example of this would be LED, the WDM scheme (wavelength division multiplexing) uses micro-optical elements.

We go through and describe and discuss printer bar where each pixel is imaged onto the detector. See Fig. 1.2.

Optical waveguide devices—lens to input and output light from single mode fibers, or arrays thereof. There are a

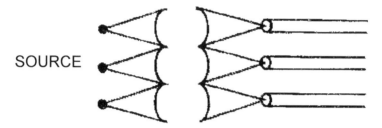

Figure 1   Sources to fiber array.

variety of applications, the most important of which is the efficient coupling of light from a laser diode to a single mode fiber. See Fig. 1.3. More recently, these devices are part of the techniques people have used to produce imaging structures, one will see that by virtue of the desired small dimensions, a wide variety of methods become viable which would not be otherwise practical on a larger scale. Examples are binary-optic structures, and to a large extent, gradient index structures. In the case of gradient index, this is easily understood. The focal length of a lens with a parabolically shaped radial profile is proportional to $\Delta n/\Delta R$, not to Δn itself. Here, $\Delta n/\Delta R$ represents the gradient of the refractive index change with the radius of the lens. Whereas it might be difficult to produce large index differences, it is not required if one can achieve small index differences over small distances. For diffractive lenses which are constructed from periodic structures of the order of wavelengths of light, the resolution capability limits one to 4–5 periods. For a typical case in the visible wavelength range, this limits lens sizes to $<100\,\mu$m.

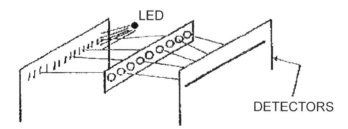

Figure 2

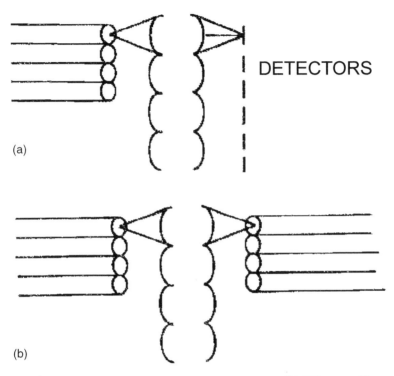

(a)

(b)

DETECTORS

Figure 3   (a) Fiber array to detector array; (b) Fiber-to-fiber array connector.

In the third part of the book, we single out and discuss two important integrated micro-optic areas. In these chapters, one starts to feel how the new devices utilize and integrate many different types of optical elements as well as share, to a large extent, similarity of fabrication methods. These two chapters will deal with, respectively, the fabrication of microdiffraction gratings, the elements, and properties of optical isolators.

Microgratings and optical isolators have come to the fore, to a large extent, because of the emergence of optical communication systems. As an example, the importance of fiber gratings has come about from the application called "wavelength division multiplexing", or WDM for short. This has to do with the optical communication scheme where more than one wavelength is propagating in the same optical fiber. The need to separate, and otherwise act independently act on different propagating wavelengths has produced the need for diffraction grating. What is new is that these gratings have to be implemented on a microscale, mated and integrated to an optical system. We will cover all the important methods that have arisen over the last few years to satisfy this new need. Also, we will cover the representative applications.

Chapter 8 will be devoted to an optical device whose need has also arisen from the new "photonic" technology. This technology has to do with the next stage of the optical communication network. This will deal with using light itself to perform many of the functions that are now done electronically. The initial example of this is the optical amplifier. Whatever the particular scheme, it will always need an optical element to prevent light from going backwards in the train that would prematurely de-excite the inverted population. This optical element is the optical isolator. It is included not only to explain another important optical element but also because it provides an excellent example of how many different micro-optic elements are brought together to produce an important new optical device.

The fourth part of the book will be devoted to three new developments in the micro-optic area. The first is that of "photonic crystals" which is covered in the Chapter 9. It is relatively new, even by the standards of the micro-optics field in general, but has made great strides in its short existence. It is an excellent example of the way new innovative technology requires the emergence and push of new methods mixed with new theories, or ways of thinking of old theories. In this case, the new way of thinking about old theories was the realization that the mathematical formulation that is used to determine the behavior of electrons in solids, often called band theory, could also be applied to the way light propagates in periodic structures [5]. The link to the technology of micro-optics is that the length of the periodic scale of the structures for the important case of visible to near infrared wavelengths, corresponds to the wavelength itself. This means that fabrication methods appropriate for many of the elements discussed in the previous chapters are again needed here. To make the cycle complete, the devices that one could imagine using this photonic crystal approach for, are the very same optoelectronic applications discussed in the context of other approaches, which, in turn share the same fabrication techniques. The specific area that has made the greatest progress is that of photonic crystal fibers. From a mere suggestion of the concept four short years ago, there have been demonstrated fibers with losses rivaling the standard optical fiber.

The last two chapters are dedicated to even newer microoptical phenomena. In Chapter 10, we review the phenomenon of the interaction of short pulse lasers with solids, in particular glass. A number of unique consequences result because of the shortness of the duration of the pulse relative to the physical response such as the thermalization of the photoelectrons. Moreover, one is dealing with the unusual combination of very high peak intensity $>10^{11}\,\text{W/cm}^2$ at relatively low energy per pulse. The former drives nonlinear processes quite efficiently, but the thermal contribution is relatively small. In spite of its short existence a number of applications have already been suggested [6] and a few have been demonstrated.

The last chapter is both the newest and in some sense the oldest since it was first suggested back in the late fifties. Vesalago [7] was the first to investigate the behavior of light in a medium that has a negative refractive index together with a negative magnetic permeability. It was purely a theoretical analysis. However, its reality and demonstration is quite new [8], and has struck an interesting chord since the behavior of light acquires some counter intuitive properties. We will discuss this in Chapter 11.

REFERENCES

1. *Integrated Optical Circuits and Components*; Hutcheson, L.D., Ed.; Marcel Dekker Inc.: New York, 1987.

2. Iga, K.; Kokubun, Y.; Oikawa, M. *Fundamentals of Microoptics*; Academic Press: New York, 1984.

3. *Handbook of Laser Science and Technology, Supplement 2; Optical Materials*; CRC Press: Boca Raton, FL, 1995.

4. Handbook of optics. In: *Devices, Measurements and Properties*; Bass, M., Ed.; McGraw-Hill: New York, 1995.

5. Yablonovitch, E. Photonic band-gap structures. JOSA B **1993**, *10* (2), 283–295.

6. Liu, X.; Mourou, G. Ultrashort laser pulses tackle precision machining. Laser Focus World 101–118, August 1997.

7. Vesalago, V.G. The electrodynamics of substances with simultaneous negative values of ε and μ. Sov. Phys. (Uspekhi) **1958**, *10* (4), 509.

8. Pendry, J.B. Optics: positively negative. Nature **2003**, *423*, 22.

2

Refractive Elements

2.1. OPTICS REVIEW

2.1.1. Basics

In order to follow the development and performance of refractive microlens elements, some rudimentary understanding of geometric optics is required. A number of references are given for the reader to consider. What we need here, at the very least, is an understanding of the basic terminology. It might be useful to bear in mind that there is really no conceptual difference in the optical principles relating to small refractive lenses as compared with large lenses; however, there is in the way one formulates the optical design. In the simplest case of paraxial ray tracing for large-diameter lenses, the so-called thin-lens approximation [1] is used, which simply means that the deviation of the rays through the lens thickness can be ignored. The paraxial specification means that Snell's law can be written as $\phi_i = n\phi_r$. Here ϕ_i and ϕ_r are the angles of incidence and refraction measured from the surface normal as shown in Fig. 2.1. When the lens thickness is comparable, or exceeds, the lens diameter, one has to resort to what is

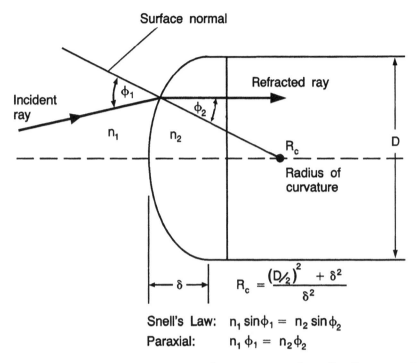

Figure 2.1 Definition of the salient terms to describe the ray path through a spherical lens. Ray incident from left in medium of index n_1 making an angle with respect to the normal at that point of ϕ_1, refracting to an angle ϕ_2 in the lens medium of refractive index n_2. The exact form of Snell's law is given relating the incident angle to the refracted angle, as well as the paraxial approximation.

often called the thick-lens formulation. Referring to Fig. 2.2, one has a set of definitions that derive from the desire to maintain the simple lens maker's formula relating the focal length of the lens to the object and image distances

$$\frac{1}{f} = \frac{1}{s_o} + \frac{1}{s_i} \tag{2.1}$$

Here, f represents the focal length of the lens, s_i is the image, and s_o is the object distance. In the thick-lens formulation, one must use a new set of definitions of the various common lens terms. These definitions are summarized in Table 2.1 based

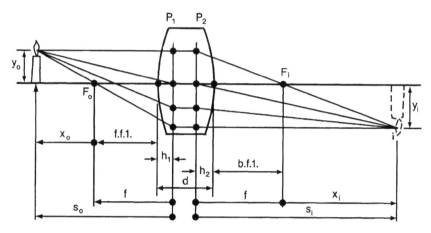

Figure 2.2 Diagram defining the various terms and distances relating to a thick lens which are included in Table 2.1.

on the description of Figs. 2.1 and 2.2. One can see that the principal planes, P_1, P_2, are defined in such a way as to make Eq. (2.1) valid if the object and image distances are measured from them and not the lens boundaries. One should realize from the expression in Table 2.1 that the position of the principal planes can lie outside the physical lens. We can see from the drawing of Fig. 2.2 how the principal plane is defined. The intersection of a parallel ray with the extension backward of a ray from the focus must lie on the principal plane.

With the aid of the Table 2.1 and Fig. 2.2, the reader should be able to begin to understand the interplay of the various parameters as they impact a given desired optical application. For example, the distinction between the effective focal length (EFL), and the respective front and back focal distances is important. From a practical point of view, by way of an example of a trade-off, the distance most usually specified is the total conjugate distance. This is the distance from the object to the image, and usually represents a physical limitation in a design. Within this distance, the lens thickness and working distances can be varied, but not independently from the EFL. One can see all this interdependence from the expression for TC given in the Table 2.1. We will refer back to this table at various times.

Table 2.1 Definition of Optical Terms in Thick Lens Formulation (Referenced to Figure 2.1)

Term	Definition	Mathematical expression
R_1	Radius of curvature of first surface	
R_2	Radius of curvature of second surface	
s_o	Object distance (centerline distance from principle plane to object)	
s_i	Image distance	
P_1	First principle plane	
h_1	Centerline distance from first lens surface to first principle plane	$h_1 = \frac{T(n-1)}{n}\frac{\text{EFL}}{R_2}$
P_2	Second principle plane	
h_2	Centerline distance from second lens surface to second principle plane	$h_1 = \frac{T(n-1)}{n}\frac{\text{EFL}}{R_1}$
EFL	Effective focal length	$[(n-1)\{1/R_1 + 1/R_2 - (n-1)/nR_1R_2\}]^{-1}$
NA	Numerical aperture	r_L/EFL
ffL	Front focal length (front working distance)	$f - h_1$
bfL	Back focal length (back working distance)	$f - h_2$
T	Lens thickness measured at lens centerline	
r_L	Lens radius	
TC	Total conjugate (total distance from object to image)	$(s_0 - h_1) + (s_i - h_2) + T$
$f\#$		$\frac{\text{EFL}}{2r_1} = 1/2\text{NA}$

2.1.2. Performance Criteria

The literature is rich in describing the characterization of lens performance within the geometrical optics regime [1–3]. In general, the analysis is built up in terms of aberration theory which describes mathematically the distortion of a wavefront from the perfect spherical form. This is done in terms of an expansion of the difference between the emergent phase front and a perfect spherical wave, as shown in Fig. 2.3a. The ray aberration as a consequence of the wavefront aberration is also indicated. One expands the difference in the expression for the wave from a perfect spherical wave in terms of the parameters described in the blowup of a portion of the wavefront shown in Fig. 2.3b. The parameters are the height off the axis, ρ, and the polar coordinates, r and ϕ of the reference point on the wavefront. One expands the difference in a power series in ρ, ϕ, and r, the terms other than the pure r^2 term representing the deviation from the spherical wave, or aberrations. The convenient aspect of this approach is that one can relate certain terms in the expansion with specific types of common optical image distortions. For example, consider the following terms of the expansion:

$$Br^4 + Fr^3 \rho \cos \phi - Cr^2 \rho^2 \cos^2 \phi + E \rho^2 \cos \phi + \cdots \quad (2.2)$$

The first term represents what is termed *spherical aberration*. Note that it is present on-axis ($\rho = 0$) The second term accounts for the optical distortion called *coma*. One can see that it is an off-axis effect and has an angular term. The third term is called *astigmatism*, and the fourth term is called *distortion* which can be positive or negative. All of these produce the commonly known distinct optical distortion effects. However, one can continue in the series to higher and higher order terms where the physical meaning becomes more obscure, but nonetheless represents terms of importance. The present analytical approach is to expand the difference in terms of Zernike polynomials, $Z_{nm}(r,\phi)$ One then can extend the analysis to as many terms as desired. In general, the number used is 35. We will see this later on when we discuss the performance of specific microlenses.

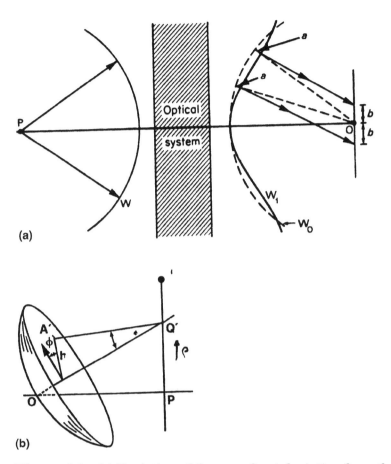

Figure 2.3 (a) Depiction of the wavefront deviation from the perfect spherical wavefront. Both the wavefront distortion, angle ϕ, and the ray distortion, b, are indicated. (b) Reference system for expansion of the deviation.

One often encounters the term *paraxial approximation*. It essentially means that one is confined to on-axis rays, $\rho = 0$, making a very small angle with the axis; that is, r is very small. We shall see that this approximation is hardly ever met in microlens applications. As an obvious example, the assumption that ρ is 0 is difficult when one considers that the microlens radius is often not that much larger than the object, which could be an LED or an optical fiber.

Regardless of the size of the lens, one still has recourse to aberration treatment of lens to determine the lens performance. Clearly, diffraction ultimately defines the limiting performance of a lens, but before that limit is reached, one must face all the limitations afforded from geometric optics. The geometric limitations for microlenses are more severe than for the conventional for the reason that with microlens applications only a single lens must supply the function rather than a multiple train where aberrations can be mitigated. For example, the effects of the geometric aberrations can be lessened by using many refracting surfaces. The other alternative is to use aspheric surfaces. This is the case where the lens figure is altered in such a way as to compensate for the various aberration terms in Eq. (2.2). For microlenses, the former is not practical because of space and alignment constraints, and the latter presents a challenging fabrication problem. As we will see, for single element microlenses, an aspheric molding operation can be used, but it is not practical for arrays. We will briefly mention the molded aspheric process in Sec. 2.3.3. Because most of the refractive microlens applications will be simple planoconvex or biconvex spherical lenses, we will be dealing in some detail on the performance of these cases.

2.1.3. Ray Tracing

One of the simplest and most useful ways to deal with the design, function, and to some extent, performance of a simple thick lens is to utilize a ray-trace analysis. Simply put, this means generating the algorithm to allow one to plot the ray trajectories. There are a number of relatively simple paraxial techniques, one of which we will describe below in detail, as well as the accurate ray-trace program with no assumptions. Today with computers, it is relatively easy to generate one of your own, or to buy one of the commercial software packages. The simple paraxial approach is useful in determining the approximate relationships between the lens parameters for a given application. For example, one wants to collimate a source with a certain NA with a lens of diameter D. This will involve the interplay of the lens curvatures, lens thickness, and total conjugate distance. It gives aberration-free performance and thus cannot be used to evaluate actual optical performance. In the next sections, a paraxial method will be described and an example of its use will be worked out. In the following section, we will use a more accurate ray-trace program in which one can begin to estimate the actual performance of the lens. For this case, in contrast to the paraxial situation, we will see the onset and consequence of the various aberrations that derive from refraction from a purely spherical surface.

2.1.3.1. Paraxial Approximation

As mentioned above, there are a number of relatively simple ways to do paraxial ray tracing. For thick lenses, a particularly useful one makes use of what is called the transfer matrix [3]. A two-component vector is formed with the first component representing the distance above the optic axis y and the second the slope of the ray m through the point. See Fig. 2.4

$$\begin{pmatrix} y_2 \\ m_2 \end{pmatrix} = \mathbf{M} \begin{pmatrix} y_1 \\ m_1 \end{pmatrix} \tag{2.3}$$

where $\mathbf{M}$ is defined as

$$\mathbf{M} = \begin{pmatrix} 1 - T/nf_1 & T/n \\ T/nf_1f_2 - 1/f_1 - 1/f_2 & 1 - T/nf_2 \end{pmatrix} \tag{2.4}$$

Here T is the lens thickness, n is the refractive index of the lens material, and $f_i = R_i/(n-1)$; R_i is the radius of curvature of its surface. One can understand this from the point of view that $\mathbf{M}$ is made up of the product of three operations, the operation corresponding to the refraction of the first surface, followed by a translation operation through the thickness of the lens, followed by the refraction operation corresponding to the second surface

$$\mathbf{M} = \mathbf{R}_2 \mathbf{T} \mathbf{R}_1 = \begin{pmatrix} 1 & 0 \\ -1/f_2 & n \end{pmatrix} \begin{pmatrix} 1 & T \\ 0 & 1 \end{pmatrix} \begin{pmatrix} 1 & 0 \\ -1/nf_1 & 1/n \end{pmatrix} \tag{2.5}$$

(a)

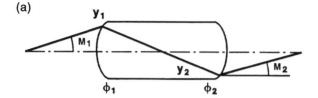

(b)

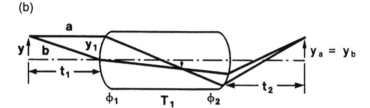

Figure 2.4 Definition of the ray parameters, the ray height h, and the ray slope m. In the lower portion of the figure are shown the test a and b rays.

In all cases, Snell's law is taken with $\sin m = m$, that is, the paraxial approximation.

By way of a demonstration of how the simple ray trace aids in an evaluation of a microlens for a given example, consider the case where we want a microlens to collimate a source. Let the properties of the lens be given by the following set of values:

$$D = 250\,\mu m, \quad R_{c1} = 200\,\mu m, \quad R_{c2} = \text{infinite}, \quad T = 589\,\mu m$$

One sets up the matrix equation in the form

$$\begin{pmatrix} y_2 \\ 0 \end{pmatrix} = \mathbf{M} \begin{pmatrix} y_1 \\ m_1 \end{pmatrix} \mathbf{M} = \begin{pmatrix} 1 - T/nf_1 & T/n \\ -1/f_1 & 1 \end{pmatrix} \qquad (2.6)$$

where $f_i = R_{ci}/(n-1)$ and m_2 is taken to be 0 since we want the light to be parallel. We want to obtain the value of m_1 since this will give us the distance we should place the source relative to the lens. Since y_1 is arbitrary, we will take $y_1 = D/2$, and $m_1 = D/2s_1$. Solving the algebraic equations yields the expected result that $s_1 = f_1 = R_{c1}/(n-1)$. If the flat surface of the lens faced the source, that is, R_{c1} is infinite and $R_{c2} = 200\,\mu\text{m}$, then the object distance would by given by $s_1 = f_1 - T/n$. This suggests the interesting configuration of butting the source directly to the flat face by choosing the lens thickness to be $nR_c/(n-1)$. This is shown in Fig. 2.5b.

For imaging an object that is a distance x_0 from the lens at a height y_0, one chooses two convenient rays as shown in Fig. 2.4b. The a-ray has the initial coordinates $(y_0, 0)$ and the b-ray, $(0, m_1 = y_0/x_0)$. The simultaneous solution of the simple algebraic expressions will yield the position of the image.

2.1.3.2. Exact Method

In this case, we use the exact ray trace where the refraction at the lens surface is properly taken into account without any approximations. The consequence of this accurate treatment is the appearance of the distortions of the image that are manifestations of the aberrations mentioned above. (In Appendix B of this chapter, the algorithm is shown for a three-dimensional ray-trace program.) Here we reexamine the performance of the collimating lens treated above in the paraxial case.

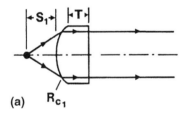

(a) R_{c_1}

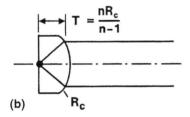

(b) R_c

Figure 2.5   Two examples of the paraxial ray trace for the case of collimating a point source with a planoconvex lens. (See text.)

We revisit the collimating example that we treated above in the paraxial approximation. We position the point source at the paraxially determined position of 0.4 mm. We will look at the resulting ray trajectory as a function of the NA of the source. In Fig. 2.6a, the NA is only 0.05 and the collimation result is equivalent to what one would get from the paraxial approximation. As we let the NA increase to 0.10 (Fig. 2.6b), we begin to see the consequence of spherical aberration. The marginal rays are no longer parallel. In the last case, Fig. 2.6c, the NA is increased to 0.20 and the consequence of the spherical aberration is dramatically evident. Often in the face of this type of aberration, one moves the object closer than that indicated by the paraxial estimate. This has the effect of reducing the spread of the outermost rays at the expense of the inner rays no longer being parallel.

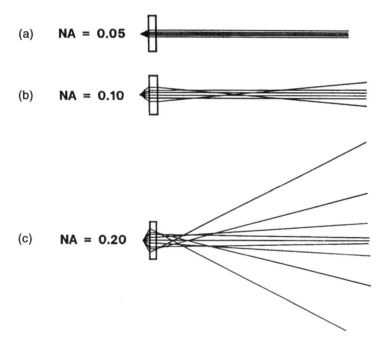

Figure 2.6 Exact ray trace of collimating function planoconvex lens specified in the text, for three different numerical apertures.

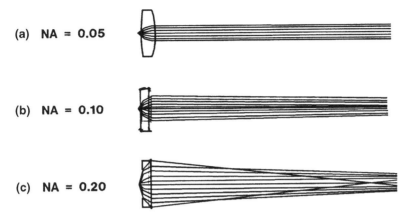

(a) **NA = 0.05**

(b) **NA = 0.10**

(c) **NA = 0.20**

Figure 2.7 Exact ray trace of collimating function for biconvex lens with the same EFL as that shown in Fig. 2.6.

For the situation where one can use a biconvex lens, one can see the significant reduction in the aberration. We repeat the three situations given in Fig. 2.6a–c for the case where $R_{c1} = R_{c2} = 500\,\mu m$ in Fig. 2.7a–c. In this case, because the refraction is less at each surface, the overall spherical aberration is reduced for comparable NA.

The performance of certain designs, and the consequence of aberrations, will be dealt with in Chapter 5, when we deal with specific applications.

2.2. FABRICATION METHODS

2.2.1. Introductory Comments

A relative large number of ways to make small refractive-type lenses has been employed over the last 10 years. Each has had some perceived advantage. Some are relatively inexpensive and utilize an existing technology, while others require new materials with special properties and unusual fabrication techniques. The particular application must determine the best material and fabrication technique according to the choices, glass/plastic, single element/array, aspheric/ spherical. For microoptical applications involving single elements, that is, for lenses with diameter <1 mm, conventional grinding and polishing are impractical. The same is true for monolithic arrays of lenses. We will try to cover the important methods available for refractive lenses, keeping in mind that there are many variations. As we move to the next chapters, the reader will see more lens fabrication methods based on other types of lenses than refractive. In this section, we will start with the most traditional and proceed to the more exotic.

2.2.2. Molding

2.2.2.1. Plastics

This is the classic method of preparation where a mold of the desired surface is made in a nonreactive material by some precise method. This provides the form which is translated to the end element by placing it in intimate contact with the mold. This is a quite straightforward process for the case of plastic elements where injection molding is used. See Fig. 2.8 [4]. The technique, however, has the disadvantage of the high initial cost of the mold, especially if single-point diamond turning technology is required. There have been plastic lens arrays made by molding, examples of which made by USPL [4] are shown in Fig. 2.9.

There are somewhat simpler ways to produce plastic lenses, or lens arrays, for example, hot pressing. In this method, a sheet of thermoplastic (polycarbonate, PMMA, or polystyrene) is heated so that it softens sufficiently to allow it to be permanently deformed. The heated sheet is then pressed against the mold. The softened lens material forms a protrusion at each aperture, as shown in Fig. 2.10a. The mold may be made of stainless steel and the pattern formed by any convenient method such as chemical etching. This particular method has the advantage that the lens surface does not come in contact with the mold. A more complete drawing of the lens-forming operation is shown in Fig. 2.10b.

The example results [5] indicate that lens arrays with individual lens diameters of the order of 1 mm are possible, with good spherical surfaces limited to the central region of

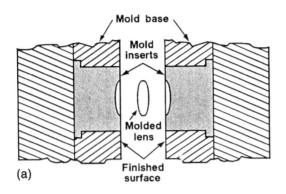

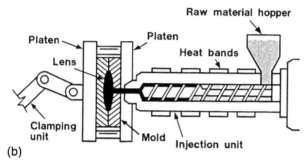

Figure 2.8 Diagram of extrusion molding process to make plastic lenses. The upper drawing shows a blow-up of the mold itself. (From Ref. [4].)

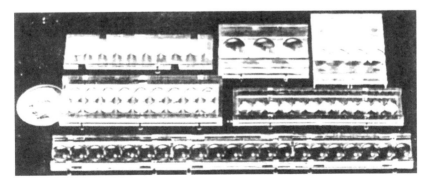

Figure 2.9 Photograph of various molded lens array configurations. (From Ref. [4].)

(a)   (b)

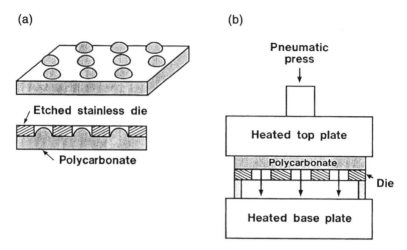

Figure 2.10 Schematic drawing of hot-pressed lens array in polycarbonate, also showing the heated press.

the lens. The radius of curvature is in the range of 1.1–2.4 mm and controllable by the pressing time. The lens performance was estimated by a spot size test that was not well described. It general, this is a poor method to use to evaluate lens performance. It is not the size of the Airy disk that is important, but how much intensity is contained in the central ring.

2.2.2.2. Plastic on Glass

To avoid some of these problems, there are hybrid methods, for example, plastic lenses on a glass substrate. Adaptive Optics Associates reports [6,7] fabricating lens arrays in this manner from a metal master. The master is made of a high-purity annealed and polished material. After the pattern is formed, a release agent is applied to the surface. Using standard replication techniques, a small amount of epoxy is placed on the surface and the optical substrate is placed on top. This is shown in Fig. 2.10. Representative types of arrays are shown in Table 2.2.

A SEM photograph of an array of 70-μm-diameter lenses made by this method is shown in Fig. 2.11, along with the images produced by these $f/12$ lenses.

Table 2.2 Representative Lens Arrays Made by Molding

Aperture (μm)	Array size	f	FL (mm)	Format	Fill factor (%)
100	120 × 120	5	0.5	Square	>98
100	68 × 68	17	1.7	Square	>98
200	60 × 60	5	1.0	Square	>98
200	36 × 36	33	6.5	Square	>98
400	20 × 36	8	3.2	Square	>98
400	21 × 21	63	25	Square	>99
500	300 × 300	6.6	3.3	Square	>97
600	13 × 13	18	6.3	Square	>94
768	8 × 8	8	65	Circle	>79
1016	12 × 12	256	260	Square	>98

Source: Selected from data sheet of Adaptive Optics Associates (United Technologies), Cambridge, MA.

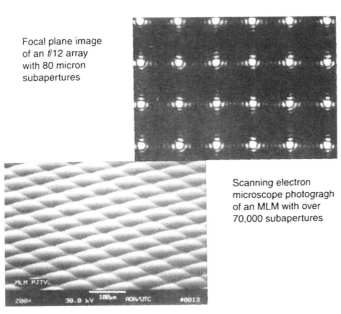

Focal plane image of an f/12 array with 80 micron subapertures

Scanning electron microscope photogragh of an MLM with over 70,000 subapertures

Figure 2.11 SEM photomicrograph of 80-μm-diameter molded plastic lenses and the images produced by the $f/12$ lenslets. (From Ref. [7].)

Using precisely controlled machines, it is possible to make mold cavities with very high surface figure accuracy. As pointed out above, the application dictates what the material must be, glass or plastic, and the choice of material determines the difficulty of the subsequent molding operation. For example, for the high temperature process required for glass, the mold life, and in general, the difficulties accompanying the higher temperature process make the process quite different from what it would be for plastic. To make this point, we will describe a process reported for molding 5-mm-diameter aspherics as an example of what is involved. The fact that it is an aspheric surface is really not an issue in the problems encountered, save may be in degree, because of the higher required accuracy of the replication process.

2.2.2.3. Glass

The molds are made to a surface accuracy of a few thousands of a wave with a mirrorlike finish. To obtain such surfaces, a technique called single-point diamond turning is used [8]. In its simplest description, these are very precise lathelike machines that can control cut depths to 0.1μ in. A typical diamond turning tool is represented in Fig. 2.12. Machines such as those manufactured by Moore Specialty Tool Co. must be operated in a temperature and humidity-controlled environment. The cost of such a machine, including the facility to place it in, can be as high as \$1.5 million. An example of a complete molding assembly is shown in Fig. 2.13. It consists of two molds (one for each lens surface), a precision drilled alignment sleeve to control centration and tilt of the molds, a ring member to form the outside diameter of the lens, and finally a glass preform. In addition to the accurate tooling for high-numerical-aperture lens applications, there are two important factors in accurate replication: forming at high viscosity and maintaining an isothermal environment. Both of these conditions are required in order to limit distortion. There is a lower limit to the diameter of a lens that can be made by a single-point diamond turning mold and it stems from the radius of curvature of the cutting tool relative to

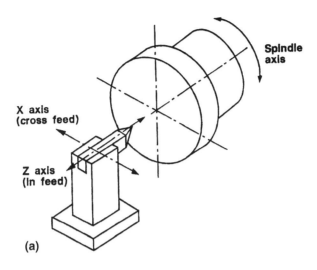

X axis
(cross feed)

Spindle
axis

Z axis
(in feed)

(a)

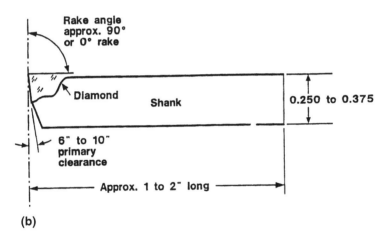

Rake angle
approx. 90°
or 0° rake

Diamond   Shank

0.250 to 0.375

6″ to 10″
primary
clearance

Approx. 1 to 2″ long

(b)

Figure 2.12 (a) Representation of a generic single-point diamond turning machine. (b) Typical cutting tool design. (From Ref. [8].)

the curve one is trying to cut. Typical curvatures are of the order of 300 μm.

The molding process described above might be more aptly called compression molding [9] of a glass preform. The glass preform is contained within the molding assembly which is heated to the molding temperature. Then pressing

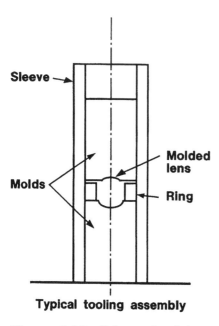

Typical tooling assembly

Figure 2.13 Schematic of the molding assembly. Consists of two molds, an alignment sleeve, and a glass preform as shown. (From Ref. [9].)

is accomplished by applying a load to one of the molds. Thereafter, the glass is cooled to below its transition temperature before removal.

The thermophysical properties of the glass play a crucial role in making the molding process a practical method. The glass has to have a lower transition temperature $T_g < 400°C$, than typical glasses, so that the forming can be done at a sufficiently low enough temperature to make the tooling and molding cost effective. This primarily means what composition the mold material must be, and how many pressings it will allow. In general, in order to soften the glass, some sacrifice of the glass durability is made, which can be a problem. A glass composition that has been used to mold glass lenses is the Pb/Zn-flouro-phosphate system [10].

Another molding process has been reported for making lens arrays, termed *contactless* molding [11]. The method is to heat the glass above T_g and then press into it with a mold

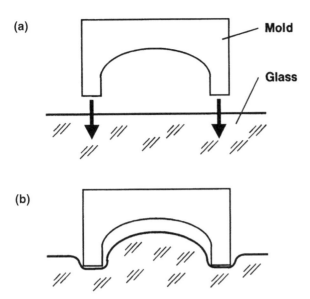

Figure 2.14 Schematic representation of contactless molding operation in glass. The ribs of the mold are pressed into the soft glass, which takes on a near-spherical surface because of surface tension. The soft glass never touches the interior of the mold. The mold is made of special alloys that do not react with the glass, and release easily. (See Ref. [11].).

made up in the form of a grid. The glass is forced up in the mold but does not come in contact with the mold. The glass is thus influenced by surface tension and forms a surface of minimum energy. In the simplest case, if the grid is made up of circles, then a spherical element is produced. In the case that the grid symmetry is something else, distortions away from a spherical shape occur at the boundaries. A picture of an array made in this way is shown in Fig. 2.14. As in any molding process, the life of the mold is the major problem.

2.2.3. Glass vs. Plastic

It is probably worthwhile to generally compare the advantages and disadvantages of glass and plastic lenses for micro-optics. A great deal of progress has been made over the years in making harder and tougher plastic materials with higher refractive indices. This has translated into improved scratch resistance and overall durability. But this is not the major issue with lens applications. The major drawback of high-performance plastic lenses has to do with the thermal stability of the lens performance. One can quantify this property by defining the temperature coefficient of optical path, nL, through the expression

$$G = \frac{\Delta(nL)}{L\,\Delta T} = \frac{(n-1)\Delta L}{L\,\Delta T} + \frac{\Delta n}{\Delta T} = (n-1)\text{CTE} + \frac{\mathrm{d}n}{\mathrm{d}T} \quad (2.7)$$

where CTE is the coefficient of thermal expansion. One wants the value of G to be as small as possible. For glasses $\mathrm{d}n/\mathrm{d}T$ is negative and partially cancels the CTE part, and the resulting value of G is relatively small. For plastics, the CTE is very large and dominates the value of G.

The overwhelming advantage of plastics is their low cost while maintaining a high performance level. Unfortunately, many of the applications of microlenses require ultra performance levels, not yet achievable by plastics. In less demanding applications, in particular for lens arrays, the plastics are more competitive. Different problems then arise which have to do with maintaining dimensions over long distances, or run-out as it is often called. In addition to being a less rigid material, the higher thermal expansion could pose a problem. Other aspects that have to be considered are how one bonds the plastic material to other materials. Plastics often absorb moisture and swell, creating problems with maintaining a secure bond. Using the plastic lenses on glass avoids all of these problems, leaving only the disadvantage of the softness of the lens surface with respect to abrasives.

2.2.4. Photoresist Based

There are a number of methods by which one can obtain reasonably good spherical microlenses that involve the use of patterned photoresist. This is the standard process where resist is spun on to a suitable substrate, then exposed to light through a mask to alter its solubility to a subsequent development solution. The convention is that a negative resist is one where after exposure it is resistant to the solvent, whereas a positive resist is one that dissolves where exposed [12]. In the following, we will see how this can be used to produce microlenses.

2.2.4.1. Melted

The simplest and most ingenius way to create a microlens is to make a resist pattern in the shape of disks. In this case, one exposes the resist of thickness L through a mask made up of clear circles with diameter D. Upon development, this turns into the disk shape shown in Fig. 2.15 [13,14]. One

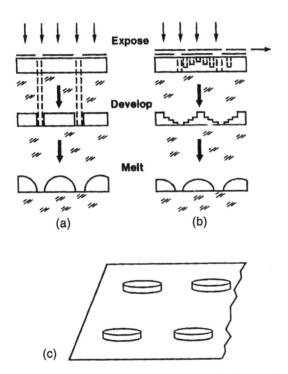

Figure 2.15 Representation of the melted photoresist method to form microlenses. Cylindrical-shaped columns or rows are patterned into the photoresist. Upon heating, the resist flows and spherical are produced. The resist can be prepatterned to obtain a better approximation to a spherical surface.

now merely heats the substrate to melt the photoresist. The surface forces control the shape and a segment of a sphere represents the minimum energy surface. The first-order relationship between the circle diameter, the resist thickness, and the resulting radius of curvature of the surface, and ultimately the focal length of the lens so formed, is given by the expression

$$R_c = (n-1)f = \frac{D^2}{4L} \tag{2.8}$$

It has been reported [14] that this method is effective in making lenses that are close to hemispherical, and correspondingly less so as one deviates to shallower lenses. The effective range of f-numbers (f/D) is in the range of 1–2.5. To increase the f-number, one can replace the ambient medium with a fluid of intermediate index between the resist and air. Five to seven hundred and fifty micrometer is the range of diameters of lenses that have been produced by this method. Another limitation is imposed by the thickness of resist that can be applied, around 50 μm. Yet another limitation is that one cannot place the lenses close enough to achieve high fill factors unless a different lens boundary shape is used.

Daly et al. [14] present a fairly comprehensive report on the fabrication process, including such facts as the actual resists that were used and some of the process steps. This is given in Table 2.3. They also describe some novel variations

Table 2.3 Photoresists Used in the Melted Method

Type	Deposition (rpm/min)	Thickness (μm)
Shipley AZ 1400-37	1500/1	5
Hoechst AZ 4620-A	2500/0.16	8
	1200/0.16	20
	800/0.1	40
	800/0.06	50

Taken from Ref. [14].

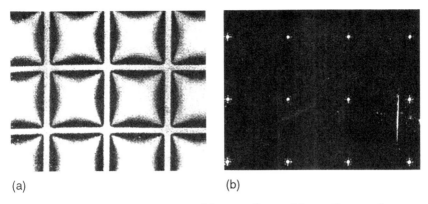

(a)   (b)

Figure 2.16 Square patterned lenses formed by orthogonal exposure. Images formed by the lenses are also shown.

such as making a preform resist pattern before melting, as
shown in Fig. 2.16a, to address the problem of making longer
focal length lenses. Another interesting idea was to place the
resist lenses on an existing metal apertured pattern to
provide a dark matrix surround. Otherwise light would come
through the nonlens portion of the array. They also showed
lens with square cross-section made by superimposing two
cylindrical arrays, as shown in Fig. 2.16a. They mention the
expected poorer performance because of the edge effects. For
close-packed designs, there appears to be a dependence on
the ultimate shape and the shape before melting. For example, see the stepwise resist structure shown in Fig. 2.16b.
They report on a variety of lenses, both cylindrical and
spherical, which we reproduce here as Table 2.4.

2.2.4.2. Etching

Following the procedure in the previous section, we can
proceed to use the shaped resist to provide a way to etch
the pattern into the substrate. For example, lenses etched
in this way are shown schematically in Fig. 2.17. Obviously,
the major issue is to choose an etching process that is one to
one in solubility. That is, the resist and glass must etch at
the same rate. The etch of choice is RIE (reactive ion etching),

Table 2.4 Representative Sizes and Numerical Aperture of Melted Resist Lenses

Focal length (µm)	Aperture (µm)	NA	Measured FWHH (µm)
Cylindrical			
78	88	0.56	1.0
105	95	0.45	1.1
147	100	0.34	1.4
200	102	0.25	1.95
300	102	0.17	5
500	100	0.1	10
Spherical			
630	750	0.6	1
570	280	0.31	3
180	125	0.34	1.7
110	65	0.3	1.8

Source: Ref. [14].

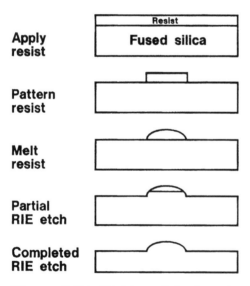

Figure 2.17 Etching of the lens into material using the melted photoresist as the graded mask. The initial steps are as shown above. It is important to maintain a one-to-one etch rate between the resist and the material.

where one can exercise some control over the relative etch rates by using a combination of oxygen-, and fluorocarbon-containing gases. Lenses with less than $\lambda/4$ waves of third-order spherical aberration have been made [15]; their diameters are comparable to those mentioned above.

Clearly, the substrate can be other than glass. There have been reports of lenses made in silicon and polymide [15].

2.2.5. Microjet Fabrication

The use of the microjet technology, developed primarily for printing, has also been adapted to making microlenses [16]. The analogous microjet system used a piezoelectric ceramic with a microchannel machined in it with a nozzle on one end connected to a reservoir on the other. An electrical pulse bends the channel and forces a droplet through the aperture. The droplet is directed to a substrate which is mounted on an xyz micropositioner (Fig. 2.18); the liquid drop solidifies on

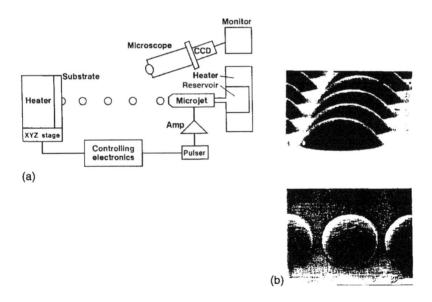

Figure 2.18 Schematic of microjet method of depositing microlenses. See text for description. Inset shows SEM photographs of lenslets on 100 μm centers.

contact with the substrate and surface tension causes a spherical surface to form. The lens diameters are in the order of 80 μm and are roughly hemispherical. This method produces much the same result as the melted photoresist technique. One forms near hemispherical lenses and the packing fraction is limited because of the lenses running together. No description or characterization of the fluid used is given.

MicroFab Technology, Inc. reports being able to deliver spheres of fluid with diameters from 25 to 100 μm at rates of 6000 per second using this piezoelectric drop-on-demand ink jet printing technology [17].

2.2.6. Photosensitive Glass

2.2.6.1. Photothermal Process

One of the more interesting methods of producing refractive microlens arrays, at least as far as the phenomenon is concerned, is that which involves the use of a special photosensitive glass [18,19]. The basis of the effect is generated by a photonucleation of a phase crystallization. (lithium meta-silicate) within the glass which produces a physical change in density. The total crystal content is of the order of 10–20 vol%. Under the appropriate exposure pattern, this density change can be used to produce surface features that ultimately act as lenses. A schematic of the process is shown in Fig. 2.19 and a typical SEM result is shown in Fig. 2.20. The photopatterning is done by conventional photolithographic masking techniques. After the glass is exposed, it is heated to about 600°C to effect the crystallization. What results are circular regions where the light is blocked, surrounded by crystallized regions of higher density. The effect of this is to squeeze the soft unexposed glass beyond the surface that then forms a minimum energy surface. The surface bumps are formed on both surfaces. We can estimate the height of the bump from a simple expression:

$$\frac{\delta}{T} = \left(\frac{2}{3}\right)\left(1 - \frac{\rho}{\rho_0}\right) \tag{2.9}$$

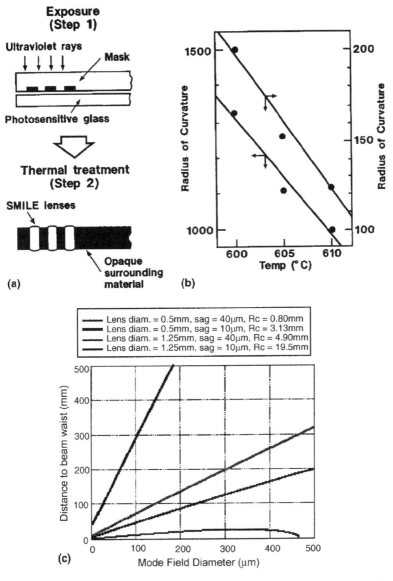

Figure 2.19 (a) a schematic of the photosensitive process. Special glass is exposed through a mask, then heated. Exposed area densifies and squeezes soft unexposed glass to form spherical proturberances. (b) is a graph of the radius of curvature of the induced lens as a function of the top temperature of the thermal schedule. It is shown for two different lens diameters. (c) distance to focus us mode field diameter for various lens paromodes.

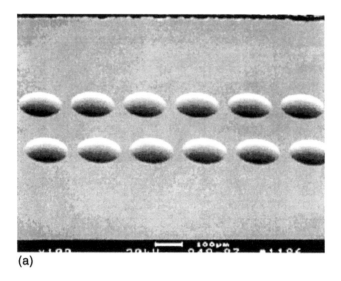

(a)

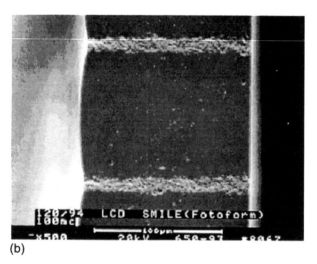

(b)

Figure 2.20 (a) Scanning electron micrograph of lenses formed by SMILE. In this case, the lens diameter was 160 μm on 195 μm centers with a focal length of 0.4 mm. (b) A cross-sectional view showing the crystallized intervening material.

where T is the thickness, ρ is the unexposed glass density, and ρ_0 is the exposed crystallized glass density. The maximum density change observed is of the order of 1.5%. The effective depth is determined by the absorption coefficient at the activating wavelength, which is 310 nm. In general, using conventional Hg–Xe exposure lamps, exposure times of 10–100 sec are required and the depth of subsequent development is 3 mm. The samples can also be simultaneously exposed from both sides by extending the maximum thickness to 6 mm and by stacking configurations [20] as shown in Fig. 2.21.

Nonspherical lenses can also be formed. The shape of the lens is determined by the geometry of the exposed region. For example, lenses that represent segments of ellipsoids of revolution are easily produced by exposure through an elliptical

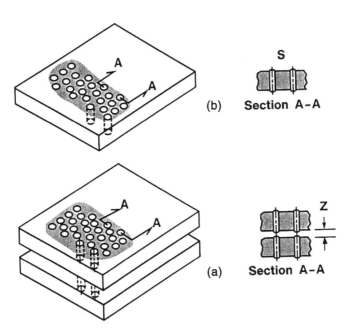

Figure 2.21   Schematic drawing of result of double side exposure of the photosensitive glass to produce a biconvex lens. Also shown is the stacking of two arrays.

mask. Thus it is possible to form an anamorphic lens by using an elliptically shaped mask. For dense packing even hexagonal-shaped lenses have been produced. A variety of shapes that have been formed are shown in Fig. 2.22. In Appendix A to this chapter, we will derive the minimum energy shape that evolved from an arbitrary closed-boundary curve. It turns out that a reasonably uniform surface evolves except near the boundaries that contain edges.

2.2.6.2. Ion-Exchange Stuffing

Another way was found to make lenses in this photosensitive glass using an ion-exchange stuffing mechanism [20]. It is shown that the ion-exchange of Na or K for Li occurred only in the unexposed glass. The substitution of the larger ion Na or K for the Li in the unexposed glass caused the glass to expand. In the exposed and developed region, the ion exchange did not occur, and moreover at the temperature of the ion-exchange, the material is rigid. As a consequence, the soft unexposed glass is squeezed beyond the surface and spherical bumps are produced. A representative result is shown in Fig. 2.23 where the radius of curvature of the lens is plotted vs. the square root of the exchange time in a KNO_3 at 550°C.

There are practical advantages of the ion-exchange method of producing lenses over the purely thermal method. The major one is the temperature uniformity afforded by the immersion of the sample in a constant temperature bath over that achievable in a furnace. This is particularly true when processing large two-dimensional arrays. Another advantage is for applications where compound structures are required. One can finish the exposed and developed patterned array to its final thickness before the ion-exchange development of the lenses. One has two pristine lens surfaces which can then be attached together as previously shown in Fig. 2.20. In the normal thermal method, one side of the sample is in contact with a substrate during the heating and this produces small defects on the lens.

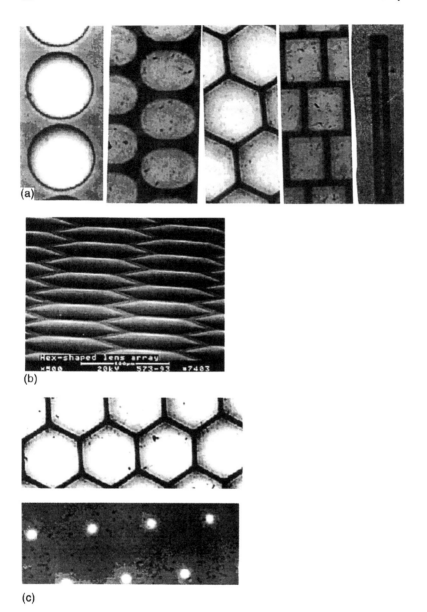

Figure 2.22 (a) Photomicrographs of various lens shapes and arrangements made from the photosensitive glass method under trade name SMILE™. (b) Regular hexagonal pattern, upper photo shows electron microscope view and the lower the geometric layout. (c) Images formed by the lens array shown in (b).

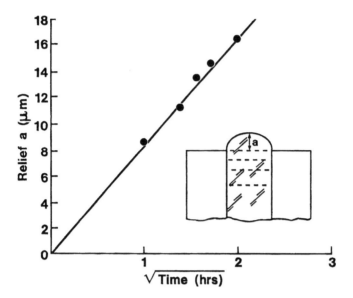

Figure 2.23 Ion-exchange stuffing to produce microlenses in photosensitive glass. Ion exchange proceeds more completely in the unexposed glass. Graph shows height of relief as a function of diffusion time at a fixed exchange temperature.

2.2.6.3. Fabricated Lenses

The lens diameters range from 80 to 1000 µm and the corresponding focal lengths are greater than 100 µm. Since the process can yield double convex lenses, the effective focal length range is extendable via the compound lens effect. The effective focal length, EFL, of a biconvex lens is given by

$$\frac{1}{\text{EFL}} = \frac{1}{f_1} + \frac{1}{f_2} - \frac{T/n}{f_1 f_2} \tag{2.10}$$

where T is the lens thickness, and f_1 and f_2 would be the focal lengths of the individual front and back lenses. Table 2.5 lists the range of lens diameters and configurations that have been fabricated. The lens arrays can be arranged in any two-dimensional pattern limited only by the minimum separation distance of 10 µm.

Table 2.5 Lenses Made by Photothermal Method

Key Optical Performance Parameters	
Lens diameter	50–1500 μm ± 8 μm
Separation between lenses	>10 μm
Numerical aperture	≤0.35
Thickness (T)	≤6 mm (2 mm typical)
Lens Sag (S)	0–120 μm ± 1 μm
Center-to-Center tolerance	±0.2 μm
Coefficient of thermal expansion *for glass microlens*	$8.4 \times 10^{-6}/°C$
Transparent	From 600 to 2700 nm
Distance to beam waist	3–500 mm

This photothermal method lends itself to the fabrication of erect one-to-one imaging arrays because the natural formation is a biconvex structure [21]. If the lenses are strong enough, a relay image can be formed midway between the two lenses. In Chapter 5, the applications of such lens bars will be explained and their performance discussed.

2.2.7. Miscellaneous Methods

2.2.7.1. Laser Heating

There have been other methods that have been described as leading to lenslike structures. One method makes use of a laser whose emission wavelength is strongly absorbed such as to produce local melting at the focal spot [22a,b]. The explanation put forward for the formation of the bump is that the glass is initially melted and then resolidified. Since the soft glass at the higher temperature has a lower density, the excess volume wells up out of the colder rigid substrate glass and reforms upon solidification into a lenslike shape.

A more recent study [23] indicates the mechanism shown in Fig. 2.24. The circular depression around the raised lens which was caused by the local heating was measured and is shown in Fig. 2.25. Typical exposure conditions leading to lens of various diameters are shown in Fig. 2.26. These data are all taken from Ref. [23].

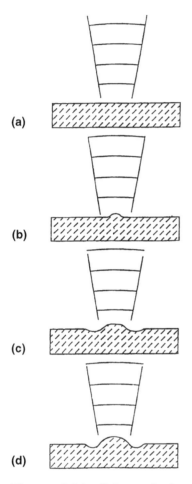

Figure 2.24 Schematic drawing of how lens forms by local laser heating. In (b) the temperature is sufficient to soften the central region; in (c) with higher temperature being achieved, the onset of the large deformation is seen. (After Ref. [23].)

Another intriguing method recently described by Lawandy et al. [24] is where the chemical etchability of certain borosilicate glasses can be altered by preexposure to high intensity light. The mechanism is not really understood but is attributed to an atomic level of damage induced by the irradiation that subsequently affects the way the material reacts

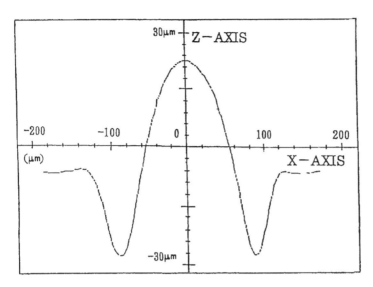

Figure 2.25 Profilometer trace over heated region. (After Ref. [23].)

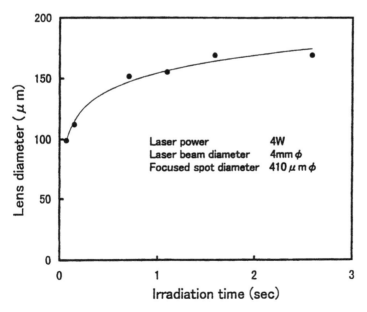

Figure 2.26 Diameter of lens as a function of irradiation time, for parameters as listed. (From Ref. [23].)

with typical glass etchants like HF. In the data reported, the glass etches preferentially where exposed.

2.3. COMPARISON AND ANALYSIS

The intention of this section is to give the reader some way to compare the various ways that microlenses can be prepared as outlined above. Clearly, the specific application defines and orders the important performance criteria; there is no overall "best way". When cost is included, the choice is even more difficult. However, we can strive to set up a method of comparison where one can get some idea of what kind of property and performance one can expect from the various ways lenses and lens arrays are made. This will be done in two ways. The first will be to summarize from the above methods what can and what cannot be done by each of the fabrication methods. We will compare the range of achievable focal lengths, lens diameters, etc., in a chart form, so we can readily see what is possible. The second comparison will be more detailed and will try to put together what data there are pertaining to the actual lens performance. Ideally, this would include a common method of characterization such as interferometry, but unfortunately little real lens characterization has been reported. Nonetheless, what has been published will be summarized.

2.3.1. Comparison of Capabilities

As mentioned above, this section deals with the type of lenses that can be made and roughly the range of parameters available. We will do this in a tabular form to permit a quick reference. We list the fabrication method vertically and the property or capability horizontally.

By asphere, we mean that such a surface can be prepared by this method. Under shapes is whether such arbitrary boundary shapes as ellipses or squares can be made. Array means whether the method lends itself to making arrays of lenses. The focal length, FL, and the diameter refer to the limiting values producible by the method. Packing means how close the lenses can be spaced, ultimately expressed as a percentage fill factor. Geometry refers to how the lenses can be arranged in a two-dimensional pattern. The size refers to the dimensions of a piece that can be processed.

2.3.2. Methods of Analysis of Lens Performance

It does not matter how ingenious and easy the lens fabrication method is if the lens performance is not adequate for the application. The question is then how to characterize lenses is such a way as to be able to obtain a reasonable estimate of how they will perform. The most common performance characterization that is given is that the lens is diffraction limited. What is usual meant is that the spot size in the image plane was somehow measured and compared to the diffraction limited number which is either $1.22\lambda/\text{NA}$, if the intensity across the field is flat, or $(2/\pi)/\text{NA}$ if the intensity is guassian. Unfortunately, the measured spot size, as interpreted as the central portion of the Airy profile, does not change even in the presence of significant spherical aberration. As we will show below, even if the surface curve is not perfectly spherical, the first ring diameter is not significantly altered.

    The irradiance distribution in the focal plane in the presence of a phase aberration, $\phi(\rho)$, is given by the following expression [25]:

$$\frac{I(r)}{I_0} = \left(\frac{k}{R}\right)^2 \left| \exp[i\phi(\rho)] J_0\left(\frac{kr\rho}{R}\right)\rho \, d\rho \right|^2 \tag{2.11}$$

where $k = 2\pi/\lambda$, R is the distance from the pupil plane to the image plane (focal length of the lens), a is the radius of the aperture in the pupil plane (lens radius), and ρ is the radial position in the image plane. This rotationally symmetric phase aberration term can be expressed in terms of the classical Zernike polynomials representing the fourth, sixth and eighth order spherical aberrations. For $\phi(\rho) = 1$ corresponding to the perfect lens, one can integrate Eq. (2.11), and obtain the well-known Airy function

$$\frac{I(r)}{I_0} = \left| \frac{2J_1(kar)/R}{kar/R} \right|^2 \tag{2.12}$$

Mahajan [25] investigated a number of informative cases with this formulation which could form a basis of an ultimate performance comparison for microlenses. He used the Strehl ratio as the parameter measuring the magnitude of the aberration. The Strehl ratio is defined as the value of Eq. (2.11) at $r = 0$, that is, $I(0)$. One can show that the Strehl ratio is very well approximated by the expression

$$S = \exp(-\sigma_\phi^2) \tag{2.13}$$

where σ_ϕ is the standard deviation of the spherical aberration. In a typical Zernike analysis of the interferometric data for a given lens, this number would correspond to RMS or peak-to-valley number under the appropriate aberration. The results are shown in Fig. 2.27. The interesting aspect of this calculation is that the central ring diameter is not changing with the degree of aberration. The main effect of the

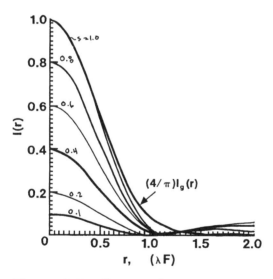

Figure 2.27 Computed point spread function vs. radius (normalized to the product of the wavelength λ and the focal ratio F). The various curves are for different degrees of aberration, as indicated by the Strehl ratio S. Also drawn in the Gaussian profile. (Graph is replotted from Ref. [24].)

Table 2.6 Comparison Table

Fabrication	Asphere	Shapes	Array	FL (μm)	Diameter (mm)	Packing/ geometry
Molded						
Plastic	Y	Any	Y	>100	>0.1	100%/any
Glass	Y	Any	N	>500	>2[a]	–
Contactless	N	Any[b]	Y		>0.1	80%/any
Resist						
As is	N	Circle	Y			
Etched	N	Circle	Y			
Microjet	N	Circle	Y		0.08–1	80%/any
Photosens glass	N	Any[b]	Y	>200	0.08–1	80%/any
Laser	N	Circle	Y	>100	>0.1	?/any

[a] Depends on radius of curvature of cutting tool.
[b] The surface figure is dependent on the boundary curve.

spherical aberration is the reduction of the intensity of the central ring and the corresponding increase in the outer ring.

2.3.3. Data Comparison

What the analysis in the above section points out is that for a true comparison of the microlens performance by any of the methods described above, either the point spread function (PSR) be measured or a Zernike analysis of interferometric data should be done. The reader is referred to Table 2.7 for a listing of the Zernike terms.

Zernike polynomial analysis of the molded glass asphere has been reported. These are relatively large lenses in the microoptic sense. These lenses are produced to provide high resolution at high numerical aperture. A typical result for a molded bisphere is shown schematically in Fig. 2.28 representing an overall rms 0.05λ distortion from a perfect spherical wavefront. The numerical aperture for this lens was 0.45 [26].

Aperture limitations imposed on interferometry make the measurement of very small lenses a problem. Nonetheless interferometric data have been reported for microlenses made by the photosensitive glass process, SMILE. The complete

Table 2.7 Zernike Polynomial Notation

n	m	Z_{nm} Polynomial		Characterization
0	0	0	1	Piston
1	+1	1	$\rho \cos \phi$	X tilt
	−1	2	$\rho \sin \phi$	Y tilt
	0	3	$2\rho^2 - 1$	Focus
2	+2	4	$\rho^2 \cos 2\phi$	Astigmatism 0° or 90°
	−2	5	$\rho^2 \sin 2\phi$	Astigmatism ±45°
	+1	6	$(3\rho^2 - 2)\rho \cos \phi$	X coma and tilt
	−1	7	$(3\rho^2 - 2)\rho \sin \phi$	Y coma and tilt
	0	8	$6\rho^4 - 6\rho^2 + 1$	Spherical and focus
3	+3	9	$\rho^3 \cos 3\phi$	
	−3	10	$\rho^3 \sin 3\phi$	
	+2	11	$(4\rho^2 - 3)\rho^2 \cos 2\phi$	
	−2	12	$(4\rho^2 - 3)\rho^2 \sin 2\phi$	
	+1	13	$(10\rho^4 - 12\rho^2 + 3)\rho \cos \phi$	
	−1	14	$(10\rho^4 - 12\rho^2 + 3)\rho \sin \phi$	
	0	15	$20\rho^6 - 30\rho^4 + 12\rho^2 - 1$	
4	+4	16	$\rho^4 \cos 4\phi$	
	−4	17	$\rho^4 \sin 4\phi$	
	+3	18	$(5\rho^2 - 4)\rho^3 \cos 3\phi$	
	−3	19	$(5\rho^2 - 4)\rho^3 \sin 3\phi$	
	+2	20	$(15\rho^4 - 20\rho^2 + 6)\rho^2 \cos 2\phi$	
	−2	21	$(15\rho^4 - 20\rho^2 + 6)\rho^2 \sin 2\phi$	
	+1	22	$(35\rho^6 - 60\rho^4 + 30\rho^2 - 4)\rho \cos \phi$	
	−1	23	$(35\rho^6 - 60\rho^4 + 30\rho^2 - 4)\rho \sin \phi$	
	0	24	$70\rho^8 - 140\rho^6 + 90\rho^4 - 20\rho^2 + 1$	
5	+5	25	$\rho^5 \cos 5\phi$	
	−5	26	$\rho^5 \sin 5\phi$	
	+4	27	$(6\rho^2 - 5)\rho^4 \cos 4\phi$	
	−4	28	$(6\rho^2 - 5)\rho^4 \sin 4\phi$	
	+3	29	$(21\rho^4 - 30\rho^2 + 10)\rho^3 \cos 3\phi$	
	−3	30	$(21\rho^4 - 30\rho^2 + 10)\rho^3 \sin 3\phi$	
	+2	31	$(56\rho^6 - 105\rho^4 + 60\rho^2 - 10)\rho^2 \cos 2\phi$	
	−2	32	$(56\rho^6 - 105\rho^4 + 60\rho^2 - 10)\rho^2 \sin 2\phi$	
	+1	33	$(126\rho^8 - 280\rho^6 + 210\rho^4 - 60\rho^2 + 5)\rho \cos \phi$	
	−1	34	$(126\rho^8 - 280\rho^6 + 210\rho^4 - 60\rho^2 + 5)\rho \sin \phi$	
	0	35	$252\rho^{10} - 630\rho^8 + 560\rho^6 - 210\rho^4 + 30\rho^2 - 1$	

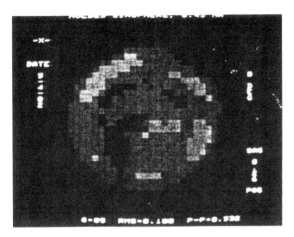

Figure 2.28 Interferometrically measured optical phase difference map of a molded aspheric bisphere. (From Ref. [25].)

tabulation of the Zernike interferometric analysis is given for a 0.4-mm diameter plano-convex lens with a radius of curvature of 1.25 mm ($f/6.25$). The coefficients for the 36 polynomial fit are listed in the left-hand column. In Table 2.8, on the right, a synopsis of the contributions is made in terms of the conventional aberration descriptions such as spherical, coma, astigmatism, etc. At the very bottom is a row which gives the overall least square deviations, listed as P–V, peak-to-valley, RMS, root-mean-square, and Strehl ratio. The RMS characterization is 0.125λ.

There are no interferometric results reported for the other microlens types, likely due to their small size. Adaptive Optics Associates claims 0.25λ deviation for lenses greater than $f/4$, although how this was obtained was not specified.

The point-spread function is a more accessible characterization method, at least in principle. There are a number of experimental techniques by which one can obtain fairly accurate 2D profiles of the intensity pattern in the focal plane of the lens. These encompass scanning slit systems, pyroelectric-based systems, and the newer CCD-based 2D arrays used in conjunction with a microscope.

Table 2.8 Zernike Analysis of SMILE Lens

Term	Coefficient	RMS (0.0001 Waves)
(*Diameter* = 0.4 mm, focal length = 1.25 mm)		
1 TILTX	0.0335	167
2 TILTY	1.6721	8360
3 FOCUS	0.2701	1559
4 AST30	0.0046	19
5 AST31	-0.0154	63
6 COM30	0.0777	275
7 COM31	-0.0785	278
8 SPH3	0.2583	1155
9 3TH50	-0.0016	6
10 3TH51	-0.0064	23
11 AST50	-0.0004	1
12 AST51	0.0040	13
13 COM50	-0.0035	10
14 COM51	-0.0029	8
15 SPH5	0.0098	37
16 4TH70	0.0048	15
17 4TH71	-0.0016	5
18 3TH70	0.0007	2
19 3TH71	-0.0003	1
20 AST70	0.0023	6
21 AST71	-0.0045	12
22 COM70	-0.0004	1
23 COM71	0.0094	24
24 SPH7	-0.0030	10

(Continued)

Table 2.8 (*Continued*)

Term	Coefficient RMS (0.0001 Waves)						
ORD	SPHE	COMA	ASTI	3THE	4THE	5THE	SUBT
RMS Error (0.0001 waves)							
3	1155	391	66				1221
5	37	13	13	23			47
7	10	24	13	2	16		33
9	15	43	26	29	9	7	61
11	13						13
SUBT	1156	394	73	37	18	7	1224
RMS Residue							112
Total RMS							1229
ORD	SPHE	COMA	ASTI	3THE	4THE	5THE	
P–V error (0.0001 waves)							
3	3875	2210	321				
5	196	91	80	132			
7	43	189	101	16	102		
9	98	388	223	230	66	45	
11	66						

Overall average values (in waves)

Peak	Valley	P–V	RMS	Strehl Ratio
0.414	−0.246	0.660	0.125	0.540

Measured PSFs for three SMILE™ lenses are shown in Fig. 2.29 [20]. They are compared to the calculated Airy curves. In all cases, one can see the wings are larger, particularly for the larger lens diameter. This clearly indicates

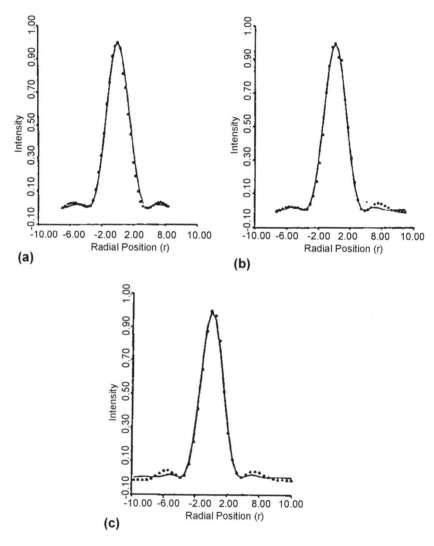

Figure 2.29 Measured point spread functions for SMILE lenses compared to best-fit Airy pattern. The lens diameter was 0.4 mm and the EFLs were 0.14, 0.19, and 0.36.

aberrations. Unfortunately, the method compared the normalized curves that did not permit a quantitative estimate of the extent of aberrated behavior. Daly et al. [14] report PSF data for the melted photoresist lenses over a range of diameters which is shown in Table 2.4. They make a comparison to the Airy curve based on the full width at half maximum. The results are not particularly consistent. In addition, it is not clear what the FWHH really means, since Majahan has shown that the normalized patterns over a wide range of aberrations have essentially the half width.

APPENDIX A. LENSES DERIVED FROM SURFACE TENSION

Three of the fabrication methods described above relied on surface tension to produce the surface figure. The methods were the melted photoresist, the contactless molding, and the photosensitive glass. In many applications (see Chapter 5), it is important for arrays of lenses to have a high fill factor, that is, to have the lenses occupy nearly 100% of the area. This requirement is a problem for the lens fabrication methods mentioned above because the lens figure is not independent of the shape of the lens. In this section, we will calculate the minimum energy surface figures for different boundary conditions which will be relevant for applying these techniques for making lenses with noncircular boundaries.

The reader is referred to Fig. A.1 for a statement of the problem. We want to determine the equation of the surface $z = f(x, y)$ bounded by the curve C, where a uniform pressure p is applied to the surface whose material has a surface tension coefficient α. This is entirely equivalent to the variational problem of finding the minimum (extremum) of the integral area:

$$A|f| = dA \sqrt{1 + \left(\frac{\delta f}{\delta x}\right)^2 + \left(\frac{\delta f}{\delta y}\right)^2} \, dx \, dy \qquad (A.1)$$

under the constraint of the constant volume. The variational problem is expressed as $\delta(A + \lambda V)$ where λ is the Lagragian

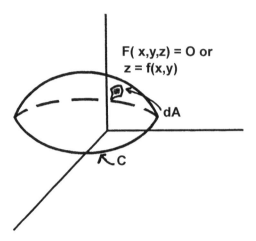

Figure A.1 Representation of the surface described by the equation $F(x, y, z) = 0$ bounded by the curve C.

mutiplier and is determined by the condition $V(\lambda) = V_0$ The Euler Lagrange solution of Eq. (A.1) is

$$\nabla \cdot \left\{ \frac{\nabla f}{\sqrt{1 + |\nabla f|^2}} \right\} - \frac{p}{\alpha} = 0 \tag{A.2}$$

For the well-known case where C is a circle, one can obtain an analytical solution to Eq. (A.2) in cylindrical coordinates assuming axial symmetry. The solution is an equation of a portion of a spherical surface, viz.

$$\frac{f(\rho)}{R_0} = \sqrt{\left(\frac{2}{p}\right)^2 - \left(\frac{\rho}{R_0}\right)^2} - \sqrt{\left(\frac{2}{p}\right)^2 - 1} \tag{A.3}$$

where R_0 is the radius of the boundary circle, and $P = R_0 p / \alpha$.

For other shapes of the boundary curves, one has to resort to computer numerical solutions as available from such software packages as PDE/Protran. We calculated the square and the hexagon since both of these would lead to high fill factors. The data are presented as isodeviation lines, where the deviation

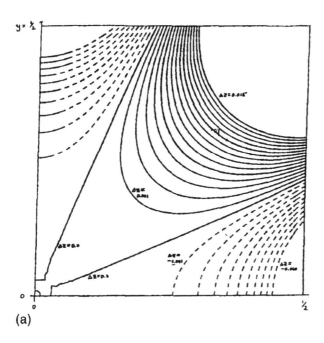

(a)

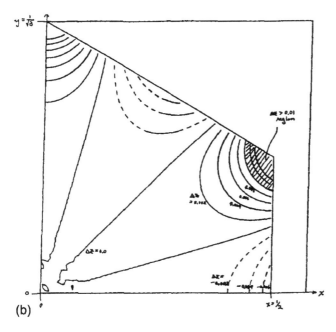

(b)

is from a perfect sphere. For the square geometry, because of the symmetry, only one quadrant is shown in Fig. A.2a. The length of the square is L and all the coordinates are normalized to L, including the deviation term Δz. One can clearly see that the distortion cuts deeply into the lens. For $L = 400\,\mu m$, a Δz of 0.001 corresponds to roughly one wave at 550 nm.

The situation is much improved for the hexagonal geometry situation where again we have shown the result for the smallest repeated section in Fig. A.2b. The distortion corresponding to $\Delta z = 0.001$ is only around the apices of the hexagon, and the major portion of the lens conforms to a sphere within one wave at 550 nm.

There is an application that relates to this concern for the nature of the curvature of microlenses. This is the application where it is desired to circularize a beam. Often the emitting regions of solid-state sources are not regular, resulting in a different beam spread in the x and y directions. If one collimates this output in the conventional way, one obtains an ellipse for the shape of the beam. It is frequently desirable to have the beam circular. One approach is to use two cylindrical lenses, as shown in Fig. A.3. By choosing the correct ratio of the focal lengths f_1/f_2, one can compensate for the different beam-spreading angles. Where space is an issue, there is a way to use a microoptic lens that has the appropriate difference in radii of curvature in the x and y directions. It turns out that the shape required is that from a section of a torus, shown in Fig. A.4. The equation for sag of the lens cut from such a section is

$$x^2 + y^2 + z^2 = a^2 + R^2 + 2R(a^2 - z^2)^{1/2} \qquad (A.4)$$

Here R is the radius of the torus and $2a$ is the circular thickness.

Figure A.2 (a) Quadrant map of isodeviation of surface height relative to a segment of a spherical surface for the case of a lens with a square cross-section of width L. All quantities x, y, Δz are normalized to L. The normalized radius of curvature is $R_c/L = 1.4$. (b) Same as in (a) except the cross-section is hexagonal with characteristic length L.

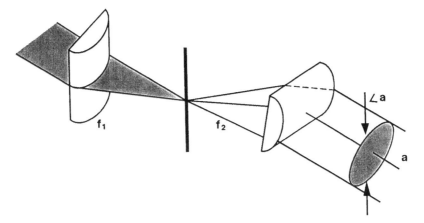

Figure A.3 Method of circularizing a beam with the use of two cylindrical lenses. The ratio of the focal lengths determines the correction.

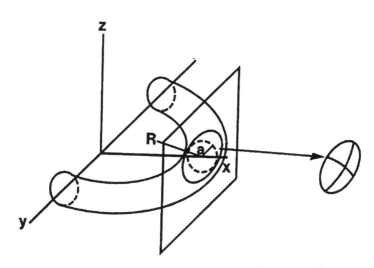

Figure A.4 Sketch of the slice of a toroid to produce an anamorphic lens with two spherical surfaces. The one radius of curvature corresponds to the radius of the toroid and the other to the radius of the thickness.

APPENDIX B. RAY-TRACE ALGORITHM

The reader is referred to Fig. B.1 for the definition of the symbols used. In addition, we define the following relationships:

$$x_1 = -\left(\frac{T}{2} + t_1\right), \quad x_4 = \frac{T}{2} + t_2$$

$$(x_2 + a_1)^2 + y^2 + z^2 = R_1^2, \quad (x_3 - a_2)^2 + y_3^2 + z_3^2 = R_2^2 \quad \text{(B.1)}$$

$$a_1 = \frac{T}{2} - R_1, \quad a_2 = \frac{T}{2} - R_2$$

where $t_1, t_2, a_1, a_2, R_1, R_2$, and T are all greater than 0. Starting at an arbitrary object point (x_1, y_1, z_1), the direction

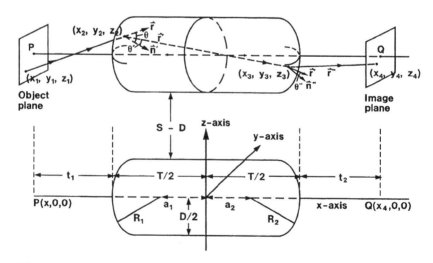

Figure B.1 Drawings showing the symbols and notation used in the ray-trace algorithm. The upper figure shows the 3D perspective showing the coordinate points at each surface consistent with the ray-trace algorithm. The lower curve indicates the dimensional variables.

cosines of the ray to a point on the lens (x_2, y, z) are given by

$$l_1 = \frac{x_2 - x_1}{B}, \quad m_1 = \frac{y - y_1}{B}$$
$$n_1 = \frac{z - z_1}{B}, \quad B = \sqrt{(x_2 - x_1)^2 + (y - y_1)^2 + (z - z_1)^2} \tag{B.2}$$

where

$$x_2 = -\sqrt{R_1^2 - y^2 - z^2} + a_1 \tag{B.2a}$$

At the point (x_2, y, z) on the lens, the direction cosines of the normal to the surface are

$$l_2 = -\frac{(a_1 + x_2)}{C}, \quad m_2 = \frac{-y}{c}, \quad n_2 = \frac{-z}{c} \tag{B.3}$$

with

$$C = \sqrt{(a_1 + x_2)^2 + y^2 + z^2}$$

The angle of incidence θ is obtained from the scalar product of the unit vector $\mathbf{r}(l_1, m_1, n_1)$ of the incident ray with the unit vector $\mathbf{n}(l_2, m_2, n_2)$ of the normal, where both $\mathbf{r}$ and $\mathbf{n}$ point toward the right, as do all other unit vectors

$$\cos\theta = l_1 l_2 + m_1 m_2 + n_1 n_2 \tag{B.4}$$

The refracted ray with unit vector $\mathbf{r}'$ must be in the plane of incidence, that is, the plane determined by $\mathbf{r}$ and $\mathbf{n}$.

Mathematically this can be expressed as

$$\mathbf{r}' = a\mathbf{r} + b\mathbf{n}$$

The value of a and b can be obtained by substracting the following equations:

$$\mathbf{r}' \cdot \mathbf{r}' = 1 = a^2 + b^2 + 2ab\cos\theta,$$
$$(\mathbf{r}' \cdot \mathbf{n})^2 = \cos^2\theta' = a^2\cos^2\theta + b^2 + 2ab\cos\theta \tag{B.5}$$

which yields

$$1 - \cos^2 \theta' = a^2(1 - \cos^2 \theta) \quad \text{or} \quad \sin^2 \theta' = a^2 \sin^2 \theta$$

where θ' is the angle that the refracted ray makes with the normal. Applying Snell's law, one obtains

$$a = \frac{1}{n} \tag{B.6a}$$

where n is the refractive index of the glass. To obtain b one substitutes the value of a into the above equation and solves for b:

$$b = \frac{1}{n}\left\{\sqrt{\cos^2 \theta - 1 + n^2} - \cos \theta\right\} \tag{B.6b}$$

The direction cosines of the refracted ray can now be obtained using the following equation:

$$l_3 = al_1 + bl_2, \quad m_3 = am_1 + bm_2, \quad n_3 = an_1 + bn_2 \tag{B.7}$$

The equation of the refracted ray is given by

$$y_3 - y = \frac{m_3}{l_3(x_3 - x_2)}, \quad z_3 - z = \frac{n_3}{l_3(x_3 - x_2)} \tag{B.8}$$

where the point (x_3, y_3, z_3) is located on the back spherical surface whose equation is

$$(x_3 - a_2)^2 + y_3^2 + z_3^2 = R_2^2 \tag{B.9}$$

Solving these two equations simultaneously yields the value of x_3

$$x_3 = D \pm \sqrt{D^2 - E} \tag{B.10}$$

where

$$D = (m_3^2 + n_3^2)x_2 + a_2 l_3^2 - m_3 y l_3 - n_3 z l_3$$
$$E = (m_3^2 + n_3^2)x_2^2 - 2x_2 l_3(m_3 y + n_3 z) + l_3^2(y^2 + z^2 - R_2^2 + a_2^2)$$

Substituting the value of x_3 (the positive root) into Eq.(B.8) yields y_3 and z_3.

The direction cosines of the unit vector $\mathbf{r}''$ normal to the surface at (x_3, y_3, z_3) are now determined to be

$$l_4 = \frac{x_3 - a_2}{F}, \quad m_4 = \frac{y_3}{F}, \quad n_4 = \frac{z_3}{F} \tag{B.11}$$

where

$$F = \sqrt{(x_3 - a_2)^2 + y_3^2 + x_3^2}$$

The incident angle on the back surface can now be obtained from

$$\cos \theta'' = l_3 l_4 + m_3 m_4 + n_3 n_4 \tag{B.12}$$

The exit ray's direction cosines are determined as before by writing

$$\mathbf{r}'' = e\mathbf{r}' + f\mathbf{n}'' \tag{B.13}$$

and following the same procedure as done on the front surface, we find that

$$e = n \quad \text{and} \quad f = -n \cos \theta'' + \sqrt{n^2 \cos^2 \theta'' - n^2 + 1}$$

The direction cosines of the exit ray are thus

$$l_5 = el_3 + fl_4, \quad m_5 = em_3 + fm_4, \quad n_5 = en_3 + fn_4 \tag{B.14}$$

If the image plane is located at $x_4 = T/2 + t_2$, the exit ray forms an image spot with coordinates (y_4, z_4) given by

$$y_4 = y_3 + \frac{(x_4 - x_3)m_5}{l_5}, \quad z_4 = z_3 + \frac{(x_4 - x_3)n_5}{l_5} \tag{B.15}$$

REFERENCES

1. Born, M.; Wolfe, E. In *Principles of Optics*; Pergamon Press: London: **1964**.

2. Ditchburn, R.W. In *Light*, Blackie and Son Ltd; Glasgow, Scotland: **1963**.

3. Willianms, C.S.; Becklund, O.A. In *Optics*, Wiley Interscience; New York: **1972**.

4. *Handbook of Plastic Optics*, 2nd Ed.; U.S. Precision Lens Inc: Cincinnati, OH, **1983**.

5. Pantelis, P.; McCartney, D.J. Third International Conference on Electrical, Optical, and Acoustic Properties of Polymers, Conference Proceeding, Plastics and Rubber Institute, London, Sept 16–18, 1992.

6. D'Amato, D.; Berletta, S.; Cone, P.; Hizny, J.; Martinsen, R.; Schumtz, L. Fabrication and testing of monolithic lenslet module (MLM) arrays. Opt. Fab./Testing. Tech. Digest June 12–14, 1990, p. 11.

7. *Data Sheet from United Technologies*; Adaptive Optics Associates: Cambridge, MA, 1986.

8. Hannah, P.; Rhorer, R. *Basics of Diamond Turning, tutorial*; Am. Soc. Precision Eng.: Atlanta, GA, October 24, 1988.

9. Fitch, M.A. Molded optics. In *Photonics Spectra*; October 1991.

10. Olszewski, A.R. assigned to Corning Inc. US Patent 4,362,819, 1982.

11. Dannoux, T.; Calderini, P.; Pujol, G.; Themont, J.-P. Procede et dispsitif de fabrication de reseaux de microlentilles optiques, French Patent 94-08420, July **1974**.

12. Sze, S.M. *VLSI Technology*, Chap. 4; McGraw-Hill: NewYork, 1988.

13. Popovic, C.D.; Sprague, R.A.; Neville Connell, G.A. Techniques for monolithic fabrication of microlens arrays. Appl. Opt. **1988**, *27*, 1281–1284.

14. Daly, D.; Stevens, R.F.; Hutley, M.C. *Manufacture of Microlenses by Melting Photoresist*; IOP Short Meeting Series 30; Institute of Physics: Teddington, May 1, 1991.

15. Mersereau, K.O.; Nijander, C.R.; Feldblum, A.Y.; Townsend, W.P. Fabrication and measurement of fused silica microlens arrays. SPIE Int. Sym. Opt. Appl. Sci. Eng., July 19–24, 1992.

16. MacFarlane, D.L.; Narayan, V.; Cox, W.R.; Chen, T.; Hayes, D.J. Microjet fabrication of microlens array. IEEE Photonics Techn. Lett. **1994**, *6*, 1112–1114.

17. Wallace, D.B.; Hayes, D.J.; Frederickson, C.J. In *High Throughput Array for Fluid Microdispensing in DNA Diagnostics, Product Information*; MicroFab Technologies, Inc.: Plano, TX, 1994.

18. Borrelli, N.F.; Morse, D.L.; Bellman, R.H.; Morgan, W.L. Photolytic technique for producing microlenses in photosensitive glass. J. Appl. Phys. **1985**, *42*, 2520–2525.

19. Borrelli, N.F.; Morse, D.L. Microlens arrays produced by a photilytic technique. Appl. Opt. **1988**, *27*, 476–479.

20. Borrelli, N.F. *Generation of Lens Arrays using Photothermal Techniques*; IOP Short Meeting, Series 10; Institute of Physics: Teddington, UK, May 1, 1991.

21. Borrelli, N.F.; Bellman, R.H.; Durbin, J.A.; Lama, W. Imaging and radiometeric properties of microlens arrays. Appl. Opt. **1991**, *30*, 3633–3642.

22a. Veiko, V.P.; Yakovlev, E.B.; Frolov, V.V.; Chujko, V.A.; Kromin, A.K.; Abbakamov, M.O.; Shakola, A.T.; Fomichov, P.A. Laser heating and evaporation of glass and glass-boring materials and its application for creating. MOC, SPIE **1991**, *1544*, 152–163.

22b. Veiko, V.P. Essence and comparison of different laser technologies for microoptics fabrication. In Optical Applied Science/Eng. Miniature/Microoptics; SPIE Int. Symp.; San Diego, CA, July 19–24, **1992**.

22c. Fritze, M.; Stern, M.B.; Wyatt, P.W. Laser-fabricated glass microlens arrays. Opt. Lett. **1998**, *25* (2), 141.

22d. Smuk, A.; Lawandy, N.M. Direct laser fabrication of dense microlens arrays in semiconductor-doped glass. J. Appl. Phys. **2000**, *90* (8), 4026.

23. Wakaki, M.; Komachi, Y.; Kanai, G. Microlens and microlens array formed on a glass plate using a CO_2 laser. Appl. Opt. **1998**, *37*, 627–631.

24. Sauvain, E.; Kyung, J.H.; Lawandy, N.M. Multiphoton micrometer-scale photetching in silicate-based glasses. Opt. Lett. **1995**, *20*, 243–245.

25. Mahajan, V.N. Aberrated point-spread functions for rotationally symmetric aberrations. Appl. Opt. **1983**, *22* (19), 3035–3041.

26. Maschmeyer, R.O.; Hujar, R.M.; Carpenter, L.L.; Nicholson, B.W.; Vozenilek, E.F. Optical performance of a diffraction-limited molded-glass biaspheric lens. Appl. Opt. **1983**, *22* (16), 2413–2145.

3

Gradient Index

3.1. INTRODUCTION

Although the binary-optic method of making microlenses may be the source of the current excitement in the micro-optics field, as we will see in Chapter 4, it was the gradient index method pioneered by Nippon Sheet Glass that paved the way, and in many ways still dominates the field. We will see in succeeding sections of this chapter how the ingenious use of an ion-exchange process in glass provided the first practical way to make microlenses. This method of making an imaging device is generically called GRIN, for *gradient index*, so you will see people use the description *GRIN lens* to mean any lens formed through refractive index profile.

The idea that an index profile could produce an image similar to that produced by a curved refracting surface goes back to Wood [1]. Intuitively, it would seem that one should mimic the phase front that occurs through a spherical lens, as a starting point. In other words, if the optical phase difference $\Delta(nL)$ is equal to $n \, \Delta L(r)$ for a refracting lens, then reproducing $L \, \Delta n(r)$ might produce the same effect. We shall see

69

that this turns out to be essentially true, although we will derive it from a more rigorous basis.

Marchand [2] presents an excellent and complete exposition of the phenomenon related to gradient index optics. We utilize his approach in our brief optics review given below. Iga et al. [3] is another valuable and extensive reference. The scope is more restricted, but it also contains applications of devices other than lenses.

What we are concerned with is the demonstration that there is an index profile $n(x, y, z)$ that images light to a given point in the manner similar to that of a refractive spherical lens. Moreover, we want to show what sort of limitations or approximations are necessarily imposed to achieve this end. We begin with the equation that governs the direction that a ray of light will take through a medium with an arbitrary index profile $n(x, y, z)$. We then look at the possible solutions and condition that lead to imaging under the assumption of a purely radial refractive index profile. We outline a simple ray-trace procedure that is useful in the paraxial approximation. The planar version of this process will be discussed separately because it presents a more difficult problem where the index is both a function of the radial distance and the axial distance. From there we proceed to the description of actual methods of fabrication, and finally to an analysis of the GRIN imaging performance.

3.2. OPTICAL THEORY

3.2.1. Path in a Rod with Gradient Index

The path a beam of light will travel in a medium with a refractive index distribution given by $n(x, y, z)$ can be treated in one of two ways. One way is to solve the wave optics problem as contained in the scalar form of Maxwell's equations with a spatially varying refractive index [4]. A second approach is to treat it from the standpoint of geometric or ray optics using the "least action" principle [2]. The former is more useful for the general case of an arbitrary profile, but for the kind and dimension of the devices that we will be discussing, it is more instructive to take the ray approach.

The least action principle dictates that light will travel a path in such a way as to minimize the time of flight. The integral representation of this statement, Fermat's principle, is that the time of flight (distance divided by the velocity)

$$\delta t = \int \frac{ds}{v} \qquad (3.1)$$

should be a minimum. One can put this in the classic differential arc element, $ds = \sqrt{dx^2 + dy^2 + dz^2}$, replacing the velocity with the variational calculus form by substituting for the expression $v = c/n(x, y, z)$. The variational problem can now be stated: for what function $n(x, y, z)$ is this integral a minimum?

At this point we anticipate the practical situation to come, which deals with rod lenses, and use cylindrical coordinates. Further, in the same vein we initially assume that n is a function of r alone consistent with their ion-exchange fabrication. The differential arc element in cylindrical coordinates is $ds = \sqrt{dr^2 + r^2\, d\theta^2 + dz^2}$, so Eq. (3.1) can be written as

$$\delta t = L\left(r, \theta, z, \frac{dr}{dz}, \frac{d\theta}{dz}\right) dz$$

$$= \left[\frac{n(r, \theta)}{c}\right] \sqrt{1 + (dr/dz)^2 + r^2(d\theta/dz)^2} = dz \qquad (3.2)$$

The solution proceeds via the use of the Euler–Lagrange differential equations of variational calculus*, which are formally expressed as

$$\frac{\partial L}{\partial r} = \frac{d}{dz}\left(\frac{\partial L}{\partial r'}\right) \qquad (3.3a)$$

$$\frac{\partial L}{\partial \theta} = \frac{d}{dz}\left(\frac{\partial L}{\partial \theta'}\right) \qquad (3.3b)$$

---

*For the reader unfamiliar with the calculus of variations, it is sufficient to know that the solution of the Euler–Lagrange differential equations yield the path, or function, which corresponds to the minimum condition expressed in the least action integral. For example, see Ref. [5].

where L is the integrand of Eq. (3.2) and $r' = dr/dz$ and $\theta' = d\theta/dz$. For the sake of simplicity in our development, we suppress the θ dependence. The reader is referred to Marchand [2] for the full derivation. As a consequence of our 2D approach, we will be looking at what would constitute meridional rays in a cylindrical coordinate system (see Fig. 3.1). These are rays that contain the z-axis. We will point out the consequence of this as we discuss skew rays.

Applying Eq. (3.3a) to L as defined by Eq. (3.2) yields

$$\sqrt{1 + r'^2}\left(\frac{dn}{dr}\right) = \left(\frac{d}{dz}\right)\left\{\frac{nr'}{\sqrt{1 + r'^2}}\right\} \tag{3.4}$$

where we use r' to signify dr/dz. Multiplying both sides by n, Eq. (3.4) can be written in the form

$$\frac{d\{n^2\}}{dz} = \frac{nr'}{\sqrt{1 + r'^2}}\frac{d}{dz}\left(\frac{nr'}{\sqrt{1 + r'^2}}\right) = \frac{d}{dz}\left\{\frac{n^2 r'^2}{1 + r'^2}\right\} \tag{3.5}$$

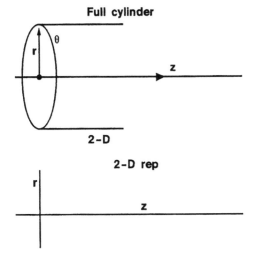

Figure 3.1 Full cylindrical coordinate system used for the complete development of ray paths; (*bottom*) reduced 2D coordinate system used in the case of assumed axial symmetry.

where we used the relation $dn/dr = (l/r')dn/dz$. We can now equate the terms in the curly brackets to obtain

$$\frac{(nr')^2}{l+r'^2} = n^2 + l^2 \tag{3.6}$$

or upon rearranging and taking the square root,

$$\frac{n}{\sqrt{l+r'^2}} = +1 \tag{3.7}$$

where l is a constant independent of z. Solving for r' in Eq. (3.7) and then integrating with respect to z yields the equation of the ray,

$$z - z(0) = \pm \int k \frac{dr}{\sqrt{n^2 - l^2}} \tag{3.8}$$

This equation defines the ray $r(z)$ once we put in a specific form for $n(r)$. We have taken the long way around to obtain the ray Eq. (3.8) to show the more general approach to the solution of the problem. There is, however, a much quicker route to Eq. (3.8) in view of the restrictions we put on the problem; viz., n is only a function of r with no θ dependence. One could have expressed the Euler–Lagrange equations in parametric form. For example, Eq. (3.4) could have been written as

$$\frac{\partial n}{\partial r} = \frac{d}{ds}\left\{\frac{n\,dr}{ds}\right\} \tag{3.9a}$$

using the identity $d/ds = (l+r'^2)^{-1/2}\,d/dz$. Similarly, the other two equations for θ and z would have been

$$\frac{\partial n}{\partial \theta} = \frac{d}{ds}\left\{\frac{n\,d\theta}{ds}\right\} \tag{3.9b}$$

and

$$\frac{\partial n}{\partial z} = \frac{d}{ds}\left\{\frac{n\,dz}{ds}\right\} \tag{3.9c}$$

Since n is not a function of z, then one could have integrated (3.9c) directly:

$$\frac{n\,\mathrm{d}z}{\mathrm{d}s} = \text{constant} \qquad\qquad (3.10)$$

This is nothing more than the statement that the z direction cosine is invariant. This equation is exactly that expressed above in (3.7), so we can immediately identify the constant as the z direction cosine which because of its invariance is identically equal to its value at $z = 0$, l_0.

3.2.2. Imaging Forms of the Radial Profile

For a specific $n(r)$, one can use Eq. (3.8) to determine the ray path. Our concern is with those $n(r)$ that lead to a formation of an image. What this means is that from a fan of rays from a given point z_0, they will pass through a given point at some other z. Consider the radial profile given by the expression [2,3,6],

$$n^2 = N^2(1 - Ar^2) \qquad\qquad (3.11)$$

Substituting this into (3.8) and integrating yields the solution

$$(z - z_0)\left(\frac{z}{l_0}\right) = \left\{ \begin{array}{c} \sin^{-1}\left(\frac{r}{r_0}\right) \\ \cos^{-1}\left(\frac{r}{r_0}\right) \end{array} \right\} \qquad\qquad (3.12)$$

or, putting the boundary conditions at $z_0 = 0$, one obtains

$$r = r(0)\cos\left(\frac{N\{A\}^{1/2}z}{l_0}\right) + \left\{\frac{(\mathrm{d}r/\mathrm{d}z)_0}{N\sqrt{A}}\right\}\sin\left(\frac{N\{A\}^{1/2}z}{l_0}\right)$$

$$(3.13)$$

Clearly, this represents ray paths that lead to an imaging condition for meridional rays as long as l_0 is constant for all rays involved. Consider the situation depicted in

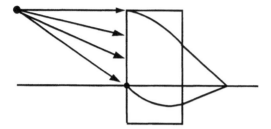

Figure 3.2 Ray trace through a body with parabolic index profile [see Eq. (3.11)] showing the imaging property.

Fig. 3.2. In this case one can easily see that all rays emanating from point P will intersect at the point z defined by

$$\tan z_{int} = \frac{r_i - r_j}{m_j - m_i} \tag{3.14}$$

We saw from above that l_0 is a constant for each ray. We are interested in the question, what would make it a constant for all rays, irrespective of initial position and slope. Considering the definition,

$$l_0 = \frac{n(r)}{\sqrt{l + r'^2}} \tag{3.15}$$

it is clear that it would remain relatively constant if $(dr/dz)^2$ is small compared to unity, and the change in n itself with r is small over the region of interest. This is tantamount to a paraxial approximation.

There is a radial profile that images meridional rays without regard to any approximation [7]. The form of the radial profile is

$$n(r) = N_0 \sec h(ar) \tag{3.16}$$

If one substitutes this into Eq. (3.8), one obtains the expression

$$\sinh(ar) = A \sin\left\{\sin^{-1}\left[\frac{\sinh(ar_0)}{A + az}\right]\right\} \tag{3.17a}$$

or,

$$\sinh (ar) = A \, \cos\left\{ \cos^{-1}\left[\frac{\sinh (ar_0)}{A + az} \right] \right\} \qquad (3.17b)$$

where $A = \sqrt{(N_0/l_0)^2 - 1}$. Equations (3.17) represent periodic functions of z and correspond to a sharply imaged condition for meridional rays which is independent of the ray or the starting point chosen [2].

In a real sense, (3.17) constitutes the ideal profile for a gradient index lens with respect to meridional rays. By ideal is meant that all rays are sharply focused irrespective of the initial ray position or slope. It is like an aspheric surface is to a refractive lens. However, if we consider the possibility of skew rays as well, then we find that there are problems. The consideration of skew rays involves maintaining the θ dependence as indicated by Eq. (3.3b) throughout the analysis. Because of the lengthy derivation, we will be content to reproduce the profile that images helical rays derived by Rawson et al. [8] of the form

$$\frac{n(r)}{N_0} = [1 + (Ar)^2]^{-1/2} \qquad (3.18)$$

The importance of the skew rays depends upon the particular lens application. We see that the best approximation with respect to all kinds of rays is the following. Consider the expansions of all three of the index profiles considered viz., Eqs. (3.11), (3.16), (3.18). A comparison of the sech to the parabolic profile is shown in Fig. 3.3. They are all expressible in the form where the lead terms are identical:

$$n(r) = N_0\left[1 - \left(\frac{A}{2}\right)r^2\right] \qquad (3.19)$$

We can see, as pointed out above, that this profile will only be good to the extent of the paraxial approximation. Inclusion of higher order terms leads to what would be equivalent to aberrations, as we will see later. The ray

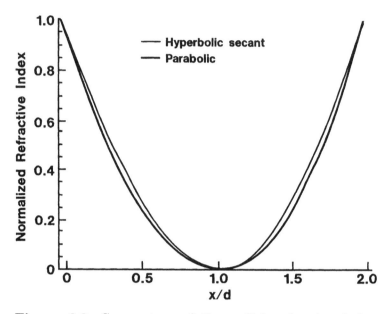

Figure 3.3 Comparison of the radial refractive index profile derived from the exact imaging sech (Ar) to that of the paraxially imaging parabolic profile, $n = n_0[1 - (A/2)r^2]$.

solutions for this profile are the same as that obtained above; see Eq. (3.13).

3.2.3. Matrix Ray Trace for Rod Lenses

Kapron [9] first suggested a matrix-based ray-trace formalism for the paraxial case. This was subsequently augmented by NSG [10] and used in their product a description/ performance publications. This has proven to be very convenient and useful for lens design. Others have published more general ray-trace methods [11], but in most cases of interest the paraxial approximation is sufficient.

The method follows the same matrix format as for the thick-lens ray tracing that was done for refractive elements in the paraxial approximation in Chapter 2 (see Sec. 2.1.3). With the use of Eq. (3.13), one can write matrix expression relating the initial ray position and slope for a lens of thickness D.

$$\begin{pmatrix} r_1 \\ m_2 \end{pmatrix} = \begin{pmatrix} \cos(\sqrt{A}D) & (1/N_0\sqrt{A})\sin(\sqrt{A}D) \\ -N_0\sqrt{A}\sin(\sqrt{A}D) & \cos(\sqrt{A}D) \end{pmatrix}$$

$$\times \begin{pmatrix} r_0 \\ m_0 \end{pmatrix} \tag{3.20}$$

From this one can write down the relationships for the focal length, working distance, and location of the principal planes. These results are shown in Fig. 3.4. The computed ray diagram for a simple case of parallel light being focused is shown in Fig. 3.5.

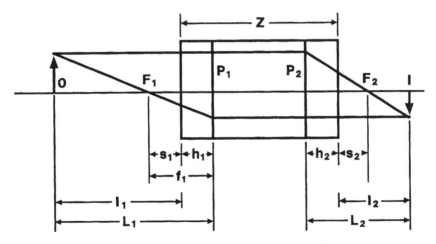

Figure 3.4 Drawing relating the commonly used parameters to describe the imaging of a GRIN lens of thickness Z. The parameters are formally defined in the same manner as shown in Fig. 2.1 and listed in Table 2.1 for conventional refracting lens. However, the actual relationships are defined through the use of Eq. (3.20). Unfortunately, the conventions for the symbols are not all the same. In particular, the symbol most commonly used for the object and images distances in the GRIN notation is l rather than s.

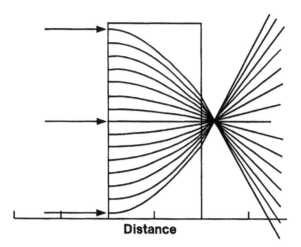

Distance

Figure 3.5 Computed ray trace through a GRIN lens with a parabolic refractive index profile. Approximate path obtained from Eq. (3.20) with $m_0 = 0$. Positions on the exit face are $r_{in} \cos(\sqrt{A}D)$ with slopes of $-N_0\sqrt{A}\sin(\sqrt{A}D)$; thus the equation of exit ray is $r = r_{ex} - m_{ex}z$ or $\ldots$, $r = r_0 \cos(D\sqrt{A}) - zN_0r_0\sqrt{A}\sin(D\sqrt{A})$.

The simple lens maker's formula can be used in the same way as it was for the conventional thick lens

$$\frac{l}{L_1} + \frac{l}{L_2} = \frac{l}{f} \tag{3.21}$$

where the L's are measured from the principal planes. Another interesting aspect of a GRIN lens with a parabolic profile is that its numerical aperture is a function of the radial position as shown in Fig. 3.6:

$$NA = \sqrt{n(r)^2 - n_0^2} \tag{3.22}$$

These expressions will be useful when we deal with the applications.

3.3. PLANAR LENS

The analysis of the imaging behavior of the so-called planar version of GRIN lenses [3,12,13] is somewhat more difficult

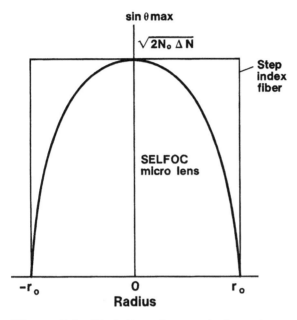

Figure 3.6 Variation of numerical aperture as a function of radial position of a GRIN lens with a parabolic profile.

to mathematically describe than that of the rod lens geometry discussed above. The way the index profile is produced is quite different, and consequently some of the simplifying assumptions made in the treatment above are not valid. In the next section we go into the fabrication methods in somewhat more detail, but for the present it is instructive to compare how the index profile is derived with a simple diagram. In Fig. 3.7a we show how the rod lens is made. In the ion-exchange method, for example, the refractive index is altered uniformly in the radial direction and constant along the axial direction. One would thus expect the refractive index profile to look as sketched in Fig. 3.7b. It is with this situation in mind, since it is the preferred way to make such lenses, that the above derivation of the ray trace was done. In the planar version of GRIN, the index is altered by ion exchange through a small opening in a mask on the surface, as shown in Fig. 3.8a. The expected index profile is shown

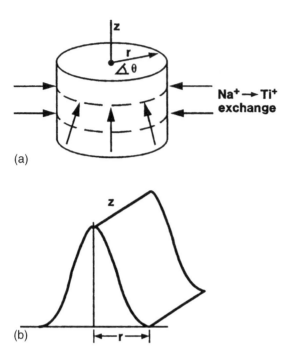

(a)

(b)

Figure 3.7 Schematic representation of the molten salt bath ion-exchange process into a glass rod to produce a radial refractive index gradient. The lower figure is a sketch of the index profile in terms of the radial and axial position.

in Fig. 3.8b. Clearly, the index is a function of both r and z. From the shape of the profile, Iga et al. [3] proposed the function

$$n(r,z)^2 = N_0^2(R(r) + Z(z)) = N_0^2\{1 - Ar^2 - \nu_n Az^n\} \quad (3.23)$$

One can easily see that this would constitute an approximation to the profile shape of Fig. 3.8b.

3.3.1. Planar Ray Trace

In this section we present a shortened approach to the development of the equation of the ray for the planar case. Iga et al. [3] and Sakamoto [13] have given complete derivations and the reader is referred to these publications for the

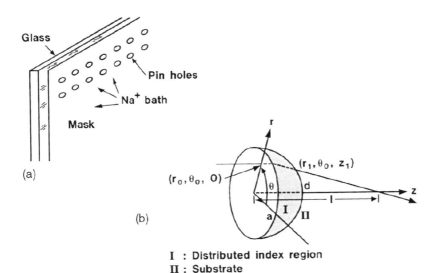

I : Distributed index region
II : Substrate

Figure 3.8 Schematic representation of planar GRIN process where molten salt bath ion exchange is done through a mask in contact with the surface, as shown. The lower figure shows the coordinate system by which the refractive index profile is characterized. One notes that the major difference produced by the planar process is that the profile is also a function of z, as well as r and θ.

details. We will try to follow the same general form used above for the rod case. One starts with Euler–Lagrange equations in parametric form; see (3.9a–c), with a change of independent parameter from the differential are length ds to it normalized to the index $n(r, z)$

$$d\tau = \frac{ds}{n} \tag{3.24}$$

Equations (3.9) now can be written as the following:

$$\frac{d^2 r}{d\tau^2} - r\left(\frac{d\theta}{d\tau}\right)^2 = \left(\frac{d}{dr}\right)\left(\frac{n^2}{2}\right) \tag{3.25a}$$

$$\left(\frac{d}{d\tau}\right)\left(\frac{r^2\, d\theta}{d\tau}\right) = \left(\frac{d}{d\theta}\right)\left(\frac{n^2}{2}\right) \tag{3.25b}$$

$$\frac{d^2z}{d\tau^2} = \left(\frac{d}{dz}\right)\left(\frac{n^2}{2}\right) \tag{3.25c}$$

To simplify matters, we again assume axial symmetry, and remove the θ dependence and return to considering meridional rays only. Using the form for $n(r, z)$ given in Eq. (3.23), one has

$$\frac{d^2r}{d\tau^2} = \left(\frac{N_0^2}{2}\right)\left(\frac{dR}{dr}\right) \tag{3.26}$$

and

$$\frac{d^2z}{d\tau^2} = \left(\frac{N_0^2}{2}\right)\left(\frac{dZ}{dz}\right) \tag{3.27}$$

In Eq. (3.26), multiply both sides by $dr/d\tau$, followed by integrating with respect to τ, and finally taking the square root, to obtain

$$\frac{dr}{d\tau} = \pm N_0\sqrt{R(r) + C_1} \tag{3.28}$$

where C_1 is a constant of integration which can be evaluated in terms of the incident r direction cosine of the ray at $z = 0$. In a similar manner, Eq. (3.27) becomes

$$\frac{dz}{d\tau} = \pm N_0\sqrt{Z(z) + C_2} \tag{3.29}$$

with C_2 defined in a similar way to C_1, viz., the z direction cosine at $z = 0$. Dividing (3.29) by (3.28) and performing the appropriate algebraic manipulations yields the expression

$$dz = \pm\sqrt{Z(z) + l_0^2R(r_0) + (l_0^2 - 1)Z(0)}$$
$$dr = \pm\sqrt{R(r) + l_0^2R(r_0) + (1 - l_0^2)Z(0)} \tag{3.30}$$

where l_0 is the incident z direction cosine, and the initial condition is given by

$$\left(\frac{dz}{d\tau}\right)_{z=0} = n(r_0,0)l_0 N_0 \sqrt{R(r_0) + Z(0)} \qquad (3.31)$$

One can then cast (3.30) as two integrals, whose equality defines the desired relation between r and z when particular forms are assumed for $R(r)$ and $Z(z)$.

$$\int \frac{dr}{\sqrt{R(r) - l_0^2 R(r_0) + (1 - l_0^2)Z(0)}}$$
$$= \pm \int \frac{dz}{\sqrt{Z(z) - l_0^2 R(r_0) + (l_0^2 - 1)Z(0)}} \qquad (3.32a)$$

We will show solutions for special cases in the next section.

3.3.2. Imaging Condition

Unlike the rod lens situation, the ideal profile is not known. What one can do is to look at the imaging from the solutions of (3.31), which is not at all an obvious procedure. A simplification can be made in order to investigate the imaging behavior. This is to consider only rays which are parallel to the z-axis. For this case, $l_0 = 1$. Further, one can assume the profile suggested by Iga expressed in (3.22), $R(r) = (1 - Ar^2)$ and $Z(z) = -\nu_n A z^n$. For the parabolic case, $n = 2$, a closed form solution (3.31) was obtained by Iga [3] and Sakamoto [13] of the form

$$\frac{r}{r_0} = \cos \frac{1}{\sqrt{\nu_2}} \sin^{-1} \frac{\sqrt{(n_2 A)}_z}{(\sqrt{1 - AR_0^2})} \qquad (3.32b)$$

(A closed-form solution was also obtained for the $n = 1$ case.) Suffice to say at this time that focusing does occur as a consequence [Eq. (3.32b)] over a reasonable portion of the lens. In the limit of the argument of the $\sin^{-1}$ term being small, one can replace it with the argument itself. Also, the cosine could then be written as $\sqrt{1 - x^2}$, could reduce to the simple form:

$$\frac{r}{r_0} = \sqrt{\frac{1 - Az^2}{1 - AR_0^2}} \tag{3.33}$$

3.4. METHODS OF FABRICATION

The fabrication of GRIN lenses has been dominated by the ion-exchange method. In this case a solid material, glass or polymer, is immersed in a bath containing an ion that will diffuse into the body, exchanging for a corresponding ion that diffuses out of the glass into the bath [14]. The net effect of the exchange of ions is to modify the refractive index. We will try to cover the details of this process. See Fig. 3.9.

An older method was the idea of building up a series of tubes of glass inside of tubes each with a prescribed different refractice index. The nest of tubes was then heated to fuse them. See Fig. 3.10. A higher temperature variation on this method was to use rings of powdered glasses [15].

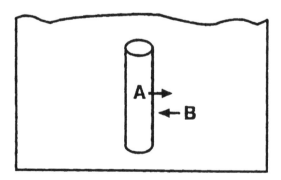

Figure 3.9 Generic representation of the ion exchange into a rod to produce a radial refractive index gradient. The bath can be a molten salt or solution. The rod will contain the counter ion. The exchange occurs through normal diffusion processes.

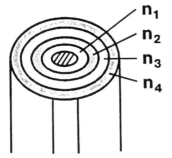

Figure 3.10 Nested rod and tube structure intended to produce a radial refractive index profile. The tubes each are of a slightly different refractive index. The idea is to heat the nested tubes until they fuse.

Finally, there are a number of photosensitive processes where the action of light on a special glass [16,17], or polymer [18], produces a chemical or structural change that leads to a refractive index change. By exposing with the appropriate intensity distribution one can produce a desired $n(r)$.

We proceed through each one of these methods in turn.

3.4.1. Ion Exchange

Ion exchange is a well-known phenomenon used in glass processing going back many years. For example, ion exchange provides strengthening of glass [19]. The idea is to utilize the diffusion of ions from a bath rich in component A into a glass poor in that component. The process is treated as a classic diffusion problem, where we have chosen the equation in cylindrical coordinates:

$$\left(\frac{1}{r}\right)\left\{\frac{d}{dr}\right\}\left\{rD_a\left(\frac{dA}{dr}\right)\right\} = \frac{dA}{dt} \tag{3.34}$$

with the following boundary conditions:

$$A = 0 \qquad 0 < r < r_0, \ t = 0$$
$$\frac{dA}{dr} = 0 \qquad r = 0, \ t > 0$$
$$A = A_0 \qquad r = r_0, \ t > 0$$

For the case where D is a constant, the solution for the concentration of A as a function of the radius and time is [20]

$$\frac{A}{A_0} - 1 = -2\sum\{\exp(-\beta_n^2 T)\}\frac{J_0(\beta_n r/r_0)}{\beta_n Jl(\beta_n)} \tag{3.35}$$

where β_n are the zeros of the zero-order Bessel function J_0 and $T = Dt/r_0^2$. The solution is shown in graphical form in Fig. 3.11.

There, of course, will be another equation involving species B, with diffusion coefficient D_b, which diffuses out of the glass. The process will be controlled by the slower of the two reactions and our diffusion constant will implicitly mean D_{ab}. The distinction being made is between the interdiffusion

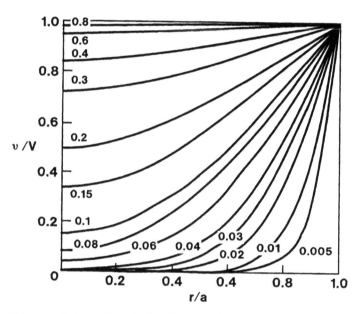

Figure 3.11 Classical solution of the diffusion equation in cylindrical coordinates, for the normalized concentration of the ion diffusing into the rod. Zero represents the center of the rod. The parameter is Dt/r_0^2, where D is the diffusion coefficient, r_0 is the radius of the rod, and t is the time.

process governed by $D(A, B)$ and the self-diffusion processes stipulated by $D(A)$ and $D(B)$. The diffusion constants are themselves concentration dependent over some range of concentration because of mutual interactions. For a good discussion of this see Ref. [21]. There is a straightforward way to measure the concentration dependence of the diffusion coefficient, usually referred to as the Boltzman method [22]. It is based on the measurement of the diffusion profile in a semi-infinite slab after some time t. From this single measurement, one can estimate $D(c)$. One can reduce the partial differential equation

$$\left(\frac{d}{dx}\right)\left(\frac{D\,dc}{dx}\right) = \frac{dc}{dt} \tag{3.36}$$

to an ordinary differential equation by the change of variable, $y = x/\sqrt{t}$. One then obtains the equation

$$\left(\frac{d}{dy}\right)\left(\frac{D\,dc}{dy}\right) = \left(\frac{-y}{2}\right)\left(\frac{dc}{dy}\right) \tag{3.37}$$

One can integrate this expression with respect to y to obtain an explicit expression for $D(c)$, after changing back to the original variables.

$$D(c) = \left(\frac{1}{2t}\right)\left(\frac{dx}{dc}\right)x\,dc \tag{3.38}$$

Referring to Fig. 3.12, one can see how the method is used. The profile $c(x)$ is measured at some time t. The integral represented by the shaded area is numerically determined. Then for each value of c, as shown for an arbitrary point on the figure, the derivative is estimated (dashed line) and the value of $D(c_1)$ is obtained from (3.38). One then proceeds to do this for any value of c to generate the full $D(c)$ relationship. We will come back to this relationship in the next section to show how it might be used to an advantage in the design of a refractive index profile.

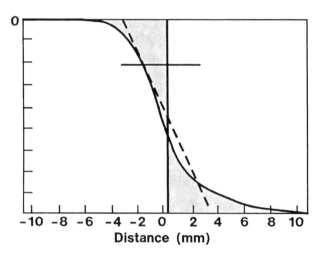

Figure 3.12 Method used to estimate the concentration dependence of the diffusion coefficient. The measured concentration of a given species is plotted vs. distance. See text and Eq. (3.38) for explanation of method.

3.4.1.1. Relation of Refractive Index to Concentration

Explicitly, one assumes that the refractive index is proportional to the concentration of species A, or more precisely, to the difference between A and B. There are equations that relate the refractive index to the concentration of various constituents with some success. One is the Lorentz–Lorenz equation which relates the refractive index to the atomic polarizability α of the constituent ions that make up the glass structure:

$$\frac{n^2 - 1}{n^2 + 2} = \left(\frac{4\pi}{3}\right) \sum N_i \alpha_i \tag{3.39a}$$

or, an often-used simpler form,

$$n^2 - 1 = \left(\frac{4\pi}{3}\right) \sum N_i \alpha_i \tag{3.39b}$$

The polarizabilities are a numerical measure of the ease of which an external electric field can influence the electron distribution about the ion. For oxide glasses, certain ions are known to dominate the sum written in Eqs. (3.39). We have listed the polarizabilities of some common ions in Table 3.1. The additional problem encountered is the ease with which these ions diffuse, which is related to how the ions are bonded in the structure. Presently, most have followed the simple rule that the singly charged ions are the most ionic and thus are the most mobile. For applications where a large index change is desired, the ions of choice have been Ag^{+1} or Tl^{+1} for alkali.

For the sake of simplicity in the single ion-exchange process, it is convenient to express the refractive index as [21]

$$n^2 = a_1 c + a_2 \qquad\qquad (3.40)$$

where c is the concentration of the exchanged constituent, and a_1 and a_2 are empirical constants.

Table 3.1 Polarizabilities of Various Ions (Cubic Angstroms)

Ion	α
Li^{+1}	0.03
Na^{+1}	0.41
K^{-1}	1.33
Rb^{+1}	1.98
Cs^{+1}	3.34
Ag^{+1}	2.4
Tl^{+1}	5.2
Ca^{+2}	1.1
S^{-2}	4.8–5.9
O^{-2}	0.5–3.2
Cl^{-1}	2.96
I^{-1}	6.43
F^{-1}	0.64
Pb^{+2}	4.9

Source: J. R. Tessman, A. H. Kahn and Wm. Schockley, Electronic polarizabilities of ions in crystals, *Phys. Rev.* 92(4).

3.4.1.2. Control of Index Profile

In the last section we described the ion-exchange process as it pertains to a way to spatially change the refractive index of a rod. In light of the mathematical description of the required refractive index given above, it is important to see how people have used the diffusion process to achieve the desired concentration gradient.

Consider a simpler form of Eq. (3.38b) where we consider only the contributions of the ions that are to be exchanged:

$$n^2 = c_0 + c_1 A + c_2 B \tag{3.41}$$

which is true at all values of r. Further, since $A + B = 1$, one can write this as

$$n^2(r) = (c_0 - c^2) + (c_1 + c_2)A = C_0 + C_1 A \tag{3.42}$$

For constituent A one substitutes expression (3.35). Consider the expansion of $J_0(\beta_n r/r_0)$ for small values of the argument:

$$J_0\left(\frac{\beta_n r}{r_0}\right) = 1 - \left(\frac{\beta_n}{2r_0}\right)^2 r^2 + \frac{(\beta_n/2r_0)^4 r^4}{4} \tag{3.43}$$

One can see that for each term of the series of (3.35) there will be an r^2, r^4, r^{2k}, component, resulting in an overall form of (3.43) of

$$n^2(r) = C_0 + C_1(b_1 + b_2 r^2 + b_4 r^4 + \cdots) \tag{3.44}$$

where $b_n = F_n(t, D, r_0\beta_n)$. This is the desired form of the profile. The constants relate to the constants of the material and of the process. So, what one has to work with are the conditions that best allow the form of (3.44) to achieve whatever level of approximation is desired. For example, in the parabolic case one would choose the parameters such that the coefficients $>b_2$ were as small as possible. Iga et al. [3] shows in a graph (Fig. 3.13) how these coefficients vary with the parameter DT/r_0^2, where D is the diffusion coefficient, T the time of the diffusion, and r_0 the rod radius. One sees that one

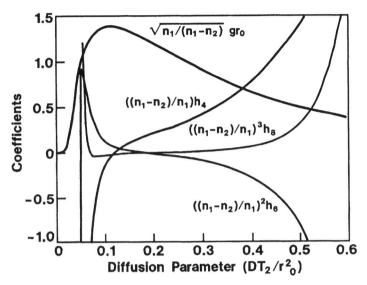

Figure 3.13 Refractive index coefficients as defined from Eq. (3.44) as a function of the diffusion parameter Dt/r_0^2. Each coefficient represents a different order in the radial expansion and the idea is to choose the appropriate value of the parameter to produce as minimum an aberration as possible. The n_1 and n_2 refer to the maximum values obtainable with constituent 1 or 2 (after Iga et al. [3]).

would choose a value of the parameter in the vicinity of 0.15. One can also use this graph to see if it were possible to achieve an approximation to the hyperbolic secant profile if one could find a value of the parameter where $b_4 = 2/3\, b_2^2$ and $b_6 = 17/45\, b_3^2$.

Another approach taken to obtain the desired parabolic or secant hyperbolic profile is that taken by Messerschmidt et al. [21]. They propose to use the glass composition through the ionic interaction energy to modify the concentration dependence of the diffusion coefficient. It is shown that it is this interaction between the two diffusing species that accounts for the concentration dependence of the diffusion constant. They show that for a parabolic profile to ensue, using the Boltzman method described above, the diffusion parameter Dt/r_0^2 would have to vary with concentration in the following way:

$$\frac{D(A)t}{r_0^2} = \frac{\sqrt{A}}{4} - \frac{A}{6} \tag{3.45}$$

The proposal is to study the concentration dependence of the diffusion coefficient in different glass systems and seek out the parameters that would allow one to approach the form indicated by Eq. (3.45).

3.4.1.3. Process and Material Considerations

There are a number of important material and process properties that bear on the performance and practicality of ion-exchange fabrication. Since these are often complex subjects in themselves, we will only briefly touch on them here to give the reader an idea of what is involved.

The first set of properties pertain to the glass composition. The specific glass composition impacts the magnitude of Δn that can be achieved through the solubility of the exchangeable ion [23], thus influencing the diffusion coefficient, the activation energy for diffusion, the dependence of the diffusion coefficient on concentration [24,25], and the chemical durability to the ion-exchange bath. All these properties of the glass can play an important role in the performance for certain applications, and therefore understanding the origins and mechanism of the particular property can lead to a better or more versatile lens.

The second set is that which involves the ion-exchange process itself, in particular that which has to do with the composition and properties of the exchange bath. In Table 3.2 we have listed the commonly used bath compositions with comments in terms of the problems, if any.

3.4.1.4. Methods of Fabrication

Rod Lenses: The manufacture of GRIN rod lenses by the ion-exchange technique is accomplished by a large batch process [14]. Meter-length or longer millimeter-diameter rods are

Table 3.2 Compositions and Properties of Ion-Exchange Bath

Ion	Radius in pm	Coordination number	Refractive index change ΔM	LD_{50} in mg/kg	Salt	Remarks
Li^+	59	4	0.02	710	Li_2CO_3	High tensile stresses
	76	6				
Na^+	99	4	−0.02 to 0.0002	1955	$NaNO_3$	Tensile stresses
	102	6				
K^+	138	6	0.009	1894	KNO_3	Compressive stresses
Rb^+	152	6	0.01	1200	RbCl	High price
Cs^+	167	6	0.04	1200	$CsNO_3$	Slow diffusion
Ag^+	126	6	0.1 (0.22)	2820	Ag_2O	Low thermal stability
Tl^+	150	6	0.1	25	Tl_2SO_4	Additional expenditure for safety

LD_{50} refers to toxicity level.

bundled together and placed into large ion-exchange baths for times as long as 120 h. They are cut and polished to length to produce the desired optical characteristic. NSG reports a typical sequence of profiles taken at various times for a Cs-exchange done at 570°C, as shown in Fig. 3.14 [26]. The particular bath was not specified. The optical characterization parameter is the pitch which means when the argument of the expressions in Eq. (3.20), viz., $\sqrt{AD}$, is equal to 2π. The ray diagram for a 0.25, 0.50, and 0.75 pitch lens is shown in Fig. 3.15. An example listing of the types of SELFOC SML (Selfoc Micro Lens) lenses that are available is shown in Table 3.3.

As we will see in Chapter 5 when we discuss applications, one of the major device areas has to do with lens bars that produce one-to-one erect imaging. We show this schematically in Fig. 3.16a. These lens bars are made by laying out the rods into rows. Then one row is placed on top of the other,

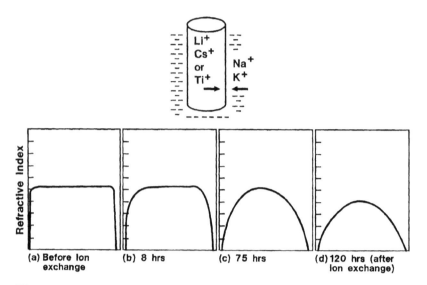

Figure 3.14 Evolution of radial profile as a function of time after ion exchange into a rod (from Ref. [26]).

as shown in Fig. 3.16b. We will discuss the application of such devices in Chapter 5.

It is not clear what limitation, if any, this process imposes on possible rod lens diameters that can be practically made. It may involve such cost features as rods of glass that are much smaller in diameter and have a greater propensity to break during the process. We will see in the next chapter when we deal with the radiometric aspects of one-to-one imaging bars that the lens diameter does play a role.

There is a similar process available in plastic rods where the ion exchange in the anion [3,18].

Planar: The fabrication technique is outlined in Fig. 3.17a. A metal mask with open holes of radius r_m is photopatterned onto the glass that is then exposed to an ion-exchange bath. The distributed index region $n(r, z)$ results from the concentration profile indicated in Fig. 3.17b. An interferometric picture of an exchanged region is shown in Fig. 3.18. NSG [26] has modeled the concentration profile and found that good fits can be obtained by taking into

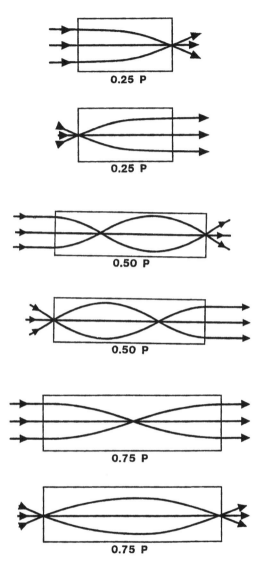

Figure 3.15 Schematic ray trace of rod lens with parabolic profile in typical use configurations as a function of pitch P which is defined is when $(\sqrt{A}D) = \pi/2$ in Eq. (3.20).

Table 3.3 Range of SELFOC GRIN Microlenses

NA	Pitch[a]	Working distance[b]	Diameter	Length
0.46	0.23	0.21 mm	1.8 mm	4.26 mm
0.46	0.25	0.21	1.0	2.58
0.60	0.25	0.21	1.8	3.65
0.46	0.29	–	1.8	5.37

[a] See text, equals $A^{1/2}D = 2\pi$.
[b] Distance from output surface to image.
Source: Newport Catalog.

account the concentration dependence of the diffusion coefficient through the expression

$$D(C_a) = D_0 \exp\frac{KC_a}{C_0} \tag{3.46}$$

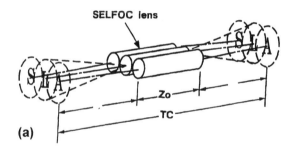

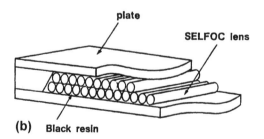

Figure 3.16 Schematic representation of an array of rod lenses used to image in an erect one-to-one manner. Lower figure shows the way the rod lenses are arranged to form the two row bar lens. See Chapter 5 for the applications of such arrays.

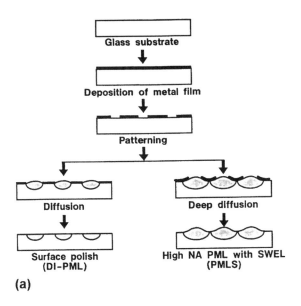

(a)

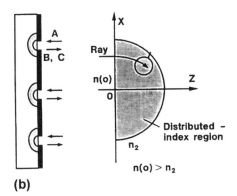

(b)

Figure 3.17 Upper figure is a representation of the planar ion-exchange process. The first stage provides a patterned mask for the subsequent exchange, As a consequence of the volume difference upon exchange, a swelling occurs as shown. This provides a refractive component to the lens, if desired. The lower figure shows an indication of the shape of the index alteration produced by this type of ion exchange.

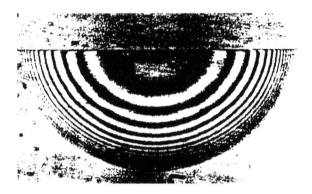

Figure 3.18 Interferogram of a thin axial slice of a plane GRIN lens (from Ref. [3]).

The effect this has on the profile, as measured by the value of the parameter K, is shown in Fig. 3.19. Values of K between 5 and 7 seem to provide a reasonable fit to the experimental measurement of the profile.

Often the ion exchange is done in a field-assisted fashion where an electrode is attached to the back side to direct the flow normal to the surface. In this case the diffusion equation is modified to account for the bias produced by the field

$$\frac{dC}{dt} = \frac{Dd^2C}{dx^2} - \frac{\mu E dC}{dx} \tag{3.47}$$

The right-hand portion of Fig. 3.17a indicates an additional process whereby one obtains a refractive contribution from the swelling effect produced by the ion-exchange and the index gradient [26a]. As noted, this swelling occurs in general but is slight when the ion exchange in not carried out to a large extent. In this case, the slight raised bumps are polished off to obtain the pure GRIN case of the left-hand depiction. Alternatively, when the exchange is carried out to a much greater depth as shown, the exchanged regions nearly overlap and the volume expansion is more significant. The advantage of this swelling feature is that it approximates a spherical shape and thus increases the effective NA of the lens.

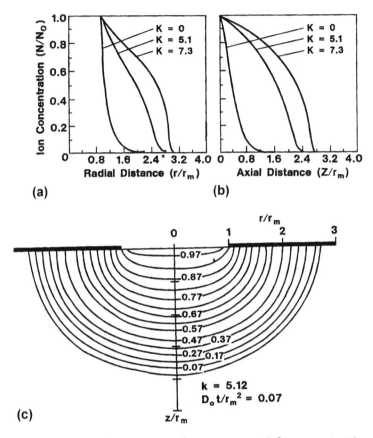

(a)

(b)

(c)

$k = 5.12$
$D_o t/r_m^2 = 0.07$

z/r_m

Figure 3.19 Upper set of curves: spatial concentration profiles calculated with concentration dependent diffusion coefficients from Eq. (3.46). The magnitude of K indicates the degree of deviation from ideal diffusion. Lower curve is the computed version of the refractive index profile shown in Fig. 3.18 (from Ref. [26]).

A typical set of properties for a single planar lens and a coupled pair of the conventional plano–plano type are shown in Table 3.4 [3]. Lens diameters that can be produced by this method range from the order of 100 µm to 1 mm. As in some of the data that follows, the spatial characterization is displayed in terms of the parameter r/r_m, where r_m is the radius of the hole through which the ion exchange is carried out. This implies equivalent performance of the lenses over this size

Table 3.4 Typical Data for Planar Microlens

	Single lens	Coupled
Lens diameter	0.9 mm	0.9 mm
Depth of index	0.45	0.45
Focal length (in air)	2.9 mm	–
(in glass)	4.0–4.5	2.5
Numerical aperture	0.15	0.25
Index of substrate	1.52	

Source: Ref. [34].

range. We will see examples of this range in Chapter 4 where we will deal with the applications of such arrays.

3.4.2. Thermal Mixing/Diffusion

The ideas here are ways to somehow provide an index profile by bringing mixtures of different materials into radial contact in a controlled way such as to subsequently be able to produce a given refractive index profile. As usual, there are high-temperature requirements for inorganic glasses, and low-temperature methods for polymeric materials. We will cover these separately.

3.4.2.1. Glass

The classical way of producing a gradient index pattern in ordinary glass is to make a cylindrical nest of glasses, as shown in Fig. 3.20, and then to heat the set to above the softening temperature. The idea is to blend the glasses together to smooth out the abrupt index changes. Ideally, one would hope to smooth the profile into something like a continuous profile. In practice, a lot of problems can and do occur. The blending at the interfaces does not extend very far in the tubular regions. This makes it difficult to achieve a reasonably continuous profile unless one uses a large number of very thin shells. It is also difficult to obtain a continuous range of compatible glass compositions within a large refractive index interval. It is doubtful this method could make very small

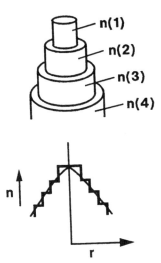

Figure 3.20 Sketch of the nested tube approach, producing a radial index gradient. As indicated, the profile will be stepwise unless the thickness of each tube is thin and there is some degree of interdiffusion at the boundaries.

lenses and it is even harder to imagine how it could be applied to an array.

Lightpath, Inc. [15] has suggested a way around the blending issue by a process combining glasses with differing refractive indices at a high temperature where the glass can more easily flow together. It appears at present that the process has been used only to produce axial gradients, and no radial gradient results have been presented.

A third method proposes to use a flame-deposition technique which is used in the manufacture of optical waveguide blanks [27]. Here, a silica precursor and a dopant precursor are burned in a flame to form a mixed composition soot which deposits onto a rotating mandrel. The mandrel is removed and the soot blank is consolidated at 1500°C. What results is a glass with almost theoretical density. This process is shown schematically in Fig. 3.21. By controlling the proportion of the two constituents as a function of time, and therefore composition as a function of radial thickness, one can produce any desired refractive index profile [27].

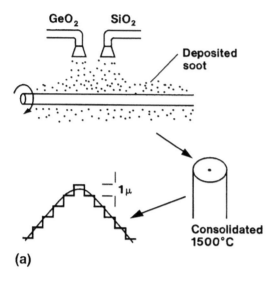

(a)

(b)

Figure 3.21 Flame-deposition method of producing radial gradient index profile. Mixture of the precursors are burned to form the soot in proportion to the desired index change with radial distance. The layer thickness per revolution of the bait rod yields the step dimension as shown in the lower sketch. A photomicrograph (100×) of a pattern produced by this method is shown. Note the resolvability of the laydown layers.

There are inherent problems in this process if it is to be utilized for the fabrication of GRIN lenses. For example, there is a layered nature to the deposition which survives the consolidation. The net effect of this is to produce what amounts to a staircase approximation to the desired profile. This structure produces undesirable features for a GRIN lens application; thus the problem is to reduce the preform size sufficiently so as to make the discontinuous structure effectively disappear. Even after a reduction of $50\times$ one can still make out the ringlike structure as seen in the photograph in Fig. 3.21.

A corresponding problem relates to the center opening of the preform where the mandrel was. Consolidation does close the hole but there is still some remnant of the hole. If one reduces the size sufficiently by redrawing the blank at an appropriate temperature, then eventually both these problems would disappear.

3.4.2.2. Polymers

The polymers present a way to do gradient index at a lower temperature. This can be done either by mixing different materials or by the use of the degree of polymerization. It should be kept in mind that the intention in most studies to date has been to produce optical fiber with a radial profile to increase the system bandwidth, not to make microlenses. It should be clear from the analysis presented above that the lens application is much more demanding of the accuracy of the profile. As a consequence, one does not see much lens characterization, but rather results pertaining to light propagation in such fibers. Because of this we will describe only the methods where accurate control of the index pattern is likely.

The most controlled method is based on a method described in a Kodak paten [28] which we show in Fig. 3.22. A mixture of monomers is injected into a centrifugal mold, the composition of the mixture being varied according to that resulting in the desired profile by essentially creating a plurality of concentric cylinders of differing refractive index. The mixture is then polymerized. Another method, sketched in Fig. 3.23, uses the idea of a rotating tube. A zoned

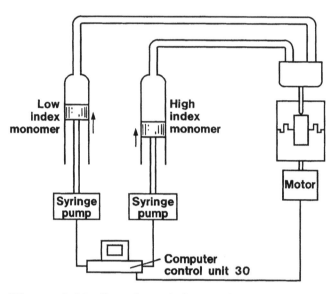

Figure 3.22 Drawing of the method used to produce a radial refractive index pattern in a polymer-based system. Each of the pumps control the relative amount of the high and low index monomers being delivered to the rotating cell.

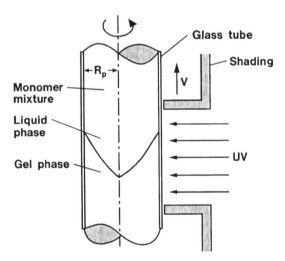

Figure 3.23 Photoinitiated method of producing radial reactive index gradient. UV exposure from the side of the rotating mixture controls the polymerization.

polymerization is effected by means of illumination from the side through a mask [18].

3.4.3. Photochemical Methods

Ideally, one would like to permanently alter the refractive index of a solid body by direct exposure to light through the initiation of some photochemical reaction. The index profile would then result as a consequence of the intensity profile of the beam. As we will see, this latter condition is not so simple to achieve. Although one can imagine a number of examples of how a reaction initiated by light might lead to a refractive index change, for example, photopolymerization, surprisingly only a few have been reported.

3.4.3.1. Photochemistry in Porous Glass

It has been reported that the introduction of photosensitive organometallic compounds into a fine porous structure of a leached borosilicate glass (Vycor, trade name of Corning, Inc., also known as Corning Code 7920) followed by exposure to light, produced sufficiently large refractive index changes that small microlenses could be made [16]. The porous glass is characterized as having about 30% porosity with an average pore size of 6 nm. The glass is essentially transparent above 250 nm.

A schematic of the process steps to form a lens is shown in Fig. 3.24. The porous sample is initially dipped into a solution containing the photosensitive agent for sufficient time to allow the solution to infiltrate the entire thickness. The sample is then air-dried to allow the solvent to evaporate. At this stage the photosensitive material is considered to be entrapped but not bound to the surface of the pores. A representative sampling of the compounds used is contained in Table 3.5 [17]. The sample is then exposed to collimated light through a mask made up of clear holes on an otherwise opaque background. The wavelength of the exposure light is determined by the absorption spectrum of the compound. In the exposed region a photoreaction proceeds, allowing the compound to bind to the glass surface. Before the exposure

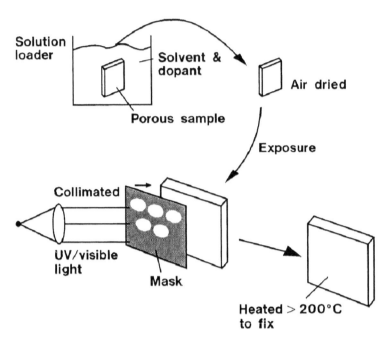

Figure 3.24 Flowchart of the steps to produce a radial refractive index pattern in porous glass. The glass is loaded with a photosensitive material and then exposed through a mask. The light produces irreversible chemical changes that lead to permanent incorporation of the dopant.

Table 3.5 Photosensitive Organometallic Compounds

Compound	Exposure wavelength
$Fe(CO)_5$	$< 500\,nm$
$Fe_3(CO)_{12}$	$< 700\,nm$
$M(CO)_{10}$	$< 500\,nm$
$\quad M = Mn, Re$	
$M(CO)_6$	$< 400\,nm$
$\quad M = Cr, W, Mo$	
$Ru_3(CO)_{12}$	No effect
$Ti(cycopentadiene)_2Cl_2$	$< 600\,nm$
$(CH_3)_3SnI$	$< 700\,nm$

Source: Ref. [17].

the organometallic can be removed by solvent extraction; after exposure it cannot. A proposed mechanism for the photolysis reaction is shown in Fig. 3.25 [29]. At this stage there is a color change in the exposed regions indicating that a photoreaction has taken place. The last step involves heating the sample to oxidize, or otherwise stabilize the metal ion, which ultimately produces the desired refractive pure index change in the absence of any absorption. The explanation is that one considers the exposed region as doped with a small concentration of metal oxide species. The image formed from lenses made by this process using $(CH_3)_3SnI$ as the photosensitive agent is shown in Fig. 3.26. In the lower photo is shown the interferogram of the lenses. The lens diameter was $150\,\mu m$.

A red-orange color is reported upon photolysis, which is attributed to the formation of I_2 from the photoreaction involving $(CH_3)SnI$. Thereupon, heating to $150°C$ returns the exposed region to a colorless state with the refractive index permanently altered. The index change has been reported to remain even into the consolidated state [29].

What is interesting about this process, and not explained in the papers dealing with the results, is how the near-parabolic profile arises. If it were simply a photoprocess, then the index profile should be relatively flat-topped across the exposed circular aperture with some rounding at the boundary from a combination of lack of parallelism of the exposure light and the near-field diffraction from the mask openings. The fit to a parabolic profile obtained from the interferometric data is shown in Fig. 3.27. It is reasonably good out to a radius of $60\,\mu m$, compared to the radius of the exposure of $75\,\mu m$. The

Figure 3.25 Proposed photoreaction in titanocene-doped porous glass that leads ultimately to incorporation of TiO_2 species in exposed regions. The extent of photoreaction is proportional to the intensity.

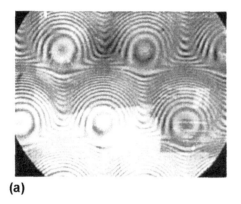

(a)

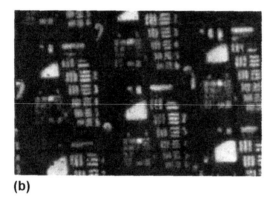

(b)

Figure 3.26 Upper photo is an interferometric trace of the induced refractive index pattern produced in a $(CH_3)_3SnI$-doped sample exposed through a mask containing 150-μm-diameter clear circles. The lower photo shows the images produced by the induced radial profile.

index change across this radius was typically of the order of 2–3×10^{-3}. Using the profile form given above,

$$\frac{Dn}{n_0} = \left(\frac{A}{2}\right) r^2 \qquad (3.48)$$

one obtains a value of $\sqrt{A}$ equal to $0.8\,\text{mm}^{-1}$. SELFOC GRIN lenses have values that range from 0.2 to $0.6\,\text{mm}^{-1}$. As a point

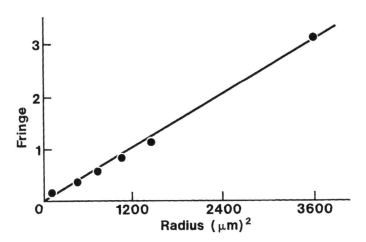

Figure 3.27 From the interferometric trace of Fig. 3.26a, the fringe displacement is plotted against the square of the radial position to ascertain the extent of a parabolic profile.

of comparison, recall that the focal length of a GRIN lens is expressed as

$$f = [n_0 \sqrt{A} \sin(\sqrt{A}Z)]^{-1} \tag{3.49}$$

where Z is the lens thickness.

There are two likely reasons that the profile is not flat. The first is that the index change proceeds layer by layer so that the initial exposed region acts on the light for deeper layer exposure. In a way, it can be thought as a weak focusing effect. One can imagine the effect of this with the aid of Fig. 3.28. Initially, the focal length is at infinity and gradually moves in as the initial layer is formed. The focus spot eventually moves into the glass and the exposure profile is not flat as a consequence. The second possibility is that there is diffusion of the photoproducts. This supported by the fact that it is primarily with the use $(CH_3)_3SnI$ that a continuous profile is produced, whereas with such compounds as titanocene and manganese carbonyl the flat-top profile is evident. An alternative process is somehow to use an exposure intensity pattern that yields a parabolic profile.

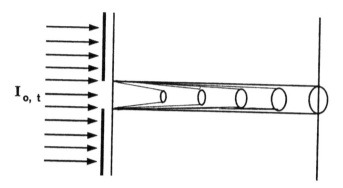

Figure 3.28 Possible explanation of how parabolic profile is produced from exposure through a circular opening mask shown at the left of the drawing. Initially, a weak lensing effect is produced, may be only at the edges near the front. This section bends the light more which produces stronger lensing, and so on.

Because of this unknown mechanism by which the gradient evolves, it is difficult, if not impossible, to estimate the lens diameter and thickness that can be effectively achieved. All of the reported data refer to lenses with diameters of the order of 150 µm [16].

There is another approach to utilizing the porous glass as a medium to support ways to produce refractive index changes. Although it has not been applied to the fabrication of lenses, but rather to the making of holograms, nonetheless it could be adapted to do so by the approach described above. The technique pioneered by Sukhanov [30] and Sukhanov and co-workers [31] was to utilize the porous glass and the material that fills it as a composite media. It should be appreciated that the difference in this approach from that discussed above is that here the porous glass is to be infiltrated with another phase or phases which are the photosensitive species. In the loading process discussed above, the doping is via a physisorbed molecular species on the pore walls.

The refractive index of the composite is then some function of the refractive index of the glass, the filler material, and the volume fraction. For example, one might express the composite index as

$$n_c = \left\{ \frac{[(1 + 2F)n^2 + 2(1 - F)n_m^2]}{[(1 - F)n^2 + (2 + F)n_m^2]} \right\}^{1/2} \qquad (3.50)$$

where n is the index of the impregnated phase with volume fraction F, and n_m is the index of the glass. What this allows you to do is to understand that if either, or both, n and F can be changed by the exposure to light, then an overall change in the composite index will occur. In the case of polymeric materials, photoinduced changes in volume can be a means to a "photorefractive" effect since it influences F. We will deal more with this material in Chapter 7.

3.4.3.2. Photosensitive Glass

There is a class of inorganic glasses containing silver, sodium, and fluorine that can be made to undergo what amounts to a phase separation of a crystalline microphase under the action of uv light and a subsequent thermal treatment. These glasses are often termed "photosensitive" glasses [32]. The mechanism involves the photoexcitation of an electron from a donor such as Ce^{+3} or Cu^{+1}, which is contained in the glass, and which is ultimately trapped by the Ag^{+1}. After a heat treatment, the silver agglomerates into a colloid that serves as a nucleus for the crystallization of a NaF phase. The schematic of the process is shown in Fig. 3.29. The crystal size can be maintained small enough to render the glass transparent. The average crystal size is in the order of 5–6 nm, and the total amount of crystal phase is in the order of 1%.

The refractive index difference between the exposed and unexposed regions arises from the differing contribution of the fluoride ion between when it is incorporated in the glass, or when it is separated into the crystalline phase. The incorporation of fluoride ions into oxide glasses decreases the refractive index [32]. In the exposed and crystallized region, the fluorine is tied up in the crystalline phase where its depressing effect on the refractive index is less.

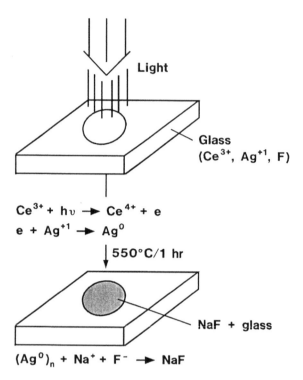

Figure 3.29 Schematic representation of way a refractive index change is produced in a photosensitive glass. After exposure and thermal treatment, a NaF microcrystalline region is produced which has a higher refractive index than the surrounding unexposed, undeveloped region.

The development of a parabolic profile is difficult to achieve with any simple masking procedure because of the need to provide an intensity profile. With a simple clear opening mask one obtains an index profile, as shown in Fig. 3.30. The Δn is 3×10^{-4} across the 150-μm diameter. The image formed in (c) is a virtual image as a consequence of the induced negative index change.

There is a photosensitive glass composition based on Ag-halide, rather than NaF as the photothermal-produced phase [32a]. In other words, the thermally developed phase after exposure is a transparent Ag-halide phase. An additional

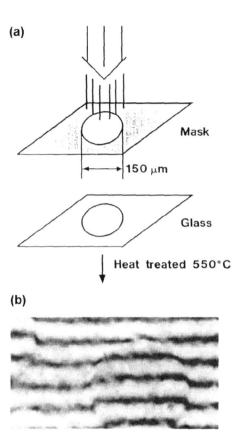

Figure 3.30 For this particular glass the exposure was 20 min under a 1000-W Hg–Xe lamp, and heated to 550°C for 1 hr (a). Interferometer trace shown in (b). Photomicrographs of images formed by gradient index lenses produced in photosensitive glass (c). Image formed by gradient index lens in a photosensitive Ag-halide-based glass (d). Exposure was through a 100-μm aperture displaced from the glass to produce a gradient exposure. The exposure conditions were 30 mJ/cm^2 at 248 nm, for 120 s at 10 Hz. The thermal development was at 560°C for 2 hr [32b].

difference is that the Ag-halide glass compositions contain no added photosensitizer such as Ce^{+3}. As outlined above, the role of Ce^{+3} is to produce the photoelectron that reduces the silver. In the Ag-halide photosensitive glass, the mechanism

(c)

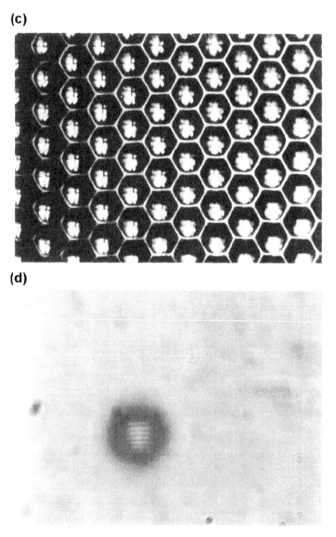

(d)

Figure 3.30 Continued.

of the development of the index change appears to be different. After exposure there is a slight color change, but no index change. The thermal development produces a broad absorption feature in the visible wavelength range indicating the formation of Ag colloids. Accompanying this absorption is a large refractive index change, the order of 0.001.

The refractive index of the produced halide phase is larger than the unexposed glass, so that a positive refractive index gradient is formed which produces a real image. An example is shown in Fig. 3.30d. In this case, the exposure condition was $30\,mJ/cm^2$ at 248 nm for 2 min at 10 Hz. The thermal development was done at 560°C for 2 hr.

3.4.3.3. Miscellaneous

Marchand in his 1978 book [2] refers very briefly to a number of methods, some of which are not included in the above. Since these methods have not ever moved forward, we merely mention the reference here and offer no further comment.

3.5. ANALYSIS AND COMPARISON

As we did in Chapter 2 on refractive lenses, we would like to evaluate the performance of GRIN lenses made by the various methods discussed above in the same way, and to indicate how certain features of the GRIN lens performance might be related to a particular fabrication method. The methods used to evaluate and characterize performance of conventional refractive lenses discussed in Chapter 2, such as interferometric Zernike polynomial analysis, or the intensity profile in the image plane, are every bit as useful and important for the characterization of GRIN lenses as they are for refractive lenses. We will present this data as it is found in the literature. The SELFOC product line manufactured by Nippon Sheet Glass has been the source of the bulk of the published quantitative data.

As we had to be concerned with the true surface figure of the microptic refractive lenses and how it depended on the method of fabrication, here we have to be concerned with the actual refractive index profile and its dependence on how it is made. One cannot decouple in all cases the effects of profile from those arising from pure aberrations. In what follows, to the extent that it can be done, we will first deal with the consequence of the profile and then to the limitations set by aberrations.

3.5.1. Profile

We have seen from the analysis given above in Section 3.2.2 that there are certain limitations in imaging performance of a GRIN lens based on the deviation from the ideal index profile. In the case of the parabolic profile of Eq. (3.19) we see that there the existence of a common focus is based on the assumption of the constancy of the l_0 ray parameter defined in Eq. (3.15). As pointed out above, this parameter is essentially constant when Δn is small, and when the square of the input ray slope is small compared to unity (equivalent to a paraxial approximation). To show this graphically, we show a case of a GRIN lens imaging in a one-to-one erect mode in Fig. 3.31. Here, to effect the righting of the image,

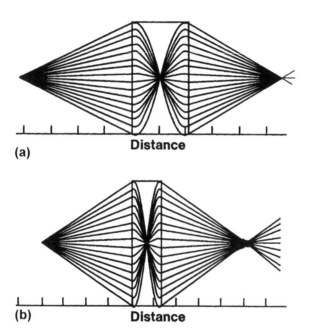

(a) **Distance**

(b) **Distance**

Figure 3.31 Computed ray trace for lens with a parabolic profile, upper curve for the case of $\sqrt{A} = 0.44\,\text{mm}^{-1}$, and the lower $\sqrt{A} = 0.66\,\text{mm}^{-1}$. The image formed by the stronger gradient lens shows evidence of spherical aberration in that the image is somewhat blurred.

a small relay image is formed at the midplane of the lens. (A two-dimensional array of lenses functioning in this mode is of significant practical utility as bar lenses for a number of paper reading devices. These applications will be discussed in Chapter 5. The computed ray trace shown in Fig. 3.31a is for a lens with a $\Delta n/n$ of 0.001, whereas in Fig. 3.31b the value is 0.0044. The working distance is the same in both cases (distance from lens surface to object), so that the external NA is constant. The ray trace was obtained from a numerical solution to Eq. (3.4) assuming an index profile of the form given in Eq. (3.19). The values of $\sqrt{A}$, which is a measure of the gradient, are obtained from the radius of the lens and the value of Dn/N_0

$$\sqrt{A} = \left(\frac{\sqrt{2Dn}}{N_0}\right)\left(\frac{1}{r_0}\right) \tag{3.51}$$

It is clear from the ray trace that there is evidence of aberration for the lens with the steeper gradient, $\sqrt{A} = 0.44\,\text{mm}^{-1}$, curve b, compared to curve a with a value of $\sqrt{A} = 0.66\,\text{mm}^{-1}$. One has a situation entirely equivalent to that of spherical lenses. We can also use this case to look at the consequence of a deviation of the profile from a parabola. This is done by assuming the equation for the refractive index profile as

$$n = N_0\left[1 - \left(\frac{A}{2}\right)r^2 + \text{B}r^4\right] \tag{3.52}$$

for the profile and letting the value of B be such that the contribution of the quartic term represents 1% of the maximum value of Dn/N_0. The ray trace of this case is shown in Fig. 3.32, where the upper curve represents the pure parabola, $B = 0$. The blurring of the image spot is clearly evident. It should be pointed out that the example case used is quite a bit more sensitive to the profile distortion than others might be, but it is, nonetheless, a case of a major application of GRIN lens arrays.

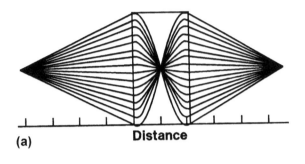

(a) **Distance**

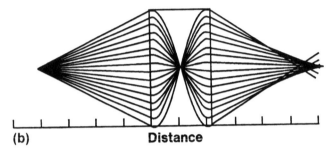

(b) **Distance**

Figure 3.32 Computed ray trace for imaging for a lens with a pure parabolic profile compared to one with a nonzero quartic term. See Eq. (3.46).

Experimentally, the situation is somewhat more complicated because the measurement of the spot size, by whatever technique, would include diffraction as well as the contribution from the profile distortion. In the example given above, one is provided with some advantage in that the diffraction-limited spot size is the same for the two cases, so any difference in the resolution could be attributed to the profile effect.

In Section 3.4.1, we briefly discussed how the process parameters are used to optimize the index profile. A good example of the nature of the empirical optimization method was shown in Fig. 3.13 where the time in the ion-exchange bath was the parameter of interest. In addition, as discussed more extensively in Iga et al. [3], the process is further refined by a heat treatment that follows the ion exchange. This adds an additional step in the diffusion process where one can now solve the diffusion equation using essentially the adiabatic boundary conditions (no material flowing out, so derivative of concentration at boundary is zero), with the initial distribution that of the rod after the ion exchange. The performance parameters are optimized with respect to the heat treatment which is characterized by the same parameter as used in Fig. 3.13, DT/r_0^2 but now T refers to the heat treatment time rather than the ion-exchange time.

3.5.2. Aberrations

As mentioned above, the standard Zernike interferometric analysis for the GRIN lenses has only been reported for the

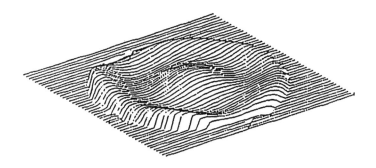

```
*/.******* Zernike Coefficients *********

A00= 2.218
A10= 2.926  A11= 0.100
A20= 0.025  A21=-0.088  A22= 0.003
A30= 0.004  A31=-0.039  A32= 0.032  A33=-0.001
A40=-0.015  A41= 0.005  A42=-0.036  A43=-0.022  A44=-0.012
A50=-0.006  A51=-0.004  A52=-0.015  A53=-0.012  A54=-0.020  A55= 0.009
                                    A63= 0.010


********* Aberration Coefficients *********

RMS              =  0.046   P-V  =  0.27
Spherical  3rd  = -0.22
Coma       3rd  =  0.15   ANGL = 310
Astigma.   3rd  =  0.05   ANGL =  41
```

Figure 3.33 Zernike profile analysis of SELFOC GRIN lens as reported by NSG.

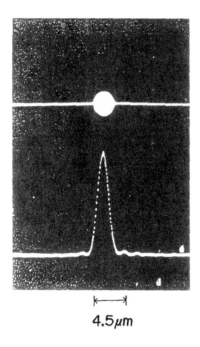

$$\lambda = 0.63 \ \mu m$$

Diffraction Limited Spot Size

$$2a = \frac{1.22\lambda}{N.A.} = 4.29 \ (\mu m)$$

Figure 3.34 Irradiance profile in the focal plane of a SELFOC GRIN lens, as reported by NSG.

SELFOC. Their reported typical rod-type lens result is shown in Fig. 3.33. A reported irradiance profile is shown in Fig. 3.34. The same mistake in the conclusion is made in this case as is generally made. As we pointed out in Chapter 2, the dimension of the first Airy ring is not particularly sensitive to the amount of aberration.

For the planar lenses, the performance is indicated in two ways. The image plane irradiance function is shown in Fig. 3.35. Here, one clearly sees the serious aberration/profile problem from the large wing contributions compared to the

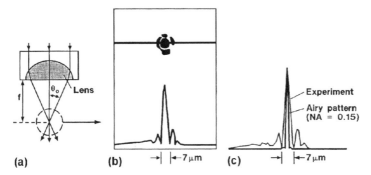

Figure 3.35 Irradiance profile in the focal plane of a stacked planar GRIN lens (as reported in Ref. [32]).

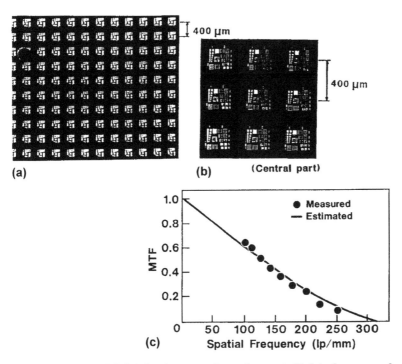

Figure 3.36 Multiple images in coherent light from a planar array of SELFOC GRIN lenses, along with the contrast vs. spatial frequency for an individual lens as reported by NSG.

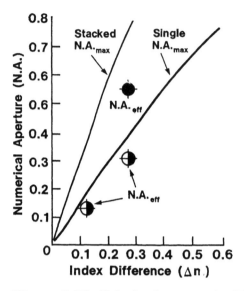

Figure 3.37 Relation between the NA and the index difference of the GRIN lens for an individual planar GRIN lens compared to the stacked configuration (see Ref. [31]).

theoretical Airy profile. Another performance criterion is the resolution as measured by the standard MTF method of a multiple image array which is shown in Fig. 3.36. Another important aspect is the achievable NA. Misawa et al. [33] report data on the NA as a function of the value of Δn for both the single and stacked lens configuration, which we show in Fig. 3.37.

The reader is referred to Iga et al. [3] for a fuller description of the aberration testing of the SELFOC lenses, both rod-like and planar. As far as the results on lenses made by the other methods, there is little quantitative data reported in the literature. Resolution results that have been presented for the photosensitive glass method are plagued with what seems to be severe profile problems. One obtains images from different portions of the lens profile, as that portion approximates a parabola. This is schematically shown in Fig. 3.38. As a consequence the object stays in focus for a long distance,

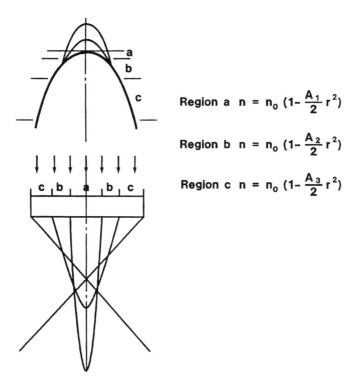

Region a $n = n_0 (1- \frac{A_1}{2} r^2)$

Region b $n = n_0 (1- \frac{A_2}{2} r^2)$

Region c $n = n_0 (1- \frac{A_3}{2} r^2)$

Figure 3.38 Imaging from a lens with a refractive index profile that deviates from a parabola. Each portion of the profile provides imaging over its parabolic range.

as each region of the lens contributes. However, the image is always of poor contrast because of the light coming from the ill-focused regions.

REFERENCES

1. Wood, R.W. *Physical Optics*; Macmillan: New York, 1905.

2. Marchand, E.W. *Gradient Index Optics*; Academic Press: New York, 1978.

3. Iga, K.; Kokubin, Y.; Oikawa, M. *Fundamentals of Microoptics*; Academic Press: Tokyo, 1984.

4. Kawakami, S.; Nishizawa, J.-I. An optical waveguide with the optimum distribution of the refractive index. IEEE Trans. Microwave Techniques **1968**, *MTT-16* (10), 814.

5. Bliss, G.A. Calculus of variations. In *Math. Soc. Am.*; Open Court Publ. Co.: LaSalle, IL, 1923.

6. Iga, K. Theory for gradient-index imaging. Appl. Opt. **1980**, *19* (7), 1039.

7. Kornhayser, E.T.; Yaghjian, A.D. Modal solution of a point source in a strongly focusing medium. Radio Sci. **1967**, *2* (3), 299.

8. Rawson, E.G.; Herriot, D.R.; McKenna, J. Analysis of refractive index distributions in cylindrical GRIN rods used in image relays. Appl. Opt. **1970**, *9* (3), 753.

9. Kapron, F.P. Geometric optics of parabolic index-gradient cylindrical lenses. J. Opt. Soc. Am. **1970**, *60* (11), 1433.

10. Nippon Sheet Glass (NSG) product sheets for SELFOC rod lenses, Clark, NJ.

11. Gregoris, D.; Iizuka, K. Ray tracing method for refractive index profiling. Appl. Opt. **1983**, *22* (12), 1820.

12. Oikawa, M.; Tanaka, K.; Yamasaki, T. Highly integrated distributed-index planar microlens and its characteristics, SPIE, 1985; Vol. 554, International Design Conference 314.

13. Sakamoto, T. Ray tracing in a planar lens with spherically symmetric quadratic index profile. Appl. Opt. **1983**, *22* (10), 1598.

14. Sono, K. SELFOC technology, *IFOC Handbook* 1980–1981 edition.

15. LightPath Technologies, Inc. Albuquerque, NM, 87109.

16. Borrelli, N.F.; Morse, D.L. Planar gradient optics Fourth Topical meeting on Gradient-Index Optical Imaging Systems, paper Dl, Kobe, Japan, 1983.

17. Borrelli, N.F.; Morse, D.L. Photosensitive impregnated porous glass. Appl. Phys. Lett. **1983**, *43* (11), 992.

18. Koike, Y.; Ohtsuka, Y. Studies on the light-focusing plastic rod prepared by photocopolymerization of a ternary monomer system. Appl. Opt. **1983**, *22* (3), 418.

19. Varshneya, A.K. *Fundamentals of Inorganic Glasses*; Academic Press: San Diego, CA, 1994; 339–344.

20. Carslaw, H.S.; Jaeger, J.C. *Conduction of Heat in Solids*; Oxford Press: Oxford, 1959.

21. Messerschmidt, B.; McIntyre, B.L.; Houde-Walter, S.N. Desired concentration-dependent ion exchange for microoptic lenses. Appl. Opt. **1996**, *35* (28), 5670.

22. Jost, W. *Diffusion*; Academic Press: New York, 1952.

23. Araujo, R.J. Colorless glasses containing ion-exchange silver. Appl. Opt. **1992**, *31*, 5221.

24. Crank, J. *The Mathematics of Diffusion*; Clarendon Press, 1975, 231 pp.

25. Houde-Walter, S.N.; Inman, J.M.; Dent, A.J.; Greaves, G.N. Sodium and silver exchange environments and ion-exchange process in silicate and aluminosilicate glasses. J. Phys. Chem. **1993**, *97*, 9330.

26. Nishizawa, K. Planar optical elements on glass substrate, Nippon Sheet Glass, 1994.

26a. Misawa, S.; Oikawa, M.; Iga, K. Maximum and effective numerical apertures of a planar microlens. Appl. Opt. **1984**, *23* (11), 1784.

27. Optical properties of glass. In Uhlmann, D.R., Kreidl, N.J., Eds.; *Optical Fibers*; M. A. Newhouse, American Ceramic Society, 1991; 185 pp.

28. Bello, J.A.; Hamblen, D.P. US Patent 5,122,314, 1992.

29. Borrelli, N.F.; Cotter, M.D.; Luong, J.C. Photochemical method to produce waveguiding in glass. IEEE J. Quant. Electron. **1986**, *QE-22* (6), 896.

30. Sukhanov, V.I. Porous glass as a storage medium. Opt. Appl. **1994**, *XXIV* (1–2), 13–26.

31. Kuchinskii, S.A.; Sukanov, V.I.; Khazova, M.V. Principles of hologram formation in capillary composites. Optics Spectrosc. **1992**, *72*, 383–391.

32a. Borrelli, N.F.; Seward, T.P. Photosensitive glasses and glass-ceremics. Engineered Materials Handbook, 1991; Vol. 4, Ceramics and Glasses in ASM, 439 pp.

32b. Borrelli, N.F. Corning Incorporated, private data.

33. Misawa, S.; Oikawa, M.; Iga, K. Maximum and effective numerical apertures of a planar microlens. Appl. Opt. **1984**, *23*, 1784–1786.

34. Iga, K.; Oikawa, M.; Banno, J.; Kokubun, Y. Stacked planar optics: an application of a planar microlens. Appl. Opt. **1992**, *21* (19), 3456–3460.

4

Diffractive Element Lenses

4.1. DIFFRACTION THEORY REVIEW

The development of methods to fabricate diffractive elements to function as microoptic lenses and lens arrays, as well as other optical elements, has become an active and productive area of research in recent years. For an excellent introductory review, the reader is referred to a *Scientific American* article by Veldkamp and McHugh [1]. The reason this has occurred is the availability of high-resolution photolithography techniques and equipment developed by the IC industry. There are other diffraction-based elements used in fiber optic devices and computer back plane applications [2,3]. Holographic elements to direct light or separate wavelengths are examples of such devices. We will cover these diffraction gratings in Chapter 7. In this chapter, we will be concerned only with those elements that act as lenses.

This is not intended to be an optics textbook, by any stretch of the imagination. However, in order to understand the design and performance characteristics of diffractive element lenses, we must devote some space to review the basic principles of diffraction theory as it pertains to the lens function.

4.1.1. Fresnel–Kirchhoff Diffraction Integral

Consider Fig. 4.1 where light in the vicinity of the point $P(x, y)$ in the x–y plane reaches point $P'(x', y')$ after passing through an obstacle located in the x_0–y_0 plane a distance r from P and s from P'. The expression for the complex amplitude of the wave from P to a point Q in the plane of the screen is the spherical wave [4]

$$E(Q) = \left(\frac{E_0}{r}\right) \exp(-ikr) \tag{4.1}$$

where we have suppressed the time dependence term, $i\omega t$, and $k = 2\pi/\lambda$, λ being the wavelength of the light; the refractive index in which the wave is traveling is taken to be unity. Utilizing Huygens' principle, every point in the x_0–y_0 plane is a source of a new spherical wave propagating toward the

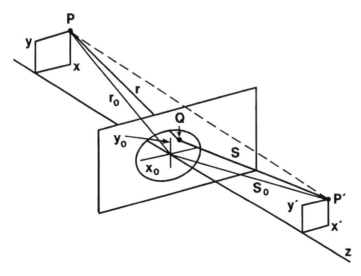

Figure 4.1 Reference diagram for the development in the text: x–y, the plane of the object; x_0–y_0, the plane of the obstacle; x'–y', the image plane.

point P', which in differential form can be expressed as

$$d\{E(P')\} = (i/\lambda)E(Q)\frac{e^{-iks}}{s}\,d\sigma \tag{4.2}$$

Here, $d\sigma$ is a differential element in the x_0–y_0 plane near the point Q. For a derivation of this equation, the reader is referred to Born and Wolfe [3]. After the substitution of (4.1) into (4.2), the integration of Eq. (4.2) over that part of the wavefront passing through the obstacle, designated as S, gives us the desired diffraction result.

$$E(P') = \left(\frac{iE_0}{\lambda}\right)\int_s \frac{\exp[-ik(r+s)]}{(rs)}\,d\sigma \tag{4.3}$$

This is a general expression except for the simplification of taking the angle between the normal to the wavefront and the direction s to be 0. In the more rigorous development, this need not be so, but little generality is lost in taking the special case. The intensity at the point P' is the product of E with its complex conjugate

$$I(P') = E(P')E^*(P') \tag{4.4}$$

4.1.2. Fraunhofer Diffraction

The availability of solutions to Eq. (4.3) depends on making certain assumptions. For example, consider the expansion of the distances r and s in a power series of the variables x_0 and y_0

$$r = r_0\left[1 + \frac{x_0^2 + y_0^2}{2r_0^2} - \frac{x_0x + y_0y}{r_0^2} + \cdots\right] \tag{4.5a}$$

$$s = s_0\left[1 + \frac{x_0^2 + y_0^2}{2s_0^2} - \frac{x_0x' + y_0y'}{s_0^2} + \cdots\right] \tag{4.5b}$$

It is easy to see that if the spatial extent of the obstacle (x_0, y_0) is much smaller than the distances r_0 and s_0, one need keep only the first two terms in Eqs. (4.5). Similarly, one can assume that r and s in the denominator of Eq. (4.3) can be taken outside the integral since they do not vary significantly over the region S in the x_0–y_0 plane, and can be taken as both equal to z_0. Further, if the source is small and essentially on the z axis, that is, $P(x, y)$ is confined to small values, then one can further ignore the second terms of Eqs. (4.5). Hence, we can now utilize all the above approximations and write Eq. (4.3) as follows:

$$E(P') = C \int_s \exp[-i(k_x x_0 + k_y y_0)] \, d\sigma \qquad (4.6)$$

Here, C is a constant phase term, and we have adopted the more useful notation, as we shall see shortly, $k_x = (2\pi/\lambda)(x'/z_0)$ and $k_y = (2\pi/\lambda)(y'/z_0)$. One can specify the region S in terms of a mathematical function which is called an "aperture" or "transmittance" function. We shall express this function in the following way:

$$t(x_0, y_0) = T(x_0, y_0) \exp i\phi(x_0, y_0)] \qquad (4.7)$$

where $T(x_0, y_0)$ represents the amplitude variation, and $\phi(x_0, y_0)$ the phase variation over the region S. Incorporation of Eq. (4.7) into Eq. (4.6) appropriately allows extension of the limits of integration to plus and minus infinity. Equation (4.6) is recognized as the 2D spatial Fourier Transform of the transmittance function, $t(x_0, y_0)$:

$$E(P') = c \int\limits_{-\infty}^{\infty} \int T(x_0, y_0) \exp(i\phi)$$

$$\times \exp[-i(k_k x_0 + k_y y_0)] \, dx_0 \, dy_0 \qquad (4.8)$$

This transform representation will prove to be of considerable use when we extend this treatment to important diffraction geometries where the transmittance function can be easily written.

Just to be complete, when the approximations leading to the form of Eq. (4.6) are made, this is referred to as the far-field diffraction regime. More commonly, it is referred to as Fraunhofer diffraction. With less restrictive assumptions about the relative magnitude of the distance from the object relative to the size of the disturbance, one refers to this regime as near-field or Fresnel diffraction. For our subsequent development of diffractive microlenses, the far-field approximation is appropriate. We will be considering diffractive elements with a numerical aperture (approximately the lens radius divided by the focal length) of the order of 0.1. This justifies the assumptions made in ignoring the higher order terms in Eqs. (4.5).

4.1.3. Diffraction Gratings

As an example of the utility of Eq. (4.8), the diffraction from a narrow slit in an otherwise opaque screen is easily obtained by writing the aperture function $t(x_0, y_0)$ by a function that is termed the "rectangular" function [5]. It is defined as follows:

$$R(x_0, y_0) = 1, \quad -a < x < a, \quad -b < y < b$$
$$= 0, \quad x > |a|, \quad y > |b| \tag{4.9}$$

where a, b are the dimensions of the slit. The utility of the rectangular function should be appreciated in that it allows us to mathematically represent a sharply defined region, which constitutes the essence of the diffraction phenomenon. The Fourier transform of R is

$$F\{R(x_0, y_0)\} = 4ab[\text{sinc}(k_x a)][\text{sinc}(k_y b)] \tag{4.10}$$

This expression, aside from being a constant, represents the diffracted amplitude obtained by integrating Eq. (4.8). We have used the common notation that sinc $x = \sin x/x$. The diffracted intensity is just the square of Eq. (4.10).

A transmission grating can be considered as an N-fold periodic extension of a slit with a separation distance of T. Mathematically, one can express this by use of the Dirac δ function for the case of a slit in the x direction:

$$t(x_0) = \sum \delta\left[\left(\frac{x_0}{T}\right) - n\right]^* R\left[\frac{x_0}{T}\right] \tag{4.11}$$

where n takes on the values, 0, ±1, ±2,....

The asterisk is written to signify a convolution. (Strictly speaking, the convolution of two functions, $f(x)$ and $g(x)$, is defined as the integral of $f(\tau)g(x - \tau)\,d\tau$.) The action of the δ function is to shift the center of the rectangular function by $T, 2T, \ldots, nT, \ldots, NT$. This is a useful way to write the transmission function since one can use the fact that the Fourier transform of a convolution is just the product of the individual transforms [6]; that is to say,

$$F\{t(x_0)\} = F\left\{\sum \delta\left[\left(\frac{x_0}{T}\right) - n\right]\right\} F\left[R\left(\frac{x_0}{T}\right)\right] \tag{4.12}$$

It can be shown [5] that the Fourier transform of an N-sum of δ functions of the form $\delta(x_0/n - T)$ is

$$F\left\{\sum_N \delta\left(\frac{x_0}{n - T}\right)\right\} = \frac{\sin[2(N + 1)(\pi T/\lambda z_0)x']}{\sin(\pi T/\lambda z_0)x']} \tag{4.13}$$

We have the Fourier transform of $R(x_0/T)$ from above, so the diffracted amplitude from a transmission grating is the product of Eq. (4.13) and the one-dimensional form of Eq. (4.10), viz., sinc $k_x a$. For an example of an intensity pattern that would be obtained from Eq. (4.13) for an array of slits of width a, arrayed a distance T apart, see Fig. 4.2.

Of more relevance to the understanding of how a diffractive element can act as a lens is the behavior of a phase grating. In the above, we talked about an array of slits where the amplitude varies periodically, and now we want to investigate a similar structure where the phase is varied. The interest in the spatial variation in phase is because a refractive element acts as a lens because it produces a phase difference that varies parabolically with the radial distance. We will come back to this point in more detail in the next section. The phase grating represents a special case of the general spatial variation of the optical path length in the x_0–y_0 plane, as represented by $\phi(x_0, y_0)$ in Eq. (4.7). The aperture or transmittance function for a phase grating with a blaze (see Fig. 4.3) can be written as

$$t(x_0) = \sum \delta\left(\frac{X_0}{T} - N\right) R\left(\frac{x_0}{T}\right) \exp(i\alpha x_0) \tag{4.14}$$

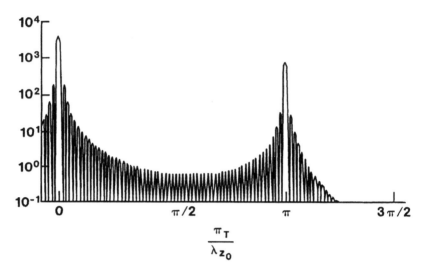

Figure 4.2 Diffracted intensity from a transmission grating according to Eq. (4.13) in the text. The individual aperture opening is x_0 and the separation distance is T. The intensity is measured at a distance z_0 from the grating, and x' measures the position in the image plane.

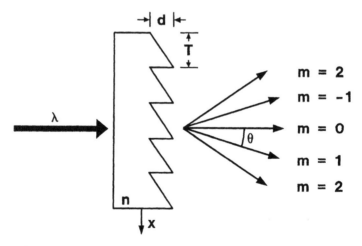

Figure 4.3 Blazed diffraction grating with an indication of the direction of the various diffracted orders.

Here, $R(x_0/T)$ is a rectangular function over the interval T. The tangent of the blaze angle is d/T, d being the depth of the grating, and the corresponding phase shift produced in the groove is $2\pi(n-1)(d/T\lambda)$, where n is the refractive index of the material. The Fourier transform of this transmittance function represented by Eq. (4.14) is

$$\sum \delta\left\{\frac{2\pi N}{T} - k_x\right\} \mathrm{sinc}\left\{\left(\frac{T}{2}\right)(\alpha - k_x)\right\} \tag{4.15}$$

Each term in N represents the successive diffracted order. In general, the nth-order efficiency would be given by the expression

$$\left[\mathrm{sinc}\left\{\left(\frac{T}{2}\right)\left(\alpha - \frac{2\pi n}{T}\right)\right\}\right]^2 \tag{4.16}$$

From Eq. (4.15) one sees that the first-order efficiency will be 100% when the value of $\alpha = 2\pi/T$, which corresponds to a groove depth d of $\lambda/(n-1)$.

4.1.4. Diffractive Elements in Relation to Refractive Elements

As we mentioned above in introducing the phase term $\phi(x_0, y_0)$ into the transmittance function of Eq. (4.7), this could be any arbitrary function. We looked at a special case of a linear function for the blazed grating. (In Chapter 7, we will analyze the case for gratings with different forms, rectangular, sinusoidal, etc.) One should realize for this case that the refractive analogue to the blazed grating is a prism. Moreover, the diffractive phase is periodic in the refractive phase. This correspondence suggests that if we take the phase function that describes a refractive lens, then a diffractive element that acts like a lens should ensue if certain conditions are met [7].

What we are trying to do is find a way to write the diffractive transmittance,

$$t_d(r) = \exp\{2\pi i \phi_d(r)\} \tag{4.17}$$

in terms of the refractive phase term, $\phi(r)$. One proceeds by comparing an arbitrary refractive phase function to the diffractive phase modulated with a limiter function [7] as shown in Fig. 4.4. The limiter function maintains the phase between the values of $-a/2$ and $a/2$ as shown. This function was done in the grating case by the depth of the groove.

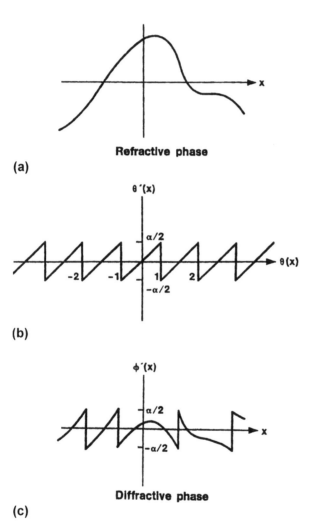

(a)

(b)

(c)

Figure 4.4 Upper curve, arbitrary refractive phase function. Middle curve represents the limiter function which limits the phase between $-a/2$ and $a/2$. Lower curve is the resulting diffractive phase function. (From Ref. [7].)

Again, one sees that the diffracted phase is periodic in the refractive phase which means that $t_d(r)$ is also periodic in $\phi(r)$. This suggests that one can write $t_d(r)$ in terms of a generalized Fourier series in $\phi(r)$ as follows [6]:

$$\exp\{2\pi i \phi_d(r)\} = \sum A_n \exp\{2\pi i n \phi(r)\} \tag{4.18}$$

The coefficient A_n is determined in the usual way for a Fourier expansion:

$$A_n = \int \exp\{2\pi i (\phi_d - n\phi)\}\,d\phi \tag{4.19}$$

Further, from Fig. 4.4, we see that $\phi_d = a\phi$ within the periodic interval, so that we can integrate Eq. (4.18) to obtain the expression

$$A_n = \text{sinc}\{\pi(a - n)\} \tag{4.20}$$

It should be clear that when $a = 1$, A_1 is equal to unity, and all other A_n's are zero. That is to say, we have found a phase function $\phi_d(r)$ that produces the same transmittance function as the corresponding refractive element. Whatever phase front is produced by the refractive element characterized by $\phi(r)$ will be exactly produced by the corresponding diffractive element characterized by $\phi_d(r)$. Putting it another way, we know that the Fourier transform of $\exp\{-2\pi i t(r)\}$ will produce a given far-field intensity pattern. Since we have shown that the same transmittance function is produced by a given diffracting structure, then the same far-field pattern will ensue.

4.1.5. Diffractive Lenses

For a spherical lens, the axial dependence of the phase follows the lens thickness; the thickness is given by the equation

$$h_r(r) = \left(\frac{1}{2R_c}\right)(r_0^2 - r^2) \tag{4.21}$$

Here, r_0 is the lens radius, and R_c is the radius of curvature. One can equally express this as referenced to zero at the center

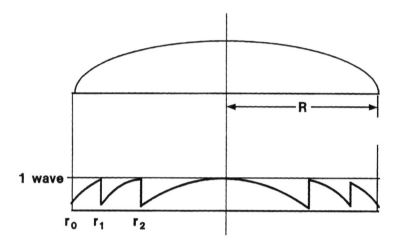

Figure 4.5 Upper curve, phase function of a refractive lens. Lower curve, corresponding phase function of equivalent diffractive lens.

$$h(r) = \frac{-r^2}{2R_c} \tag{4.22}$$

The phase, expressed as a transmittance function, is written as follows:

$$\phi(r) = \frac{-2\pi r^2}{2R_c\lambda} \tag{4.23}$$

We have shown in the previous section that this also represents the required phase function for the diffractive lens if we limit the phase difference in each interval, as shown in Fig. 4.5. We will go into considerable detail on how one produces such structures in the following sections.

4.1.5.1. Corrected Lenses

It is worthwhile to point out at this stage, before we get into describing the fabrication techniques, another potential advantage of the diffractive approach. This is that one can make corrected lenses with no additional difficulty of fabrication [7]. In the above, we talked about making a spherical lens, that is, writing the appropriate phase term that would make a diffractive element produce the same phase front as that of the corresponding spherical lens. We pointed out in Chapter 2 the various aberrations that accompany the performance of a spherical lens. If one considers, in a simple example, the surface that would produce a perfect spherical wavefront

$$\frac{-2\pi}{\lambda\sqrt{r^2 - R^2}} \qquad (4.24)$$

and one expands Eq. (4.24) in a power series:

$$\left(\frac{\pi}{\lambda_0}\right)\left[\frac{-r^2}{R} + \frac{r^4}{4R^3} + \cdots\right] \qquad (4.25)$$

One can see the spherical approximation (r^2 term) and the higer order terms that would amount to aberrations terms. In this case, we specifically show a third-order spherical aberration term. However, in principle, one can incorporate this correction into the diffractive element design. For example, the grating period would now come from solving Eq. (4.25) with as many higher order terms as one desires, or Eq. (4.24) directly. We shall see that this can be incorporated into the fabrication without difficulty.

4.2. METHODS OF FABRICATION

There are essentially two approaches to the fabrication of diffractive lenses. The first involves methods of making Fresnel-like structures, that is, trying to reproduce the refractive surface feature as a perodic grating, while the second is a relatively new approach spawned from the IC microfabrication technology, which is called binary optics. The word stems from the use of discrete levels to approximate the desired surface as shown in Fig. 4.6. There are some trade-offs, for example, the efficiency vs. the ease and/or cost which we will try to point out as we cover each method.

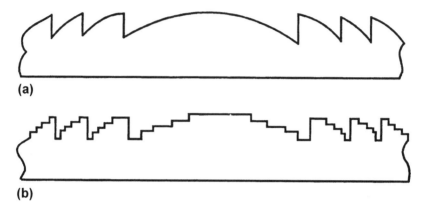

(a)

(b)

Figure 4.6 Upper curve, conventional diffractive lens with continuously varying phase. Lower curve, the binary approximation where the continuous phase is approximated in a stepwise fashion.

4.2.1. Micro-Fresnel Lenses

The methods for making micro-Fresnel lenses and lens arrays are all based on being able to provide a gray scale of exposure. This coupled with a resist that has a linear response over a reasonable exposure range forms the basis for producing a prescribed surface curve. There are essentially two ways this can be done, which we cover below.

4.2.1.1. Direct Write

In this method, a beam, laser or e-beam, writes directly onto the resist. There is no mask used. The resist-coated substrate is scanned, usually on air bearing stages, and exposed using an intensity modulated beam. The developed resist is then subjected to an etching, usually RIE [8], or is used to form a master for replication [9,10]. A focused HeCd laser is a typical source used as the exposure beam. A schematic of the process is shown in Fig. 4.7. The resolution or pixel size is reported as $0.5 \times 0.5 \,\mu\text{m}$ [8].

   This method is an attractive way of making a master from which one can replicate the desired lens array. For those materials that would be difficult to mold or emboss, each

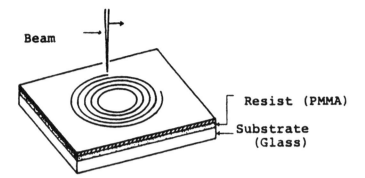

Figure 4.7 Schematic representation of direct laser beam writing onto resist to produce continuous relief pattern.

array would have to be individually produced. This will be true for any process where direct writing is used.

4.2.1.2. Gray Scale Mask

In this case, the desired graded exposure function is built into the exposure mask itself. As an example, the radial transmittance on the mask would look like that shown in Fig. 4.8a. After exposure to this mask, the pattern in the developed positive resist would be a shown in Fig. 4.8b. One way to do this is to utilize the gray scale capability of current e-beam writing systems. For one example, consider the use of a process which patterns clear dots of varying size on a chrome mask. Because the dot pitch is sufficiently small, it cannot be resolved by the projection photolithographic system. Thus the amount of light that reaches the resist-coated substrate is a slowly varying intensity pattern proportional to the average dot size in any portion of the mask. An example of such a mask pattern is shown in Fig. 4.9 where the resist used was EBR-9. Recall that this method is intended for projection systems. The optical pattern of the regions of Fig. 4.9 is thus $5\times$ smaller than shown when exposed on the resist. In this pattern, the dot pitch on the wafer was the order $0.3\,\mu m$ and hence not resolvable. At the highest optical magnification, one can actually see the dots, and the eight regions. This is not the only way one can manipulate the e-beam exposure

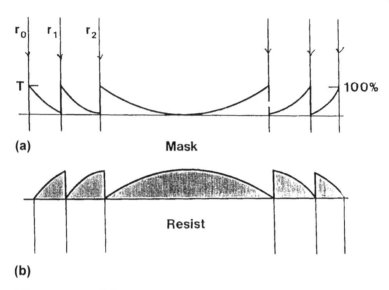

(a)   Mask

(b)

Figure 4.8 Schematic representation of an exposure through a variable transmittance mask. The upper drawing represents the mask where the transmittance is plotted in the vertical direction as a function of the horizontal position. The lower figure represents how a positive resist would develop in response to such an exposure.

Figure 4.9 Photomicrograph of a section of a variable transmittance mask made by controlling the areal density in the radial direction of an e-beam writing pattern, 800×.

to produce the desired result. One could equally use the same dot size and let the exposure vary to enlarge or reduce the dot size, or any method which produces a quasi-continuous exposure. These methods tend to be iterative in nature because one cannot measure the actual transmission pattern on the mask. One can develop, however, over a number of trials, methods that essentially calibrate the process.

Assuming one has a positive resist that has a linear response over a reasonable exposure level, one then has to fabricate the transmittance function of the mask to correspond to the desired surface figure. A positive resist is one that is subsequently removed in development where exposed. For a spherical surface, the transmittance function should follow the lens thickness

$$h(r) = R_c \left\{ \sqrt{1 - \left(\frac{r}{R_c}\right)^2} - \sqrt{1 - \left(\frac{r_0}{R_c}\right)^2} \right\} \tag{4.26}$$

where R_c is the radius of curvature of the surface and r_0 is the radius of the lens. For most cases, and to simplify the example calculation, we can assume that $R_c \gg r_0$, so we can use the familiar approximation to Eq. (4.26):

$$h(r) = \frac{r_0^2 - r^2}{2R_c} \tag{4.27}$$

One wants more light where the lens is the thinnest, so the mask transmittance function should follow the following expression:

$$T_m(r) = K\left\{1 - \left(r_0^2 - r_2\right)\right\} \tag{4.28}$$

The value of K scales the actual transmittance, so we can take it to be equal to unity for this example. From Eq. (4.22), we calculate the radii of the rings for the desired radius of curvature. We will number the rings from the radius r_0 inward. As we start in from the lens radius, we want the transmittance to vary in the outermost ring as

$$T = ar^2 + b \tag{4.29}$$

and be unity at $r = r_0$ and zero at $r = r_1$. This means the function in the first interval is

$$T(r) = \frac{r^2 - r_1^2}{r_0 - r_1^2}, \quad r_0 < r < r_1 \tag{4.30}$$

In the next ring, one wants the transmittance to continue to follow Eq. (4.27), but swing from unity at r_1 to zero at r_2. In general, the form for the transmittance of the mask in region k is given in the equation below

$$T(r) = \frac{r^2 - r_k^2}{(r_{k-1}^2 - r_k^2)}, \quad r_{k-1} < r < r_k \tag{4.31}$$

An atomic force microscope picture of the result of the exposure of a resist to a mask fabricated in the way just described is shown in Fig. 4.10 [13]. The lower curve shows a quantitative measure of the profile.

Another interesting and unique method of making a gray scale mask involves the utilization of a special glass, termed HEBS [14]. This glass develops optical absorption in a thin layer near the surface upon exposure to an e-beam. A schematic of the process is shown in Fig. 4.11. This patented material is used by Canyon Materials to produce masks. The reported induced optical density at 436 nm as a function of the indicated dose rates is shown in Fig. 4.12a. An optical micrograph (100×) of an e-beam written test pattern is shown in Fig. 4.13. Canyon further reports the required exposure level that was used to write on the glass in order to expose two representative photoresists as reproduced here in Fig. 4.12b. The beam-writing method is ideally suited for gray scale patterning since most e-beam writing packages allow the control of the current over different areas. The fact that the sensitivity of this glass is 10–20× less than typical e-beam resists is not too important since one only has to write the photomask once, whereas in the resist case one writes for each copy.

4.2.2. Binary Optics

This approach of digitizing the desired surface figure, or binary optics as it has been called, has produced the biggest

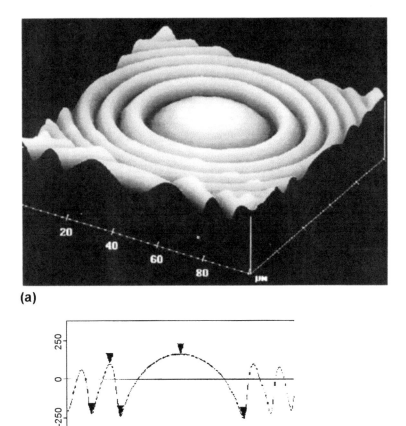

(a)

(b)

Figure 4.10 AFM image of the resist pattern formed after exposure to variable transmittance mask. The lower figure shows the trace of the pattern.

breakthrough in the fabrication of diffractive element devices, and in particular lens arrays [15–17]; see Fig. 4.6b. The reason for this is because it follows the same method that is used in the multilevel fabrication of integrated circuits. There is a price to be paid, however, in that the diffraction efficiency depends on the number of levels used to approximate the continuous surface.

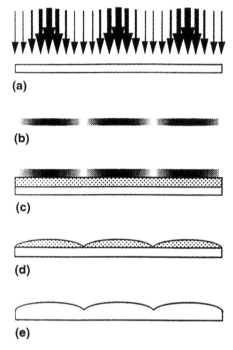

(a)

(b)

(c)

(d)

(e)

Figure 4.11 Canyon Materials, Inc. representation of the exposure process using their high-energy *e*-beam sensitive glass. The glass photodarkens when exposed to *e*-beam radiation as shown.

To the first order, one can estimate the diffraction efficiency of an N-level system by realizing that the discrete level approximation is really two structures superimposed. We will use a linear grating diagram, as shown in Fig. 4.14 for simplicity's sake. If we had a blazed grating, the efficiency was shown above to be given by the expression

$$\left| \operatorname{sinc}\left\{ \pi \left(\alpha - \frac{N}{T} \right) \right\} \right|^2 \tag{4.32}$$

where $\alpha = (n-1)d/\lambda T$. Superimposed, we have another grating of pitch T/N and a depth d/N. The diffraction efficiency of such a grating would be

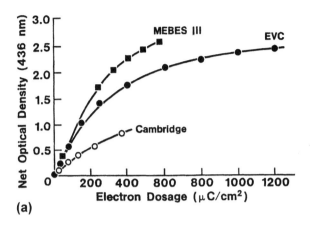

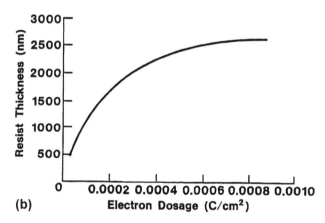

Figure 4.12   Information supplied by Canyon Materials, Inc. as to the optical density produced by given electron beam dosages, for three different machines. Also, the remaining resist thickness after the listed e-beam exposure is given in the lower figure. In this case, the resist was Shiply S1650. (From Ref. [14].)

$$\left| \sin c \left\{ \frac{\pi a}{N} \right\} \right|^{2} \tag{4.33}$$

The product of the two expressions represents the overall efficiency of the N-order binary approximation. For the lens case we can apply the same idea and assume that the

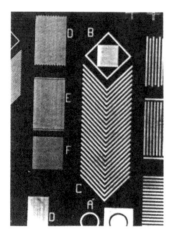

Figure 4.13 Photomicrograph of a portion of the e-beam produced pattern on the HEBS^{TM} glass.

efficiency of the Fresnel-like structure is 100%. Thus the efficiency of the N-fold binary approximation is simply

$$\text{Eff}(N) = \left| \sin c \left(\frac{\pi}{N} \right) \right|^2 \tag{4.34}$$

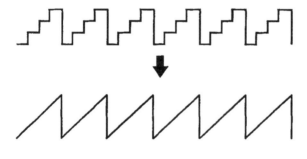

Figure 4.14 Diffraction grating representation of the behavior of a binary-optic approximation to a blazed grating. There is the contribution of the higher-order grating coming from the period produced by the steps. (From Ref. [7].)

Here, we have assumed that the condition that $d = \lambda/(n-1)N$ has been met.

How the efficiency expression relates to fabrication will become obvious in the next section.

4.2.2.1. Photolithography

For those readers unfamiliar with the photolithographic technique, we will try to incorporate into the following description some of the details. The general idea is to coat the substrate to be processed with a photosensitive organic film called a photoresist or resist for short. Typical thickness is in the tenths of micrometers range, although thicker layers are also possible when needed. The resist is usually spun on, that is, the resist is dispensed onto a spinning substrate to distribute the resist over the substrate. The resist can also be vacuum deposited. The resist when exposed to light either becomes soluble to a development solvent (positive resist), or insoluble (negative resist). Commercial resists exhibit a wide range of properties aimed at specific applications. For example, there are resists that are very sensitive: those with linear development behavior, those that are sensitive to shorter wavelengths, etc. The sensitivities are measured in terms of energy per unit area, typically in the range of mJ/cm^2 on up [18].

As an example, we will use Fig. 4.15 where the intention is to make the binary approximation to the Fresnel structure shown in Fig. 4.15b. In the first case shown in Fig. 4.15c, we will make what is called a two-level structure. A mask is made to conform to the desired geometric structure. An example is sketched in Fig. 4.16. The pattern is usually made in a chrome film to provide the optically opaque regions. In most cases, this pattern is made through the use of e-beam writing because of the required resolution. The e-beam writes onto an e-beam resist which responds in the same manner as described above. The exposed resist is developed and the Cr is removed where left unprotected by the resist. The mask is then used to expose the resist-coated substrate as shown in Fig. 4.17a. The resist provides a protection for the subsequent etching step. The method of choice is called reactive

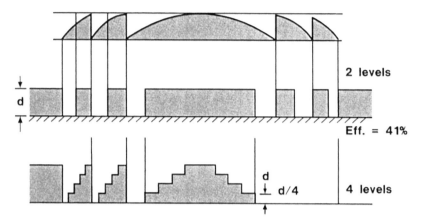

Figure 4.15 Schematic representation of a two-level and four-level approximation to a Fresnel structure.

ion etching, more commonly referred to as RIE [19]. This is a plasma-based process where the fluoride ion is produced from a precursor like CF_4 which then reacts with substrate to produce a volatile fluoride compound. The etch rates vary

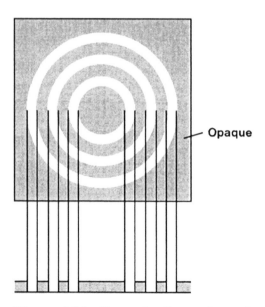

Figure 4.16 Example of a mask required for a two-level exposure.

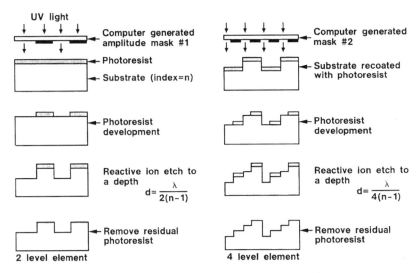

Figure 4.17 Schematic representation of what the exposure steps would be to produce a four-level binary approximation utilizing two masks as compared to a single exposure two-level.

for different materials. As a typical number for fused silica, it is of the order of $15 \, \text{min}/\mu\text{m}$. The required depth of $d = \lambda/2(n-1)$ is controlled by the etching time. The maximum efficiency that could be obtained from this two-level grating is calculated from Eq. (4.34) as 41%.

If one uses two masks, one can achieve a four-level approximation as shown Fig. 4.15c. The sequence of exposures is shown in Fig. 4.17b. The first mask exposure and etching produces the coarse structure. The etch depth here is $d/4$. This structure is coated with resist and exposed to the second mask which represents a higher resolution. Essentially every ring in the mask is split into two. The next $d/4$ step is etched as shown. The result is a four-step approximation with a maximum efficiency of 81%.

One can go to three masks, each one doubling the resolution of the preceding one which would produce eight levels with an efficiency of 95%. It is easy to see that there is a simple relationship between the number of masks used and the number of levels $L = 2^N$. One can begin to appreciate how the existing IC fabrication facility, which uses many separate masking steps to create the desired structures, lends itself to the multistep binary-optic approach. An SEM photomicrograph of a hexagonally packed array of eight-level lenses is shown in Fig. 4.18.

In a related approach [20], one can produce a two-level lens by direct e-beam writing onto an *e*-beam resist. In this reported case, the 1.0-mm diameter lenses reside in the resist. The minimum feature size was 6.4 μm. The measured efficiency was 30%.

There have been a large number of publications reporting on lens arrays fabricated by the binary-optic approach [21]. Many have been fabricated in silicon for integrated optoelectronic devices. We will try to cover these in Chapter 6 which deals with applications.

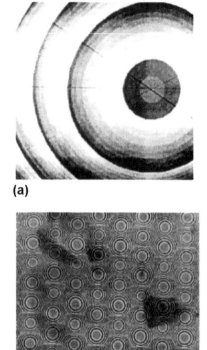

(a)

(b)

Figure 4.18 Photomicrograph of an eight-level exposure pattern.

4.2.2.2. Molding

Molding or replication into plastic from a master has been demonstrated for multilevel structures. A recent publication [3] described an eight-level 2×2 mm lenslet with a 6.5-mm focal length replicated onto a PMMA-based film pressed between the silica master and an optical flat. The master consisted of a diffractive lenslet array pattern which was fabricated with the opposite phase onto a 3-mm thick quartz glass substrate. The minimum feature size was $2 \, \mu$m, and the step depth was the order of $1 \, \mu$m. See Fig. 4.19. Although this lens diameter was relatively large, it is the feature size resolution that is important for microoptic applications. In the next section, we will estimate the required resolution for given applications. There was a brief discussion given of the thermal characteristics of PMMA, such as thermal expansion and shrinkage, and how these properties might affect the lens performance. The measured diffraction efficiency of the molded plastic lens was 91%, compared to the theoretical value for an eight-level structure of 95%. One major issue that was mentioned was the optical quality of the back side, and the wedge that is created in the relatively thick layer of PMMA. An excellent review of the status of replication in polymer-based materials is given in Ref. [22].

The molding of diffractive elements into inorganic glasses represents a more challenging problem. In addition to the mold life issues discussed in molded aspherics in Chapter 2, one has the issues associated with small submicrometer features. All this notwithstanding, there have been reports of molding diffractive elements such as gratings and holograms in glass, which will be covered in Chapter 7.

4.2.2.3. Laser Ablation

Another technique is to precisely remove the material by laser ablation [23],[24]. In this case, a focused excimer laser (both 248 and 193 nm have been used) is directed to a mask and then focused onto the substrate as shown in Fig. 4.20. The reported smallest feature that can be written is $<1 \, \mu$m, and to a depth to $0.1 \, \mu$m. The position accuracy is given as $3 \, \mu$m.

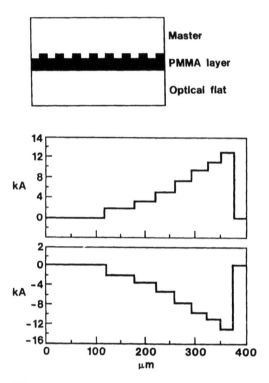

Figure 4.19 Representation of the molding of a binary-optic lens. The upper figure schematically depicts the molding, whereas the bottom figures show the relationship of the mold pattern to the material to which it is pressed into. (From Ref. [22].)

However, in a 10× system this is correspondingly reduced. The high-speed stage allows movement of centimeters in tenths of seconds. There is not much discussion of the materials that have been used although polyimide films are mentioned as an ideal material. The method, however, is not limited to this material.

4.3. ANALYSIS

The analysis of the performance of microlenses made by any of the diffraction grating processes in terms of what would be equivalent to refractive lenses is not straightforward.

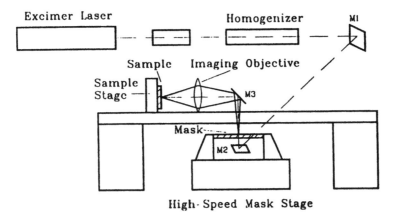

Figure 4.20 Drawing of laser ablation method of forming diffractive elements. (From Ref. [24].)

Whereas in the case of refractive lenses the distortion produced in the transmitted phase front derives from the surface figure, it is no less critical than other geometric issues involved in the diffractive lenses. For the Fresnel type, the surface profile is still an issue, and the consequence of its deviation from the desired shape can be considered in the same way as one would for a refractive lens. Perhaps a simpler way to look at this effect is to go back to Eq. (4.18). Here we expressed the diffractive phase function ϕ_d in terms of the desired refractive phase function ϕ. To the extent that the condition ϕ_d is not proportional to ϕ, more than one terms in Eq. (4.18) will have nonzero coefficients, as seen from Eq. (4.19). The practical result of this is that light will be diffracted to higher orders, reducing the efficiency in the desired first order. These deviations in the surface profile become serious when they correspond to magnitudes comparable to those of refractive lenses.

For the binary-optic lenses, one must consider the effect on the performance, both resolution and efficiency, of small errors in the depth and shape of each step. This is true for the Fresnel-type lens as well, although to a lesser extent since it is a two-level structure. For example, see Fig. 4.21. This type of analysis is difficult to portray in any simple way,

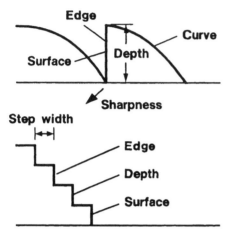

Figure 4.21 Schematic depiction of common geometric imperfections on a diffractive element that could impact the performance of a Fresnel structure, upper, and a binary optic, lower.

and is usually done by some sort of computer modeling code. This type of analysis is clearly outside the scope of this discussion [25].

What we will consider here is the issue of chromatic aberration which is severe enough to limit the use of diffractive lenses to relatively narrow band widths. Another consideration of consequence will be that of the limitation that the present fabrication techniques impose on the lens numerical aperture.

4.3.1. Chromatic Aberration

The fact that the diffraction grating spacing is derived from a single wavelength presents a problem for a diffractive lens in terms of what would ordinarily be called chromatic aberration. The focal length of the diffractive lens was given above in Eq. (4.23), where it is related to the grating spacings. One can easily see that the focal length times the product is invariant with respect to the grating spacing. Thus the focal length R_{c0} at wavelength λ_0, which corresponds to the desired wavelength, would change at any other wavelength λ by the expression

$$R_c = R_{c0} \left(\frac{\lambda_0}{\lambda} \right) \tag{4.35}$$

This represents considerably more change than a corresponding refractive element where the chromatic aberration would be solely determined by the wavelength dispersion of the refractive index. Consider the change that would occur from 400 to 700 nm. For a refractive element, a 10% change would be considered large, whereas for a diffractive element, the focal length would nearly halve. This problem limits the application of microoptic diffractive lenses to narrow wavelength sources such as lasers, although this does not seriously diminish the breadth of applications.

As far as the diffraction efficiency is concerned, the binary-optic lens is strongly dependent on wavelength whereas the Fresnel lens is not. For the binary-optic lens, one can use the grating analog, the efficiency expressed in Eq. (4.32), at a wavelength other than λ_0 $[d = \lambda_0/(n-1)]$, can be written as

$$\text{Eff} = \left| \text{sinc} \left[\pi \left\{ \frac{\lambda_0}{\lambda} - 1 \right\} \right] \right|^2 \tag{4.36}$$

which we show graphically in Fig. 4.22. The efficiency of a Fresnel lens will have essentially the same dependence on wavelength as that of a refractive element, so although its focal length is subject to the monochromatic condition, the efficiency is not.

4.3.2. Fabrication Limitations on Lens Design

In this section, we will try to relate the limitations imposed by the photolithographic fabrication techniques to the range of lens performance that can be achieved. What this means is that the photolithographic method has built-in resolution limitations, currently the order of 0.5 μm depending somewhat on how sophisticated the approach that is used. These resolution issues circumscribe the nature of lens one can make.

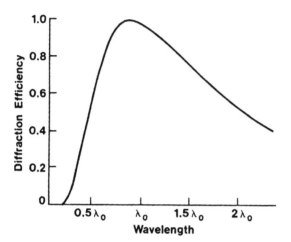

Figure 4.22 Computed diffraction efficiency of a binary-optic lens; see Eq. (4.35).

There are essentially three lens parameters of consequence, any two of which are independent: the lens diameter, the focal length, and the numerical aperture. Consider the numerical aperture, and what value might be achieved. For the moment, since we are concerned with microoptic applications, let us take the lens radius to be in the order of $100\,\mu m$. One must consider the resolution issue slightly differently for the cases of lens fabrication using a mask from that of the case where the direct-write e-beam or laser is used. This is because in the mask method, a 5–1 or 10–1 reduction is commonly used. This means the features are 5–10 times larger on the mask than they will be in the resist plane. The ultimate resolution still is determined in the wafer plane, but there are aspects unique to each method.

4.3.2.1. Resolution and Numerical Aperture

The simplest way to consider this relationship is to think of how finely one can subdivide a given ring portion of the lens pattern to make the desired feature on the lens. Consider the graded transmittance mask method mentioned above and depicted in Fig. 4.8, and actually shown in Fig. 4.9.

One has effectively taken the given grating region Δr_i and divided them up into p subregions. Essentially, the same approach is taken to produce the binary-optic structure; see Fig. 4.17. Each region corresponding to a one-wavelength phase shift is to be broken up into subregions corresponding to smaller phase shifts. The system resolution will ultimately determine the smallest number this ratio $\Delta r/p$ can assume.

The values of Δr_i, and in particular in the outermost ring, will be determined by the desired extent of the lens, which is the diameter, and the desired focal length, or the numerical aperture. To take an example, consider the case of a 50-μm-radius lens. We show in Fig. 4.23 the width of the outermost ring as a function of the numerical aperature, for each of two wavelengths, 550 and 1100 nm. One should now be able to see just how the lens design is limited by how far one can subdivide the rings considering the 0.5 μm resolution limit provided by the photolith method.

Let us consider the binary-optic approach first. If one assumes that >90% efficiency is necessary for most applications then an eight-level structure is required. This means

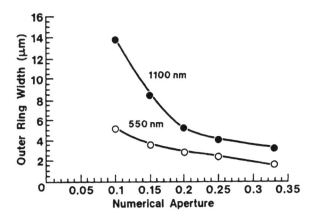

Figure 4.23 Outer ring width in micrometers vs. the numerical aperture for two wavelengths. The lens radius is 50 μm. The intention is to show the dimensional limitation that is imposed at high NA and short wavelength.

that the minimum width of the ring element must be larger than 4 µm. From Fig. 4.23, this would put an upper limit on the NA of a lens intended for visible wavelength use to about 0.15, and that for near-IR use at about 0.25. For many applications, these correspond to slow lenses.

The direct-beam writing techniques suffer similar overall resolution limitation. The direct *e*-beam write could produce somewhat higher resolution, although this is the least practical method.

REFERENCES

1. Veldkamp, W.B.; McHugh, T.J. Binary optics. Scientific American. **1992**, May, 92–97.

2. Feldman, M.R. Diffractive optics move into the commercial arena. Laser Focus World, Oct 1994.

3. Jahns, J. System design for planar optics, optical society of America, topical mtg. *Diffractive Optics: Design Fabrication, and Applications*; New Orleans, LA, April 13–15, 1992; Series Vol. 9.

4. Born, M.; Wolf, E. *Principle of Optics;* Pergamon Press: London, 1956.

5. Williams, C.S.; Becklund, O.A. *Optics*; Wiley-Interscience: New York, 1972.

6. Sneddon, I.N. *Fourier Transforms*; McGraw-Hill: New York, 1951.

7. Swanson, G.J. Binary optics technology: the theory and design of multilevel diffractive optical elements. Technical Report 854. MIT Lincoln Laboratory, 1989.

8. RPC (Rochester Photonics Corp.), Rochester, NY.

9. Kunz, R.E.; Rossi, M. Phase-matched Fresnel elements. Opt. Comm. **1993**, *97*, 6–10.

10. Gale, M.T.; Rossi, M.; Schtz, H. Fabrication of continuous-relief microoptical elements by direct laser writing in photoresist. Proc. SPIE **1994**, *2045*.

11. Daschner, W.; Stein, R.; Long, P.; Wu, C.; Leo, S.H. One-step lithography for mass production of multilevel diffractive optical elements using HEBS gray level mask. SPIE **1994**, *2689*, 153–155.

12. Wu, C. Method of Making High Energy Beam Sensitive Glasses. US Patent 5,078,771, 1992.

13. Borrelli, N.F. Corning Inc., Unpublished data.

14. Canyon Materials Inc. Hebs-glass photomask blanks, CMI product information 96-18, San Diego, CA.

15. SPIE 1991 International Symposium, Miniature and Micro-optics: Fabrication and System Applications, Tech. Conf 1544, San Diego, CA, July 21–26, 1991.

16. SPIE 1992 International Symposium, Miniature and Micro-optics: Fabrication and System Applications II, Tech. Conf 1751, San Diego, CA, July 19–24, 1992.

17. Diffractive optics: Design, fabrication and applications, Technical digest. *Opt. Soc. Am.*; Rochester, NY, June 6–9, 1994; Series Vol. 11.

18. Sze, S.M. *VLSI Technology*, Chap. 4; McGraw Hill: New York, 1983.

19. Sze, S.M. *VLSI Technology*, Chap. 5; McGraw Hill: New York, 1993.

20. Fujita, T.; Nishihara, H.; Koyama, J. Fabrication of micro lenses using electron beam lithography. Opt. Lett. **1981**, *6* (12), 613.

21. *Diffractive and Miniatrized Optics*; Lee, S.H., Ed.; SPIE Optical Engineering Press CR49: Bellingham, WA, 1993, and papers contained. (see Ref. 22 & 25)

22. Shvartsman, F.P. Replication of diffractive optics. *Diffractive and Miniatrized Optics*; Lee, S.H., Ed.; SPIE Optical Engineering Press CR49: Bellingham, WA, 1993; 165–186 pp.

23. Behrmann, G.P.; Duignan, M.T. Excimer laser Micromachining for rapid fabrication of diffractive optical elements. Appl. Opt. **1997**, *36* (20), 4666–4674.

24. Wang, X.; Leger, J.R.; Rediker, R.H. Rapid fabrication of diffractive optical elements by use of image-based excimer laser ablation. Appl. Opt. **1997**, *36* (20), 4660–4665.

25. Ricks, D.W. Scattering from diffractive optics. *Diffractive and Miniatrized Optics;* SPIE Optical Engineering Press CR49: Bellingham, WA, 1993, 189–195.

5

Erect One-to-One Imaging

5.1. INTRODUCTION

The applications of microoptic lenses and arrays can be conveniently classified into three categories. The classification is somewhat arbitrary, but it provides a useful way to compare and contrast the advantages of lenses fabricated by one method over that of another. We could discuss the applications for each of the fabrication methods separately, but in doing so we would lose some of the comparative aspects, as well as have the same device discussed in more than one place. For example, for some of the application categories GRIN lenses dominate the other available methods. This means that some applications may be better suited to a particular fabrication method. Often, such factors as achievable lens diameter or ease of producing a given geometrical pattern or overall cost are as important in the choice as the optical performance. In the way we have chosen to deal with the applications, it will be easier to point out these advantages/disadvantages as they appear.

In this chapter, we deal with the application category which involves erect one-to-one imaging. These are arrays of lenses designed to produce collectively a single image, as shown in Fig. 5.1. The upper diagram shows the overlapping

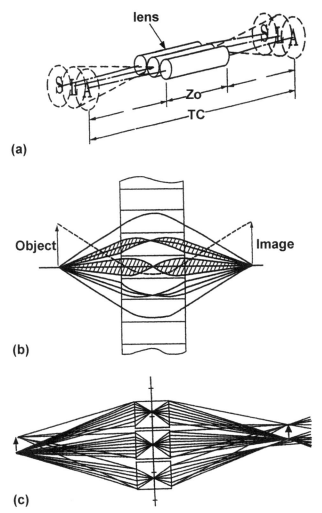

(a)

(b)

(c)

Figure 5.1 (a) Schematic of linear array of lenses operating in the erect one-to-one mode. Note overlapping of fields in the object and image planes. (b) Ray trace showing how erect one-to-one imaging occurs in a GRIN rod lens. (c) Ray trace showing how erect one-to-one imaging occurs in a thick biconvex lens.

fields and the lower two show the ray diagram of erect one-to-one imaging, (b) for the GRIN lens, and (c) for the refractive lens. The overlap of all of the individual image fields constitutes a single image. As we will see, they are used primarily for document scanning where the printed information is imaged onto a photosensor of some kind, as shown in Fig. 5.2a.

In Chapter 6, we cover the second broader category where precisely positioned two-dimensional arrays of individual lenses provide functions for individual detectors, light

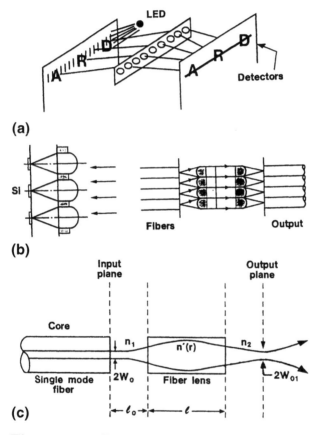

Figure 5.2 Representations of the microlens application categories: (a) one-to-one reading or scanning area; (b) 2D array for coupling light from laser arrays or fiber arrays; (c) single-element lenses used with single-fiber applications.

sources, or apertures. These applications essentially provide an optical interface for other microelectronic fabricated array devices, such as those based on CCDs (charge coupled devices), laser diodes, optical fiber arrays as indicated in Fig. 5.2b, and optical fiber-based devices as in Fig. 5.2c.

5.2. OPTICS OF ONE-TO-ONE ERECT IMAGING

As mentioned above, the one-to-one erect imaging lens array application derives primarily from the need for a spatially compact way to electronically read or scan a conventional 8.5×11 in. print document. From the simplest thin lens standpoint, it is straightforward to produce an erect one-to-one image. One places the object at a distance twice that of the focal length which would form a one-to-one inverted image at a distance $2f$ on the opposite side. By placing another identical lens $2f$ beyond this image, it would form the erect image as shown in Fig. 5.3. There are more elegant ways to do this with compound lenses, but the problem is still essentially the same, viz., the total optical path length is long: greater than four times the focal length of a typical imaging lens whose focal lengths are in the range 150–300 mm. This becomes even more of a problem when a faster lens is required. Consider again our simple case with the total conjugate distance of $8f$; for an $f/5$ lens the total distance from the object to the image is 40 times the lens diameter. This distance would be impractical for a compact document reader. What is normally done in most photocopiers is to use a folded path as shown in Fig. 5.4a. What the lens bar approach

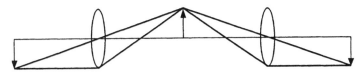

Figure 5.3 Conventional lens approach to forming one-to-one erect image.

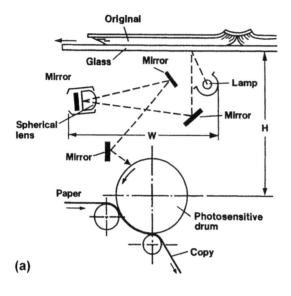

(a)

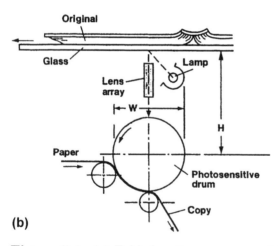

(b)

Figure 5.4 (a) Folded path arrangement for lens in photocopier application; (b) one-to-one lens array arrangement. (From Ref. [13].)

accomplishes is to allow the use of an array of small diameter lenses, which reduce the value of the total conjugate distance for the same *f* number. The imaging is accomplished through the overlapping of field of each lens in the array, as indicated in Fig. 5.1, so that the area covered is comparable to that of a larger single lens.

In the next section, we will derive the appropriate expressions for the one-to-one imaging situation for both the GRIN and conventional thick refractive lens cases. Diffractive element lenses are not particularly suited to this application primarily because of their poorer imaging capability and speed limitation in the visible wavelength region.

5.2.1. Refracting Lens Array

To derive the expression for the erect one-to-one imaging condition for a conventional thick lens, it is convenient to use the matrix method suggested in Sec. 2.1.1 along with the diagram shown in Fig. 5.5a [1,2]. We will use two rays emanating from the same point on the object a distance t_1 away from the lens: the a ray with initial position y_1 and slope of 0, represented by the vector $(y_1, 0)$, and the b ray which passes through zero at the lens with a slope of y_1/t, represented by the vector $(0, -y_1/t_1)$ [1]. Using each of these as input rays, one obtains

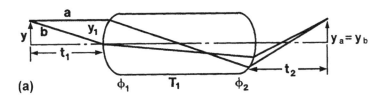

(a)

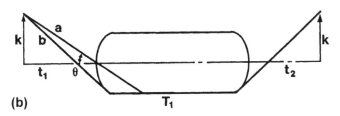

(b)

Figure 5.5 Diagrams showing the notation for the ray-trace method: (a) a and b rays defined for use with the matrix method; (b) maximum field height defined.

the exit ray vectors $\mathbf{r}_a$ and $\mathbf{r}_b$. One translated rays to the position t_2 behind the lens by applying the translation matrix

$$\mathbf{T} = \begin{pmatrix} 1 & t_2 \\ 0 & 1 \end{pmatrix} \tag{5.1}$$

The total operation on the two input vectors in then

$$\begin{pmatrix} y_0 \\ m_0 \end{pmatrix} = \mathbf{TM} \begin{pmatrix} y_{in} \\ m_{in} \end{pmatrix} \tag{5.2}$$

where $\mathbf{M}$ is defined from Eq. (2.4) and reproduced here for convenience:

$$\mathbf{M} = \begin{pmatrix} 1 - T/nf_1 & T/n \\ t/nf_1f_2 - 1/f_1 - 1/f_2 & 1 - T/nf_2 \end{pmatrix} \tag{5.3}$$

The parameters contained in $\mathbf{M}$ are the front and back curvatures defined as $1/f_1$ and $1/f_2$ and the lens thickness T. By equating the y value obtained from Eq. (5.2) for the a and b rays, one obtains the expression for the working distance t_1:

$$t_1 = \frac{T/nf_2}{T/nf_1f_2 - l/f_1 - l/f_2} \tag{5.4}$$

$$\frac{t_1}{t_2} = \frac{f_1}{f_2} \tag{5.5}$$

For the symmetric case, $f_1 = f_2 = f$, Eq. (5.3) reduces to the simple form

$$t = \frac{Tf}{T - 2nf} \tag{5.6}$$

Another expression that will be used later when the radiometric performance is discussed is the so-called maximum object field height for image-forming rays. At the erect one-to-one distance t, it corresponds to the maximum height above the axis from which light will be captured. A corresponding condition that occurs is when the position of the relay image, the small inverted image formed inside the lens, appears at the lens radius as shown; that is, $y = -R$. This is shown in Fig. 5.5b. To calculate the position of the relay image, one uses a portion of the matrix represented in Eq. (2.5). The matrix $\mathbf{TR_1}$ traces the ray after the first refraction, $\mathbf{R_1}$, and then translates the ray to a distance t within the lens. Using the a and b rays as shown, and setting $y_a = y_b = -R$, one can show that

$$k = \frac{2nRt}{T} \tag{5.7}$$

and the corresponding field angle of

$$\tan \Theta = \frac{2nR}{T} \tag{5.8}$$

One can also use two lens arrays stacked together as shown in Fig. 5.6. The one-to-one distance formula of (5.6) is still true with the condition that T is twice the thickness of the individual elements labeled T_2 in the figure [1]. The internal lenses act as field lenses and have the effect of increasing the light throughput. A field lens [3] is a lens that is placed at the field stop, which in this case is at the position of the relay image (Fig. 5.7). One can see from the ray diagram how the field lenses at the midpoint of the lens redirect the rays that otherwise would be lost to the back lens. It has no other effect on the optical system performance.

5.2.2. GRIN Lens Arrays

We can obtain the one-to-one imaging condition for a GRIN lens by using the paraxial matrix approach outlined in Chapter 3 (Sec. 3.2.3) [4–6]. We want to know what the distance l_0 should be to produce a one-to-one image for a GRIN

Figure 5.6 Stacked array to form erect one-to-one image. Intermediate lenses act as field lenses.

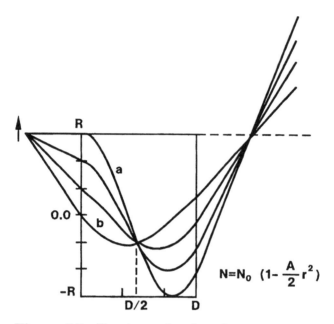

Figure 5.7 Ray trace showing the erect one-to-one imaging of a GRIN lens with parabolic profile. Also shown is the definition of a and b ray for use with the matrix method.

lens of thickness D with a given parabolic index profile characterized by the parameter A in the form of Eq. (3.19). We will use two rays eminating from a point R above the axis at a distance l_0 from the front of the lens. The transformation of the a ray $(R, 0)$ and the b ray $(0, -R/l_0)$ through the lens is obtained through the matrix equation (3.20), reproduced here for convenience:

$$\begin{pmatrix} y' \\ m' \end{pmatrix} = \mathbf{M} \begin{pmatrix} y \\ m \end{pmatrix} \tag{5.9}$$

Here, the primes are the exit values and $\mathbf{M}$ is written in terms of the distance traveled, z, in the graded index medium,

$$\mathbf{M} = \begin{pmatrix} \cos(\sqrt{A}z) & (1/N_0\sqrt{A})\sin(\sqrt{A}x) \\ (-N_0\sqrt{A})\sin(\sqrt{A}z) & \cos(\sqrt{A}z) \end{pmatrix}$$

$$\tag{5.10}$$

Equating the y values for the two rays yields the equation that the rays will come together at a distance z_0 inside the lens:

$$l_0 = \frac{-1}{N_0\sqrt{A}}\tan\sqrt{A}z_0 \tag{5.11}$$

Realizing that the action of the lens must be symmetric, one can immediately equate z_0 to $D/2$. Further, one can see that the value of the angle of $AD/2$ must fall in the second quadrant, that is, $\Pi/2 < (\sqrt{AD}/2) < \Pi$. Similar to the refracting case, the maximum field height occurs when the relay image forms at the lens boundary as shown in Fig. 5.8. One can show this from either of the a or b ray equation by setting the position of the relay image equal to $-R$. One obtains the maximum expression

$$k = -R\ \sec\frac{(A)^{1/2}D}{2} \tag{5.12}$$

The field angle is essentially k/l_0, which is given by the following expression:

$$\tan\theta = \frac{N_0 R(A)^{1/2}}{\sin[(A)^{1/2}D/2]} \tag{5.13}$$

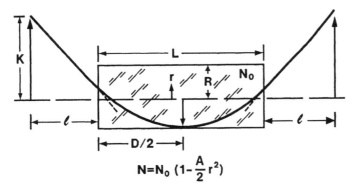

$$N = N_0\left(1 - \frac{A}{2}r^2\right)$$

Figure 5.8 Diagram showing the ray that determines the maximum field height for a GRIN lens with a parabolic index profile.

It is worthwhile noting that the quantity $N_0 R(A)^{1/2}$ is the normal definition of the maximum $\text{NA} = (A)^{1/2}(2N_0 \, \Delta n)$

5.3. DEVICE APPLICATIONS

One can catergorize the mode of operation of one-to-one erect imaging arrays into two classes, line scan and field scan [7,8]. Schematically, one can represent these two modes as shown in Fig. 5.9. The essential difference of these two modes of operation is in the radiometric aspects. In the line scan, as the name implies, the amount of light captured by the array

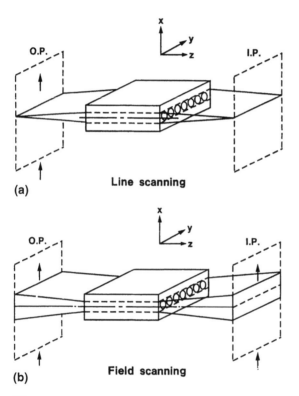

Figure 5.9 Two types of scanning applications. (a) The top is the line scan mode where each line is directly imaged to detectors. (b) The bottom is the field scan mode where the photoreceptor moves with the object for some distance.

and delivered to the image plane from any point in the object plane is simply governed by the NA of the array. This is the common way the speed of a lens is measured, array or otherwise. Alternatively, one often uses the f number or radiometric efficiency which are related as follows: $NA = (\varepsilon)^{1/2}$, $f/N = 1/2NA$. Later on in this section, we will derive the explicit expressions for the efficiencies.

For the field scan mode, the image position is maintained on the photoreceptor over some fixed interval of time, since both the object and image are moving together over some distance. For a measure of the sensitivity, one integrates the irradiance function, which is proportional to the NA as above, over the distance of travel corresponding to the time of exposure. In Fig. 5.9b, this is the x direction. This can be made somewhat more clear from Fig. 5.4a, where it is seen that both the object and the photosensitive drum move together. The maximum distance that the image remains stationary on the drum is limited by either the width of the array in the direction of movement, or the drum curvature. Field scan application requires high resolution, with speed a secondary attribute.

Common applications are listed in Table 5.1, along with important features and properties. We deal with each of these devices in the following sections.

5.3.1. Field Scan/Copiers

For the application of lens arrays to copiers, one has to be concerned with both resolution and radiometric issues [2,5]. For copiers, the image quality must be sharp which is usually quantitatively measured by an MTF (modulation transfer

Table 5.1 Applications of Erect One-to-One Imaging Lens Bars

Device	Type	Important properties
Copying machines, printers, scanners	Field scan	Compactness, resolution, depth of focus, illumination uniformity
Fax readers	Line scan	Compactness, speed, resolution, depth of focus

function) measurement. The MTF is a measure of the contrast of the image as a function of the spatial frequency. One is also concerned about how sensitive this function is when one moves away from the optimum object or image distances. Copiers have to be relatively forgiving to various copying conditions without serious loss of quality. For example, when one is copying a book with a thick binding, it is often difficult to maintain the printed page in intimate contact with the glass platten. One would still hope to make a decent copy under these circumstances.

From the radiometric aspect, what is equally important as the speed of the array is the uniformity of the irradiance. This is a function of the irradiance profile of the individual lens and how the lenses are spaced [2,9]. Copiers are relatively sensitive devices, and if the irradiance modulation is too great, then this is translated into unwanted lines or streaks on the copy. This is especially true when one wants to make a dark copy.

5.3.2. Line Scan/Facsimile Readers

A cross-section of typical fax reader assembly, known as a CIS (contact image sensor), is shown in Fig. 5.10 [9,10]. The speed of the unit is the most significant property, because it controls the data rate. The resolution requirement is not high since the printed word is the most common feature to be read. There are, of course, more expensive fax readers containing conventional lens systems that can read documents with resolution equal to that of copiers. The device shown here is the version that appears in many inexpensive fax machines.

What is important for these devices is that there exists a sufficient signal differential between what corresponds to black and what corresponds to white on the document. The threshold trigger voltage is set midway between the two signal values and must accommodate light level variations that occur from document to document. The typical output from a CIS unit using a Corning SMILE lens array is shown in Fig. 5.11 for two spatial frequencies. A comparison of the maximum and minimum signal levels achieved in the CIS

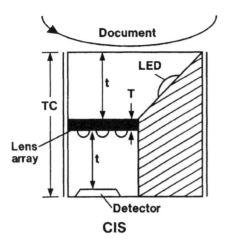

Figure 5.10 Contact image sensor (CIS) using fax reading lens for facsimile machines. The lens array tranfers the the information from the document to the detector array via the line scan mode. The typical distance for *t* is 14 mm.

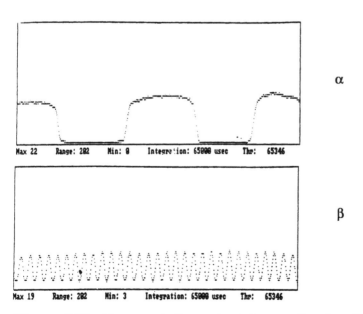

Figure 5.11 Representative output from a Corning SMILE fax lens array at two spatial frequencies. (See Ref. [10].)

unit using the Selfoc SLA-20 array and the Corning SMILE™ array is shown in Fig. 5.12a. Another comparison that is made is that of the irradiance uniformity that is shown in Fig. 5.12b. This more uniform irradiance is what would be expected because the SMILE array is made up of a larger number of closely spaced smaller lenses. We will see the quantitative explanation of this fact in the next section. Another property of interest is the change in contrast as a function of altering the object distance. This is because when the document is fed into the reader roll assembly, there is some "slop" in the object distance. A typical measurement is shown in Fig. 5.13.

5.4. RESOLUTION/CONTRAST

5.4.1. Measurement Technique and Typical Results

As mentioned above, the key property of any copying lens is the quality of the copy produced, or in optical terms, what would be referred to as the resolution. The quantification of the resolution is usually an MTF measurement where test patterns of given spacings are imaged by the lens which is then scanned to produce an intensity modulation. This measurement technique is shown in Fig. 5.14. A test target, usually chrome lines on clear glass, is illuminated with a lambertian source, here approximated by a diffuse surface integrating sphere. The target has a series of light–dark lines of progressively higher spatial frequency. The target is moved across the lens as shown. The image produced by the test lens illuminates a narrow slit in front of the detector and is displayed as a temporal modulation as a consequence of the target motion. The typical output is shown schematically in Fig. 5.15. The inset shows how the contrast would diminish with increasing spatial frequency.

An example of the contrast and depth of focus performance of a two-row GRIN-based Selfoc array, as specified by the inset to the graph shown in Fig. 5.16 [13]. The NSG-type lens is an SLA-12 signifying the field angle of 12°. The contrast results for three conditions of depth of focus are shown

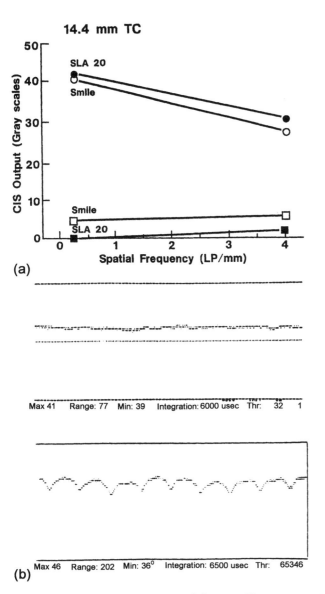

(a)

(b) Max 46   Range: 202   Min: 36⁰   Integration: 6500 usec   Thr: 65346

Figure 5.12 Comparison of fax reading contrast performance of SELFOC lens array and SMILE array. The upper line is read as white, and the lower is read as black. In the two curves below are shown the uniformity of the output to a uniform white source. The upper curve is the SMILE array and the lower in the SELFOC array. (See Ref. [10].)

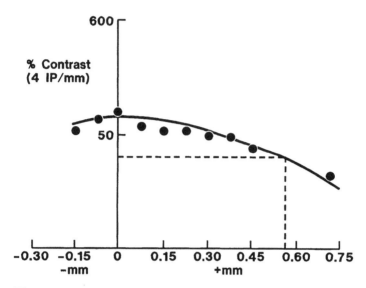

Figure 5.13 Variation in the percent contrast as a function of the offset of the object distance for fax reader. (See Ref. [10].)

in Fig. 5.17. In the first case of Fig. 5.17a, the object and image plane are symmetrically displaced; in the second, Fig. 5.17b, the image plane alone is moved; and in the third case, Fig. 5.17c, the total conjugate distance is maintained and the lens is moved. One can see that the third situation

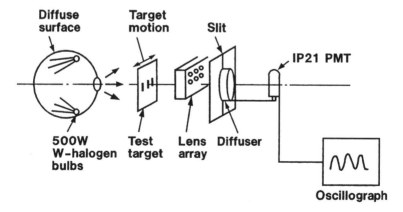

Figure 5.14 Diagram of setup to measure MTF for lens array.

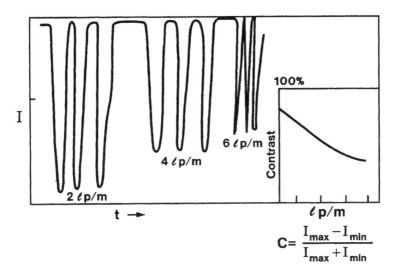

Figure 5.15 Typical MTF results at three spatial frequencies and resulting percent contrast plotted against spatial frequency.

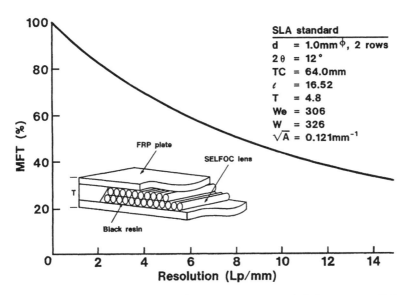

Figure 5.16 Experimental contrast vs. spatial frequency data for a SELFOC SLA lens array. Insert shows the construction of the two-row array. (From Ref. [13].)

is the most demanding in that both object and image distances are altered. The result shown in Fig. 5.17b is often called the "through focus measurement" and is of particular significance to copier applications because of the variation in the image distance due to the wobble in the photoreceptor drum as it rotates.

An example of the contrast and depth of focus performance of a refractive element-based Corning SMILE array is shown in Figs. 5.18 and 5.19 [14]. In Fig. 5.18, the contrast vs. symmetric working distance is plotted for four spatial frequencies. The array geometry is shown at the bottom. In Fig. 5.19, the through focus behavior taken at 4 lp/mm is shown. Another direct, yet somewhat qualitative test, is to observe a copy of a standard test pattern such as shown in Fig. 5.20. The lines in the pattern are marked with respect to the spatial frequency from 2 to 10 lp/mm.

5.4.2. Factors That Determine Contrast

The resolution of a one-to-one erect imaging array depends not only on the resolving capability of the individual lenses which makeup the array but also on the factors that influence the coincidence of their collective images, that is, array factors. As we have seen, depending on the maximum field height, a relatively large number of lenses are contributing to a given point of the image. Any variation in the optical parameters of these contributing lenses lead to a blurring of the collective image. To make this point, we show some data in Fig. 5.21, where the MTF was measured on a single microlens from a Corning SMILE array and then compared to the contrast measured with the array. One can see the large difference in contrast performance. One can estimate that about 12–15 lenses contribute to any given image point [2]. This estimate is obtained from the drawing in the inset where the maximum field-height radius is superimposed on the lens arrangement. Now consider the relationship between the image working distance and the lens parameters given above for the refractive lens case, viz., Eq. (5.6). If one takes the derivative of the expression with respect to the lens thickness

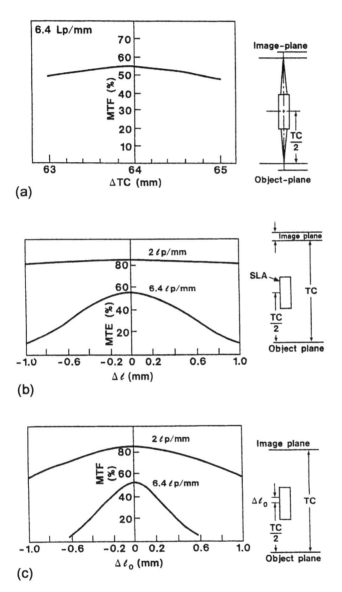

Figure 5.17 Effect of offset from optimum position on the contrast for SELFOC SLA lens array: (a) symmetric offset of total conjugate distance; (b) offset of image plane holding object distance constant; (c) unsymmetric variation of object distance and image distance holding TC constant. (From Ref. [13].)

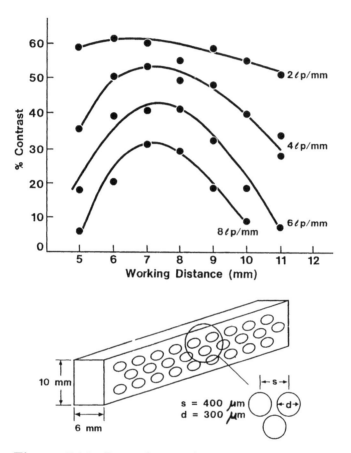

Figure 5.18 Dependence of contrast on working distance measured at four spatial frequencies for SMILE lens array. The structure of the array is shown in the inset.

T, or the lens focal length f, one obtains an estimate of the changes that would occur in the image plane distance for given variations in the these two parameters

$$\delta t = \{nf^2(T - 2nf)^2\}\delta T = \left(\frac{t^2}{T}\right)\left(\frac{\delta T}{T}\right) \tag{5.14a}$$

$$\delta t = \left\{\frac{T^2}{(T - 2nf)^2}\right\}\delta f = \left(\frac{t^2}{f}\right)\left(\frac{\delta f}{f}\right) \tag{5.14b}$$

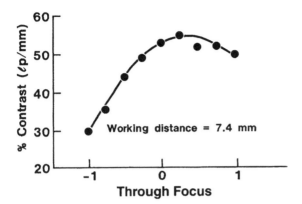

Figure 5.19 Effect of offset of object distance on the contrast for SMILE lens array.

One can appreciate how variations among the contributing lenses can cause degradation of the image, especially for designs where the total conjugate, hence the working distance t, is long. This is experimentally shown in Fig. 5.22 where the 4 lp/mm contrast is shown for arrays with lengthening optimum working distances.

For the Selfoc GRIN arrays, the problem is not as bad because the ratio of the field height to the lens radius is

Figure 5.20 Copy of test pattern formed by SMILE lens array used in a photocopier with total conjugate distance of 34 mm.

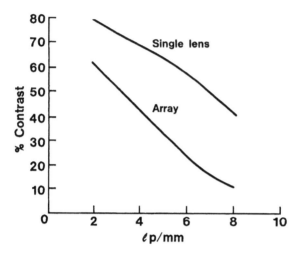

Figure 5.21 Comparison of the contrast vs. spatial frequency of a single lens from the SMILE lens array and the array as whole.

smaller and there are only two rows. Only about three or four lenses contribute to a given point in the image plane. One can go through the same anaylsis with Eq. (5.11) as we did to obtain Eqs. (5.14), with the independent variables being the

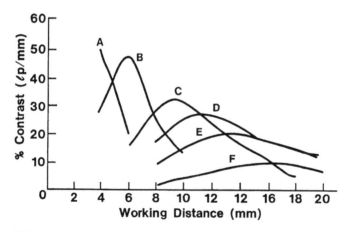

Figure 5.22 Contrast vs. spatial frequency for a number of SMILE lens arrays with different working distance. Longer working distance has poorer contrast because of greater sensitivity to image overlap.

lens thickness D and the gradient parameter $\sqrt{A}$; e.g.,

$$\delta l_0 = -\left(\frac{1}{2N_0}\right) \sec^2\left[\frac{(A)^{1/2}D}{2}\right](\delta D) \qquad (5.15)$$

One can see that this term will be large when $\sqrt{A}D/2$ is close to $\Pi/2$, which is equivalent to the refracting case when T is close to nf. It turns out that for the working distances used in devices, the latter is more likely the case than the former. The explanation for this is not mysterious; it is because the GRIN arrays were used first in scanning devices and were incorporated in the optimal way.

This discussion of what affects the contrast provides an excellent example of how the fabrication method influences the ultimate performance for a given application. The subtle point is that although the overall function is equivalent, the limitations provided by the specific method, like lens diameter, thickness, and strength, are significantly different. The GRIN method of assembling cut ion-exchanged rods of glass into two rows to form an array has certain limitations. For example, handling rods as small as 200–400 µm is impractical, or assembling more than a two-row stack is difficult. On the other hand, for the refractive array made by the photosensitive process, lens diameters above 400 µm begin to deviate from perfect sphericity, and having the thickness more than 3 mm is impractical because of the optical exposure step. Each application of one-to-one arrays has its own set of desirable and necessary properties which may not be met by any one fabrication method.

5.5. RADIOMETRY

The initial radiometric challenge is to determine the irradiance profile of a single lens. The overall profile will come from the addition of the individual profiles according to their two-dimensional layout. The method used to calculate the irradiance profile can be found in the literature. We will briefly review the method and results following that of Borrelli et al.[2]. The basic idea is to treat the exit pupil as a Lambertian source with radiance N. For a given object point, at the one-to-one image distance, the exit pupil is found by tracing a fan of rays to the edges of the entrance pupil, through the lens to the rear surface. This is shown in Fig. 5.23 for three cases: the simple refracting lens, the same lens with field lenses at the midplane, and finally the GRIN lens. The irradiance can be expressed in the paraxial approximation as

$$h(r) = h_0 \pi R^2 A(y) \tag{5.15}$$

where R is the radius of the lens, h_0 is the axial irradiance $N\Pi R_2/t_2$, and $A(y)$ is the function determined by the overlap of the projection of the front aperture to the back face, as shown in Fig. 5.23c. The expression for the shaded area is given by

$$A(y) = (2/\pi)\left\{\cos^{-1}\left(\frac{\Delta}{2R}\right) - \left(\frac{\Delta}{2R}\right)\left[1 - \left(\frac{\Delta}{2R}\right)^2\right]^{1/2}\right\} \tag{5.16}$$

One can relate the displacement $\Delta/2R$ to the lens parameters by the carrying through the method described in Fig. 5.23 by the matrix ray trace outlined above for the three cases. For example, for the refractive case, it is easy to show that the expression

$$\frac{D}{2R} = \frac{y}{k} \tag{5.17}$$

where k is the maximum field height. For the GRIN case the derivation appears in Ref. [5]. Using Eq. (5.15), one can obtain the irradiance profiles for the two lens types.

Refractive lens:

$$h(r) = \left(\frac{2}{\pi}\right)\left(\frac{N\pi R^2}{t^2}\right)\left\{\cos^{-1}\left(\frac{r}{k}\right) - \frac{r}{k}\left[1 - \left(\frac{r}{k}\right)^2\right]^{1/2}\right\} \tag{5.18}$$

where $k = 2\pi R\tau/T$.

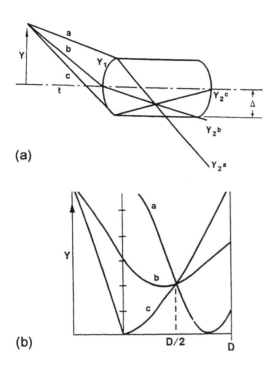

(a)

(b)

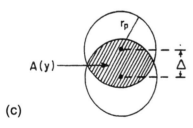

(c)

Figure 5.23   Rays showing how the radiometric efficiency is calculated. In all cases the procedure is to ray trace the two boundary rays and the central ray from a given point in the one-to-one object plane to the back lens face. The overlap is defined as shown. The refracting case is shown in (a), the refracting case with field lenses is shown in (b), and the GRIN case is shown in (c).

GRIN lens

$$h(r) = h_0 \left[1 - \left(\frac{r}{k}\right)^2 \right]^{1/2} \tag{5.19}$$

$$h_0 = \pi N n_0^2 \, AR^2 \cos\left[\frac{(A)^{1/2}D}{2} \right]$$

where $k = -R \sec(\sqrt{A}D/2)$. The profiles are shown graphically in Fig. 5.24 along with a third case of the refractive lens with field lenses. We have chosen to show the case where all the lens curvatures are equal; this is not the optimum field lens curvature case. The optimum case is when $f_{\text{int}} = T/n$, which results in all the light that enters the lens, up to the maximum field height, exits the lens. In this case, the irradiance profile is flat [2].

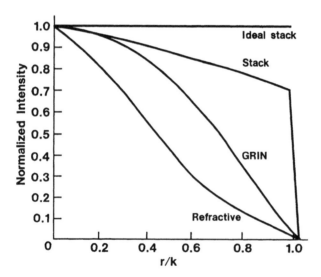

Figure 5.24 The computed irradiance profile for the three cases shown in Fig. 5.23. Actually, two cases are presented for the stacked case, one where all the four lens curvatures are the same, and the other, labeled "ideal", where the field lens curvature is optimum.

The overall power delivered by a single lens is then the integral of the irradiance profile:

$$P_s = \int h(r)r \, dr \, d\theta \tag{5.20}$$

For the refractive lens, this is given by the following expression:

$$P_s = \left(\frac{\pi}{4}\right)h_0 k^2 = NR^2 \left(\frac{\pi n_0 R}{T}\right)^2 \tag{5.21}$$

and for the case with perfect field lenses it is four times this value.

For the GRIN case, the power is given by the following:

$$P_s = h_0(\pi k^2) \tag{5.22}$$

To determine the radiometric speed and uniformity of an array of such lenses, we must consider the overlap of the irradiance profiles according to the specific geometric arrangement. For example, consider the continuous hexagonal close-packed (hcp) arrangement shown in Fig. 5.21a. The total power delivered in the area A containing n_x lenses in the x direction and n_y lenses in the y direction is simply

$$P = n_x n_y P_s \tag{5.23}$$

where the area is given by the following expression as indicated from Fig. 5.25a.

$$A = (2bRn_x)((3)^{1/2}bRn_y) \tag{5.24}$$

The average irradiance of the array $\langle h \rangle$ would be P/A and the radiometric efficiency would be this number normalized to the radiance ΠN. Thus, the average radiometric efficiency of an array of lenses arranged in a hcp structure would be

$$\varepsilon = \left(\frac{1}{\pi N}\right)\left(\frac{1}{2(3)^{1/2}b^2R^2}\right)P_s \tag{5.25}$$

where P_s is the power delivered by a single lens.

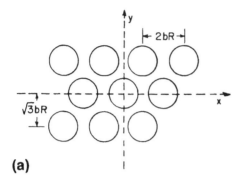

(a)

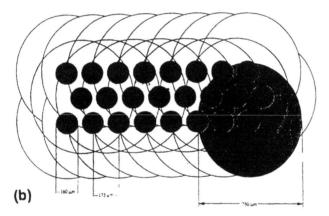

(b)

Figure 5.25 Geometric arrangement of lensets in SMILE array and the representation of the overlap of their fields. (From Ref. [2].)

For Selfoc lenses made up of p rows, Lama [15] has shown that the average irradiance is given by the following expression:

$$ph_0\left(\frac{\pi R}{4bk}\right)\left[1 - (p-1)^2\left(\frac{b^2}{4a^2}\right)\right] \tag{5.26}$$

Using the same method, but not performing the averaging, one can estimate the uniformity of the irradiance. As an example, the irradiance profile is shown in Fig. 5.26 for a two-row Selfoc array.

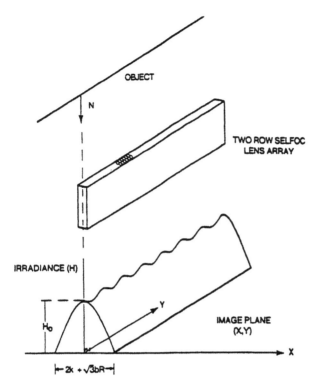

Figure 5.26 Irradiance uniformity of a two-row SELFOC lens bar. (From Ref. [9].)

5.5.1. Experimental Measurement

The radiometric efficiency of an array can be measured by the relatively simple experimental arrangement shown in Fig. 5.27. By defining the Lambertian source and detector areas, and setting the reference distance at the correct total conjugate distance, one can obtain reasonably accurate estimates of the efficiency. Some results obtained for SMILE arrays by using this method are shown in Table 5.2, where they are compared to the theoretical values obtained from Eq. (5.25).

   The radiometric efficiencies of the GRIN lens arrays can be obtained from Table 5.3, where the efficiency is equal to the square root of the numerical aperture.

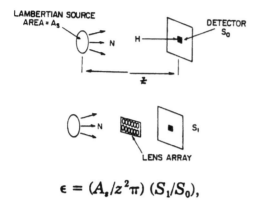

$$\epsilon = (A_s/z^2\pi)\,(S_1/S_0),$$

Figure 5.27 Representation of experimental method to measure radiometric efficiency.

5.5.2. Field Scan Exposure

We mentioned above that in the field scan mode, the image is maintained on the photorecepter for some time. This means that the one has to integrate the irradiance function over this time interval. Formally, one would write that the field scan exposure E would be expressed as

$$E = \left(\frac{1}{v}\right) h(x, \mathbf{y})\, dx \tag{5.27}$$

where x is the direction of travel with velocity v [2]. In order to deal with the geometry of the lens layout, the irradiance profiles like those of Eqs. (5.18) and (5.19) are initially averaged in the y direction, transverse to the motion direction. We will do this for the single-row GRIN case as an example following the method of Lama [15]. We have seen that the irradiance profile for a single lens is given by Eq. (5.19). To average over the y direction, we have to evaluate the integral

$$h(x, \mathbf{y}_0) = \left[1 - \frac{x^2 + y^2}{k^2}\right]^{1/2} \tag{5.28}$$

from $\pm(k^2 + x^2)^{1/2}$. (See Fig. 5.28.) This yields the following expression:

Table 5.2 Listing of Properties of SMILE Lens Arrays

Lens diameter/ spacing (μm/μm)	Thickness (mm)	TC (mm)	Measurement[a] (%)	Calculation[b] (%)
Thick single element				
400/480	6	25	0.13	0.16
350/420	6	25	0.09	0.12
350/420	6	25	0.10	0.12
450/540	6	30	0.10	0.20
450/540	6	25	0.12	0.20
Thick double element				
310/350	5.5	21	0.27	0.45
310/350	7	24	0.20	0.28
310/350	7	20	0.21	0.28
310/350	6	25	0.23	0.38
400/480	8	21	0.28	0.35
Thin single element				
160/195	0.83	5	2.7	1.4
160/195	1.85	11	0.5	0.27
200/240	2.75	13	0.40	0.21
200/240	2.75	12	0.45	0.21
Thin double element				
160/195	1.45	9	1.1	1.6
160/195	1.45	8	1.5	1.6
160/195	1.45	16	1.5	1.6
160/195	1.45	11	1.2	1.6
200/240	1.40	9	2.2	2.9
200/240	1.30	11	1.9	3.3

[a] Radiometric efficiency measured by method shown in Fig. 5.27.
[b] Calculated from Eq. (5.25)

$$h_0 = \left(\frac{\pi}{2k}\right)(k^2 - x^2) \tag{5.29}$$

Now one adds the irradiance from total number of lenses in the row, say m, and divides by the total length $m\,(2bR) + 2k$ and gets

$$\frac{h(x,\mathbf{y})}{h_0} = \frac{\pi}{4kbR}(k^2 - x^2) \tag{5.30}$$

Table 5.3 Listing of Properties of Selfoc Lens Arrays

	Symbol	SLA-06	SLA-09	SLA-20
		Type		
Lens diameter	D (mm)	1.055	1.055	1.055
Refractive index	N_0	1.538	1.606	1.563
Numerical aperture	NA	0.1	0.15	0.37
Gradient constant	A (mm^{-2})	0.016	0.032	0.201
Fiber length	Z (mm)	27–31	18–22	7–9
f Number	f/N	5	3.5	1.0
Thickness	t	4.8	4.8	4.0
Total conjugate	TC (mm)	64–75	46–55	16–20
Irradiance uniformity[a]	%	7	7	15
Number of rows		2	2	2

[a] See Fig. 5.27 for definition.
Source: From various product information sheets supplied by Nippon Sheet Glass Co.

We can now substitute Eq. (5.30) into (5.27) to obtain the expression for the single-row exposure:

$$E_1 = \frac{\pi N T n_0 A R^3}{3bv} \tag{5.31}$$

For p rows, the answer is essentially p times this result.

For the refractive lens case, where the lenses are smaller and closer spaced so that p times the interrow spacing is

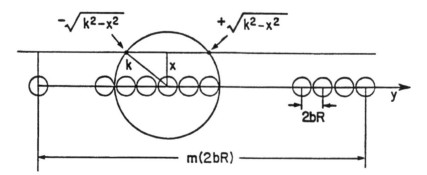

Figure 5.28 Method of determining the irradiace profile of a single row of lenses.

essentially the width, the exposure is

$$E_r = \frac{\pi^2}{2(3)^{1/2}} W(nR/bT)^2 \tag{5.32}$$

Here W is the width of the array.

5.6. FURTHER ASPECTS OF FABRICATION

In addition to the general description given in Chapter 2 of how the Selfoc GRIN and the Corning SMILE arrays were made, it is worthwhile here to discuss how the fabrication technique is optimized for the specific application, in this case one-to-one arrays. From a historical perspective, the first lens array intended for use in the one-to-one mode is attributed to Moorhusen [16], who proposed a three-piece molded plastic structure.

For the Selfoc arrays, the inset in Fig. 5.16 provides a good picture of the two-row array. A single row of lenses is laid up on a plate and one layer is placed over the other to form the two-row array in the fashion shown. The array is then potted in a black resin and ground and polished to the correct thickness. NSG makes a number of different versions of these two-row arrays. The major variable is the Δn which is controlled by the kind and amount of ion-exchanged dopant. The parameter that reflects this is the index profile constant labeled A. The range of SLA arrays made by NSG under the Selfoc trade name is shown in Table 5.3. One can see that the number following the SLA corresponds to the field angle which as we have seen above is related to the numerical aperture. In general, one can say that the mechanical fabrication of the Selfoc arrays does not significantly differ for the different applications, but the chemical part does.

For the Corning SMILE arrays, the fabrication for the field scan applications, where radiometric efficiency is not a real issue, is done by double side exposing a final thickness bar of glass, followed by the thermal development shown in Fig. 5.29a. Typically, the thickness is 6 mm, as shown in Table 5.2. One notes that radiometric efficiency is quite low,

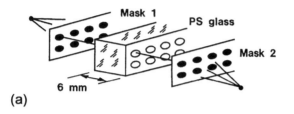

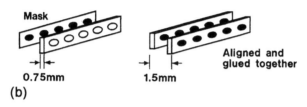

Figure 5.29 Schematic representation of the exposure method used to make the photo-sensitive SMILE™ one-to-one imaging lens arrays. (a) The double sided exposure method and (b) the structure of the stacked array.

in the order of 0.2%. However, for field scan devices, it is the exposure that counts, not the radiometric efficiency. The relative exposure number, obtained from Eq. (5.32) for a 10-mm wide, 6-mm thick, 400-μm lens-diameter array with a separation parameter of $b = 1.2$ is 43. For the Selfoc SLS-6, with a radiometric efficiency of 1%, the relative exposure is 33, obtained from Eq. (5.31).

For the line scan applications, the story is quite different. The speed of the array is paramount and the Corning SMILE arrays for this application are made from stacked thin arrays, as shown in Fig. 5.29b. From Eq. (5.25), one sees that the efficiency is increased with the inverse square of the thickness, and that the power delivered by an array with field lenses is four times that of one without. From Table 5.2, one can see that using this geometry yielded efficiencies as high as 2% compared to the SLA-20 with a 7% efficiency.

One of the disadvantages of the lens bar approach is that it only can image one-to-one. Copiers with folded conventional lenses have magnification and reduction features. Rees et al. [17] proposed to produce the reduction and magnification features by having the individual GRIN rod lenses move away from the perpendicular direction as one moves away from the center of the array. This brings the images together before (or after) the normal image plane; thus the individual images are smaller (larger) at the point of coincidence. One problem with this idea is that the rod lengths must be gradually adjusted to maintain the same image distance.

REFERENCES

1. Borrelli, N.F.; Morse, D.L.; Bellman, R.H.; Morgan, W.L. Photolytic technique for producing microlenses in photosensitive glass. J. Appl. Phys. **1985**, *24*, 2520–2525.

2. Borrelli, N.F.; Bellman, R.H.; Durbin, J.A.; Lama, W. Imaging and radiometric properties of microlens arrays. Appl. Opt. **1991**, *30* (25), 3633–3642.

3. Ditchburn, R.W. *Light*; Blackie and Sons, Ltd.: London, 1952.

4. Kapron, F.P. Geometric optics of parabolic index gradient cylindrical lenses. J. Opt. Soc. Am. **1970**, *60* (11), 1433–1436.

5. Kawazu, M.; Ogura, Y. Application of gradient-index fiber arrays to copying machine. Appl. Opt. **1980**, *19* (7), 1105–1112.

6. Rees, J.D. Non-Gaussian imaging prop. of GRIN fiber lens arrays. Appl. Opt. **1982**, *21* (6), 1009–1012.

7. Sono, K. SELFOC^{TM} Technology. *IFOC Buyers Guide Handbook*, 1980–1981 .

8. Nippon Sheet Glass (NSG). Product Sheet *SELFOC^{TM} Rod Lens Data Sheet*; NSG America: Clark, NJ 07066, 1981.

9. Matsushita, K.; Tovama, M. Unevenness of illumination caused by GRIN arrays. Appl. Opt. **1980**, *19* (7), 1070–1774.

10. Bellman, R.H.; Borrelli, N.F.; Mann, L.G.; Quintal, J.M. *Fabrication and Performance of a 1-to-1 Erect Imaging Microlens Array for FAX*, SPIE; *Miniature and Microoptics*, 1991; Vol. 1544, 209–217.

11. Matsushita, K.; Ikeda, K. *Newly Developed Glass Device for Image Transmission*. SPIE Conference, San Mateo, CA 1972.

12. Micro-optics has macro potential. *Laser Focus World*, June 1991, p. 93–95.

13. Nippon Sheet Glass Co. Selfoc™ linear lens array. *Data Sheet*; NSG: Clark, NJ 07066, September 1979.

14. Corning, Inc. Private data.

15. Lama, W.L. Optical properties of GRIN fiber lens arrays. *Appl. Opt.* **1982**, *21*, 2739–2746.

16. Moorhusen, R.W. US Patent 3,544,130, 1970.

17. Rees, J.D.; Kay, D.B.; Lama, W.L. US Patent 4,331,380, May 1982.

6

Two-Dimensional Arrays

6.1. INTRODUCTION

In this chapter, we deal with microlens array applications where the major feature will be the ability to precisely fabricate and position microlenses in two dimensions. Linear arrays are a special important case, requiring no less precision. We will try to cover the important lens array applications that include the various lens fabrication techniques discussed in preceding chapters. In the description and discussion of these applications, we shall try to make clear why one fabrication method might be better suited-over others for a particular application.

For any application, adequate lens performance will always be required, but it will be no more important than the ability to precisely pattern the lenses accordingly. In other words, one may not be able to form a lens with ideal performance, and at the same time produce thousands of them per square millimeter precisely positioned in a particular two-dimensional pattern. As we have seen in the previous chapters which dealt with the fabrication methods, some

Table 6.1 Microlens Array Applications

Application	Size[a] (μm)	Array type[b]	Key feature[c]
Linear			
Laser diode array collimators	200	GRIN	NA
Expanded beam for SMF arrays	120	GRIN, refractive	Diffraction limited
Switching 4 × 4 crossbar	>120	Refractive	Alignment, cross-talk
Computer backplane	>200	Refractive	NA
Camera autofocus	165	Refractive	Alignment, cross-talk
LED print bar	85	Refractive	NA
2D			
Shack-Hartmann	>300	Refractive	Long focal length
Parallel processing (VCSELS)	>20	Diffractive, GRIN	Lens diameter
Projection LCD-TV	100	GRIN, refractive, diffractive	Alignment, large area
Laser protection	>300	Refractive	Viewing angle
3D photography	>1000	Refractive	Large area

[a] The lens diameter represents the range used, not necessarily the optimum.
[b] Type of lens used; GRIN means gradient index.
[c] The key feature is to indicate the dominant property that the array should possess for the given application.

fabrication methods are better suited to make very small lenses, others better to be produced at precise locations, and still others with better imaging performance. It will be the trade-off of one or more of these factors that will determine the best way to accomplish the desired end. The applications that will be covered are listed in Table 6.1, along with the fabrication method.

6.2. APPLICATIONS

The applications that we cover in the next sections are intended to be representative of the wider field. In no way is this intended to be an exhaustive study. The intention is to give the reader a feeling for the areas of applications, the reasons that they are important, and the particular performance criteria required. To facilitate the process, we will use Table 6.1 as the guide to the discussion of the various applications of two-dimensional microlens arrays that we have chosen to study. The typical lens dimension is listed in the second column, although this really refers to the separation distance ultimately determined by the device that the lens is to be aligned to. For example, if it were a CCD array, it would be the center-to-center distance of the active detector areas on the silicon wafer. In the third column, we list the type of microlens that has been reported for the particular application, and finally, in the last column we list the key, or limiting property for the given application, if there is one. As an example, for the laser diode application, the numerical aperture of the lens is crucial because the NA of the laser diode is usually > 0.5. However, in a fiber-to-fiber application, the fiber NA is low, and the major issue is insertion loss. This often translates to alignment.

The first five applications that are listed are primarily aimed at optical fiber-based systems. Because of many common requirements, these will be discussed as subsections in the same overall section. The other applications are unique and will be discussed separately. The general approach will be to give a brief introduction to the particular application and the key requirements, and then to follow with the experimental results obtained from one or more of the fabrication techniques that have been reported.

6.2.1. Optical Fiber-Based Applications

6.2.1.1. Laser Diode Array Collimators

In this application, it is desired to efficiently collect light from a high NA laser diode (LD) array and provide this as an input to an array of single mode fibers (SMF), as shown in Fig. 6.1 [1–3]. The application requires two things. The first is to efficiently collect the light from the LD and the second to image it so that it is efficiently captured by the SMF. For the first, it is necessary to consider the numerical aperture

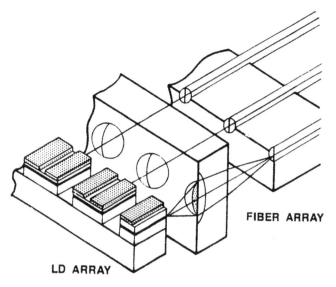

FIBER ARRAY

LD ARRAY

Figure 6.1 Schematic drawing of a linear lens array being used to couple the output from a laser diode array to a single-mode fiber array.

of the microlens, which is a measure of the amount of light that is collected. For the second, it is important to consider those factors that control the coupling efficiency into the fiber, for example, the spot size in the focal plane of the fiber.

The problem for ordinary spherical lenses, or what would be equivalent to them in GRIN microlenses, is that these two properties are in opposition. To increase the efficiency of capturing the light from the high NA source, one naturally encounters a poorer focal spot definition as a consequence of added spherical aberration. The higher the numerical aperture, the greater the contribution of spherical aberration and thus the more blurred the focus spot. This is shown from the ray traces of Fig. 6.2 for a biconvex (spherical curvature) structure having a 200-μm diameter. The extent of spherical aberration as indicated by the blurring of the focal spot is shown as a function of the increasing paraxial numerical aperture (radius of the lens divided by the effective focal length).

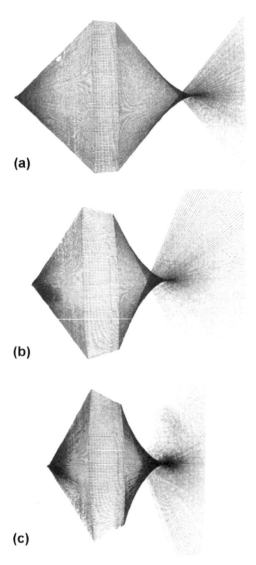

Figure 6.2 Computed ray-trace diagrams for microlenses with different numerical apertures to illustrate the effect of spherical aberration. The ray trace has a different scale for the vertical and horizontal directions. The lenses are 200 mm in diameter with a biconvex structure. The numerical aperture for the trace (a) is 0.14 with an EFL of 0.356; for (b), the numerical aperture is 0.18 and an EFL of 0.273; and for (c) the NA is 0.21 with an EFL of 0.21.

As we have seen in Chapter 4, diffractive element micro-lenses can be corrected for spherical aberration and thus would appear to be the better choice for this application. The problem is primarily in obtaining the high numerical aperture. As we have pointed out in Chapter 4, the limitation here stems from the resolution of photolithography.

Nishizawa and Oikawa [2] have shown the use of a GRIN microlens for this application. Their results are given in Fig. 6.3. They plot the insertion loss in db against the NA of the microlens for five cases of the beam spread of the source. The inset defines the asymmetric angular beam spread of the source. Their data indicate that a 1-db loss would occur for a microlens with an NA of near 0.4 when used with a 0.5-NA source.

The emitting area of the LD is in the order of $2 \times 2\,\mu m$, which yields a diffraction-limited beam spread of >0.5. This estimate comes from the diffraction-limited beam spread emanating from a 2-µm slit, yielding roughly a full angle of λ/D. This means that the microlens should have an NA of 0.5 or greater.

To deal with the situation, Oikawa et al. [2,3] used a special kind of GRIN lens. They combined the normal GRIN lens discussed in Chapter 3 with a refracting effect brought

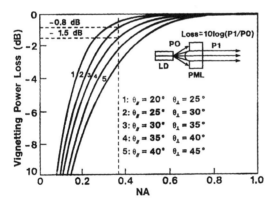

Figure 6.3 Power lost from a source with a beam spread angle indicated by the θ as a function of the numerical aperture of the microlens. (From Ref. [3].)

about by retaining the swelling effect produced by the ion exchange. This is represented in Fig. 6.4 by the sketch, together with the specific lens parameters that were used. What is most interesting and instructive is the insertion loss data. Radiometric lens performance is one thing; aligning and registration of the microoptical elements is an equally important issue. They show the loss with respect to variation of the object distance in Fig. 6.5a, the lateral displacement of the object (LD) in Fig. 6.5b compared to no lens, the effect of the variation of the image distance in Fig. 6.5c, and the lateral displacement of the image (fiber) in Fig. 6.5d. One can see the extreme sensitivity to any misalignment.

Now consider the output end where the light is to be focused onto the single mode fiber. The fiber has typically a 5–10 μm core diameter and an NA of the order of 01–02. For maximum coupling, one wants to match the NA of the exit

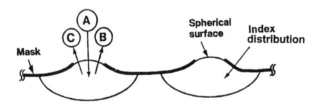

(Lens Diam 2a = 85 μm)

Diffusion Area (μm)	Diffusion Length (μm)	focal length (μm)	NA
250	85	110	0.37
400	156	78	0.48
422	168	55	0.61

Figure 6.4 Schematic drawings of the GRIN lens with "swell". The ion-exchange process produces a volume change which causes the lens to bulge. This can be used as a refracting component to the GRIN lens to significantly increase the NA. The insets indicate the effective performance numbers. (See Refs. [2] and [3].)

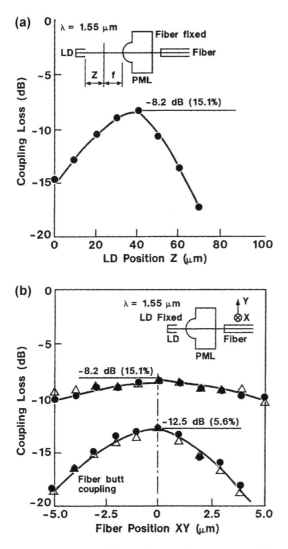

Figure 6.5 Effect of misalignment of the lens on the loss. The particular kind of offset is indicated by the drawings. (From Ref. [3].)

focus beam with the fiber. This makes sense since any light ray making an angle greater than this would not get trapped by the fiber. One way to do this is to magnify the image, that is, working with an arrangement that is not symmetric with

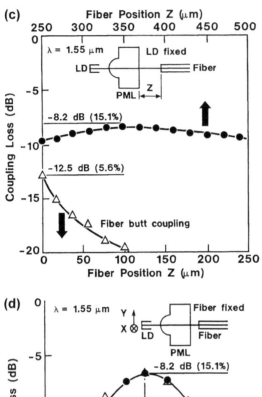

Figure 6.5 Continued

respect to object and image. We show an example with the aid of Fig. 6.6. The lens properties are listed in Table 6.2. The schematic drawing of the one-to-one arrangement is shown in Fig. 6.6a [4]. The 4–1 reduction schematic is shown in

(a) 1:1

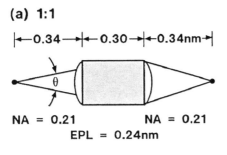

NA = 0.21   NA = 0.21
        EPL = 0.24nm

(b) 4:1

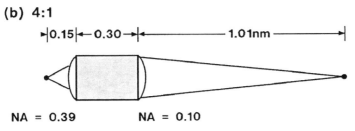

NA = 0.39   NA = 0.10

Figure 6.6 For the lens whose properties are listed in Table 6.2, A shows the schematic of the one-to-one imaging arrangment and B is the computed ray trace for this situation. In C is shown the four-to-one reduction arrangement where the NA of the fiber is matched.

Table 6.2 Lens Properties for Example Design

Lens type	Biconvex refraction
Diameter	0.2 mm
Lens radius of curvature	0.17 mm
Lens thickness	0.3 mm
EFL	0.24 mm
Principal plane	0.14
1–1 Configuration	
Object working distance	0.34 mm
Image working distance	0.34
Object and image side NA	0.21
4–1 Configuration	
Object working distance	0.15 mm
Image working distance	1.02 mm
Object side NA	0.39
Image side NA	0.10

Fig. 6.6b. The design was done by using the expression relating the object and image distances to the effective focal length

$$\frac{1}{\text{EFL}} = \frac{1}{X_0} + \frac{1}{X_i} \qquad (6.1)$$

and making $X_i = 4X_0$. The result is that $X_0 = 1.25\text{EFL}$. Recall that the distances are measured from the principal planes; thus we have $X_0 = 1.25 \ (0.24\,\text{mm}) = 0.29\,\text{mm}$ and the object working distance is then $0.29 - 0.14\,\text{mm} = 0.15\,\text{mm}$, while the image working distance is $4 \ (0.24\,\text{mm}) - 0.14 = 1.02\,\text{mm}$. The object NA is 0.39 and the exit NA is 0.10. One can see the big improvement afforded by this design. The input NA is increased by moving the object closer, while the exit NA is reduced to match that of the waveguide. One can also view this from the spot size argument as well. By lowering the NA of the exit beam, the spherical aberration is lessened; thus the blur of the image is less. All of this is from a purely geometric optic standpoint. One must also consider the diffraction aspect as well. With the lower exit NA, the diffraction limited spot size will be larger. Thus, although there is less aberration, the inherent spot size will be bigger. It does not take much misalignment to cause intolerable insertion loses.

6.2.1.2. Expanded Beam/Gaussian Optics

When one is dealing with the optics of single-mode waveguide applications, the more appropriate physical treatment is what is termed Gaussian optics [3a]. The reason for this is that the light emitted from a laser source, or the mode propagating in a fiber is almost always the lowest order single mode, or gaussian As we will discuss in more detail in Appendix A, the imaging of a gaussian beam, as opposed to a plane wave, must take into account the deviation from a plane wave, namely, a radial varying field and a curved wavefront. As a consequence of this when one focuses or collimates the output from a laser or optical waveguide, one finds the beam characteristics shown in Fig. 6.8b. The ray-method used above, although useful in an approximate way in this regime, is not adequate to deal with the actual imaging phenomenon.

Referring to Fig. 6.8b, what is important is the behavior of the beam waist (diameter) at the focus point as a function of the input beam waist, the distance to the lens, and the focal length of the lens. The expression relating these quantities shown in the figure is given below.

$$
\frac{1}{w_2^2} = \frac{(1 - d_1/f)^2}{w_1^2} + \left(\frac{\pi w_1}{\lambda f}\right)^2
$$

$$
d_2 = f + \frac{(d_1 - f)f^2}{(d_1 - f)^2 + \left(\pi w_1^2/\lambda\right)^2} \tag{6.2}
$$

A more complete description of the Gaussian optic approach is given in Appendix A. The optimization of the coupling process, either from laser to fiber, or fiber to fiber will be governed by satisfying these equations. It is interesting to note the geometric approach essentially takes the waist to be 0. One can still define a quantity that is essentially that of the ray numerical aperture, $\theta = \lambda/\pi w_0$, where w_0 is the diffraction limited beam radius. In this approximate way, one can still optimize the coupling efficiency from a geometric optic point of view by matching the numerical apertures.

The application is to collimate the output beams of an array of SMFs for a distance sufficient for the insertion of a functional optical element between them. The initial lens array collimates the light from the input array and a corresponding lens array to refocus the beams onto the output array. All this is to be done with the minimum of loss of signal. The application is represented schematically in Fig. 6.7. The optical element to be inserted can be a passive element like an optical filter, or a more complex element like an optical isolator, or an active element like a liquid crystal array that could gate the passage of light from any of the fibers. A computed ray trace of such a lens function is shown in Fig. 6.8a. The numerical aperture of single mode waveguides can range

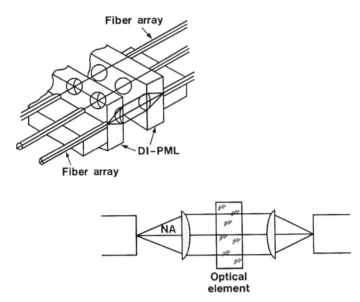

Figure 6.7 Schematic of the optical path for beam expanding application, and a proposed V-groove structure to implement the function. Single-mode fibers are placed in etched V grooves in Si to produce the precise spacing. The first lens array is aligned to collimate the light from each guide, and the second to refocus the light onto the output array. The intervening parallel beam opening is to be used to insert hybrid elements.

from values of 0.1–0.3. The ability to collimate the output as a function of the fiber NA is illustrated by way of a ray trace in Fig. 6.9. The microlens here is a biconvex structure using a 200-μm-diameter SMILE array as a prototype. The upper trace (a) is for a numerical aperture of 0.16, while the lower (b) is designed for an NA of 0.25. One can see the poor collimation at the higher NA.

Oikawa et al. [3] report the insertion loss as a function of the separation distance L, as shown in Fig. 6.10. Here they use a swelled GRIN lens as the collimator for the LD and a conventional planar GRIN to refocus the light onto the fiber array. This represents a more difficult situation than the fiber-to-fiber because of the higher NA requirement for the LD collimating lens.

(a)

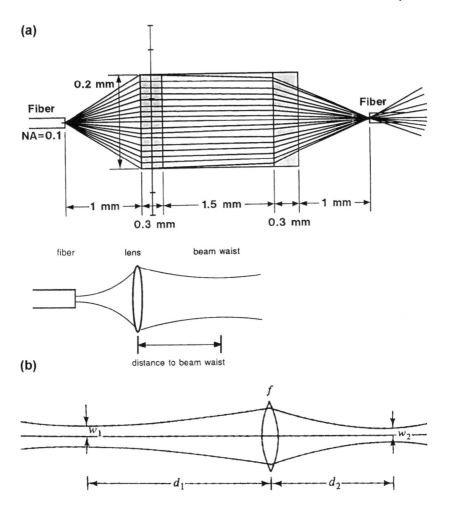

Figure 6.8 (a) Computed ray trace for lens array to perform expanded beam function. For this case, the lens array was designed to accept light from a single-mode fiber with an NA of 0.1. The lens diameter was 0.2 mm, and the opening was 1.5 mm. (b) Actual Gaussian treatment of collimation where beam waist is shown.

As a matter of comparison, in the single-element case where GRIN rod lenses are used as the collimators, the insertion loss can be quite small, <0.5 db [2]. As an example, we show in Fig. 6.11 the utilization of this structure with a

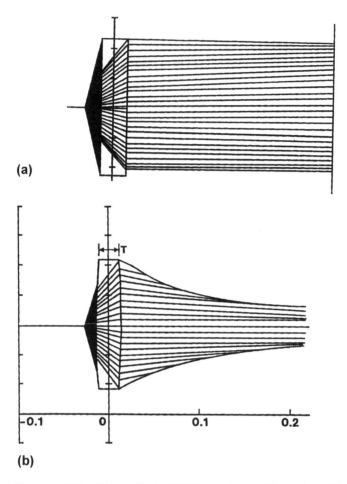

Figure 6.9 The effect of NA on the collimation. The upper ray trace is designed for an input NA of 0.16, the lower for an input NA of 0.25. One can see the effect of spherical aberration on the poorer collimation performance of the higher NA lens array.

filter which directs light to the appropriate output fiber. The alignment is a critical parameter.

6.2.1.3. Crossbar Switch

This application is to take light of wavelength λ contained in one of N input fibers and direct it to any arbitrary one of N

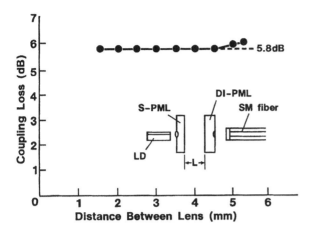

Figure 6.10 Coupling loss as a function of separation L. (From Ref. [3].)

output fibers. This is shown schematically in Fig. 6.12 for what would be called a 4×4 crossbar [5]. The input of 16 fibers is grouped into four sets of four, each fiber in the set carrying one of the wavelengths. The output fibers are also

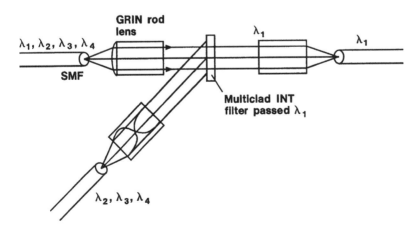

Figure 6.11 Example of the use of GRIN rod lenses in expanded beam dielectric filter-based wavelength demultiplexer. The lenses here are rod lenses, not the planar array type. Function here is to separate one wavelength and send it to another location.

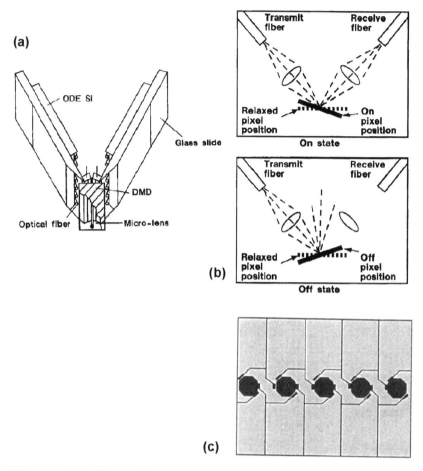

Figure 6.12 Example of crossbar switch application using a lens array. Signal brought in on the linear array of 16 fibers. The light is focused on to the micromirror as shown. Depending on the position of the mirror, it determines whether the light is transmitted or not. (See Ref. [9].)

grouped into sets of four. Any combination of output wavelengths can be made to appear from each set.

In this case, the switch is fabricated from a thin silicon film technology [6–9]. The photomicrograph of the actual silicon structure is shown in Fig. 6.12c. The silicon element is electronically addressed, which has the effect of electro-statically deforming the silicon membrane, as shown in Fig. 6.12b. The membrane has two positions, either allowing the light to get to the output fiber or not. By the appropriate switching sequence, any wavelengths can appear at any of the four outputs.

The role of the microlens array is to collimate the light from each of the 16 fibers and focus it onto the deformable mirror and then to reimage the focal spot to the output fiber as shown in the figure. The realization of this device by TI [5] was achieved using the 16-element SMILE^{TM} lens array [10] shown in Fig. 6.13. The lenses were planoconvex, 0.6 mm thick, with a focal length of 0.4 mm. The lens diameter/spacing was 160/195 μm. The effective NA of this arrangement was only 0.1 which was, however, consistent with the multimode fiber.

The reported operation of the device produced greater than 30 db contrast in all channels. The overall insertion loss

Figure 6.13 The SMILE lens array used for the crossbar switch application. The lens diameter is 160 μm on 195-μm centers with a focal length of 0.4 mm. The upper is the actual array, while the bottom is a SEM picture.

was reported as 24 db, but only 6 db was attributed to the optical system, the rest having to do with the splitting and recombining of the signals.

6.2.1.4. Computer Backplane

As computer designers seek ways to increase speed, it was suggested that the time it takes to communicate between chips, boards, and shelves could be reduced if it were done optically rather than electronically. This argument is essentially based on getting around the limitation posed by the RC time of connecting cables. The idea is to have the output of a chip or board within the computer converted to light and then have this transmitted to the detector array on the other chip or board. There are a number of ways that this transmission can occur, but we will only be concerned with the multimode fiber optic method. The other major method is called free space and involves diffractive elements to provide a directed beam pattern. (This will be covered in Chapter 7.) The fiber optic method does provide a more mechanically flexible approach, but otherwise there are advantages to both methods.

The optical application is similar to that of the collimator application discussed at the beginning of Sec. 6.2.1, in that the source will have a large NA. But here the fiber is usually multimode; thus a higher NA is required, >0.3. Otherwise, the design method is exactly the same as that presented above. An example is shown in Fig. 6.14a. The light-emitting diode array is coupled to a fiber array through the use of a stacked pair of biconvex lens arrays. A simulated ray trace of the stacked biconvex lens array (SMILE) disigned for this use is shown in Fig. 6.14b [11]. The input NA of this design is 0.34.

6.2.2. Other Related Applications of Linear Arrays

6.2.2.1. LED Print Bar

The application is to transmit the light from an array of red LEDs onto a detector array. This is a fast optical printing

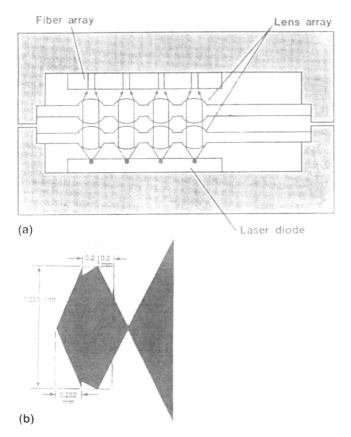

Fiber array   Lens array

(a)

(b)

                                                                Laser diode

Figure 6.14 Schematic drawing of laser to fiber array. A computed ray trace is shown in the lower figure with the use of the stacked biconvex design. The lens diameter is 0.225 mm, and the radius of curvature of each surface is 0.326 mm. The effective focal length is 0.36 mm.

head with the resolution determined by the pitch of the LEDs which in this case is 85 μm, which translates into 300 dpi (dots per inch) for the example shown in Fig. 6.15. An actual photograph of the lens array is shown above the LED array. The diameter of the microlenses was 75 μm on the 85-μm pitch [12]. A photograph of one of the individual images formed by the microlens is shown in Fig. 6.16. The numerical aperture of the microlenses was 0.1, and the

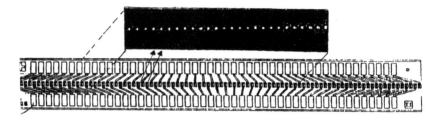

Figure 6.15 LED print bar application. The 300-DPI LED array is illustrated with the lens array shown in perspective. The lens array was made up of 75-μm lenses on 84.66-μm centers.

overall separation between the LED array and the detector array was 0.78 mm.

One of the problems with designing a lens for this type of application is that the lens diameter can only be as large as the pitch of the LED. The diameter represents the limitation of the SMILE lens array and is pushing the limit of the planar GRIN approach as well. For diffractive elements, the problem is that, although this is not a small diameter from a fabrication point of view, it becomes increasingly difficult to fabricate a lens with a high NA.

6.2.2.2. Camera Autofocus

One microlens array that has reached production is that which was used by Honeywell [13] to produce an autofocus element for camcorders and 35-mm SLR cameras. It provided a through-the-lens autofocus technique that is a desirable feature because of its better accuracy. This is done with the use of a beam splitter that takes some of the light that enters the objective lens to the sensor, as shown in Fig. 6.17. It should be noted that the distance from the lens to the film plane is equal to that to the sensor plane.

The autofocus concept is based on a radiometric rather than imaging principle. When an object is in focus the amount of light arriving at the image plane is equal from any arbitrary halves of the imaging lens. The Honeywell device uses this fact in the way that we will show with the use of the blowup of the sensor shown in Fig. 6.18. There are two

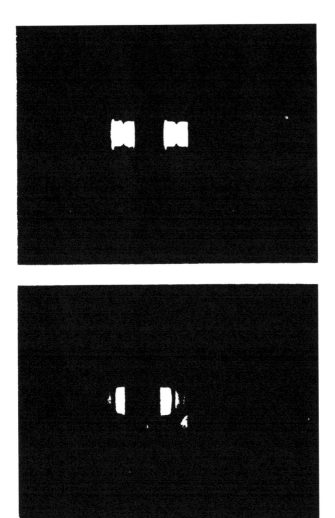

Figure 6.16   Actual photograph of the light-emitting area of one diode and the corresponding image produced by a lens of the array.

detectors behind each microlens. The method will be to compare the outputs from the string of A detectors to the outputs from the B string. When the object is in focus, the image appears at the lens plane and an equal amount of light reaches each of the two detectors, as depicted in the middle ray trace. Here, for the sake of simplicity, we are looking at

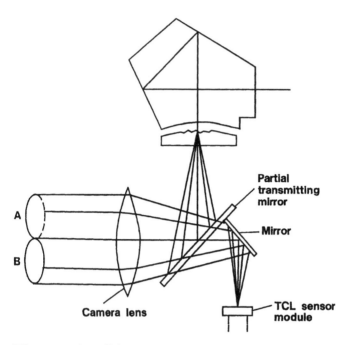

Figure 6.17 Schematic of Honeywell autofocus concept. Part of the light from the objective is directed to the sensor element located at the same distance as the film plane.

the light from only one point in the object field. When the focus position precedes the correct focus, one gets the situation depicted in the upper ray trace. For this situation, light from the upper half of objective reaches the A string, whereas light from the lower half tends toward the B string. If the focus is beyond the correct focus, the opposite behavior is observed as shown in the lower ray trace. A schematic of what the output from the respective detector would see is shown in the inset. The lack of overlap of the outputs would indicate that the object is not in focus. The nice thing about this method is that depending on the phase of the output curves, that is, the A string leading or lagging, indicates which way the objective lens should move to attain focus.

The initial lens array used for this device was a molded plastic array manufactured by USPL [14]. It was subsequently replaced with SMILE lens array for a number of reasons [15].

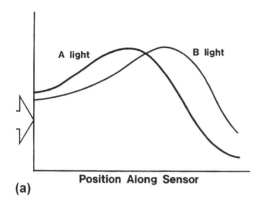

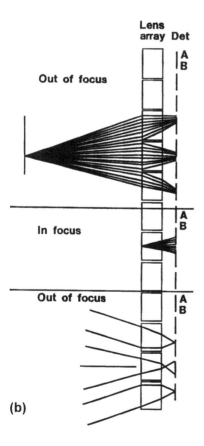

The cross-talk was less because of the opaque surround to the lenses, the uniformity and positioning of the lenses were better, and the glass lenses were more securely attachable to the detector die. The array in this case was made up of a row of 26 planoconvex microlenses, 160-μm-diameter lenses on 195-mm centers with a focal length of 400 μm.

6.3. TWO-DIMENSIONAL ARRAYS

There are a number of newer applications that require 2D arrays of lenses. The technology of producing 2D arrays does not differ in any substantial way from that to make linear arrays except that maintaining the precision of location is compounded by the added dimension. In other words, 2D microlens arrays can be made by molding GRIN, diffractive, etc.; however, when such issues as fill factor and uniformity are important, then differences in the method of fabrication can make a difference.

6.3.1. Shack-Hartman Wavefront Analyzer

This is an astronomical telescope application where an incoming wavefront is sampled by means of an $N \times N$ array of detectors, such that it can provide a basis for correcting disturbances that appear over time [16]. The prime example is that where atmospheric disturbance is to be removed from a ground-based astronomical system [17]. There are certain design conditions for optimum utilization of the device which bear on the optical properties of the microlens array. In particular, they involve the interplay between the wavelength of

Figure 6.18 (a) Basis for autofocus module. If an object is in focus, the light is focused at the lens array plane and the light is split equally by each lenslet onto the split detectors behind the array. (b) If an object is out of focus, either too close, or too far, one set of detectors sees more light than the other. The relative output of the two sets of detectors is indicated by the diagram. When they coincide, the object is in focus.

the light and the focal length of the microlens, to the optimum subimage to grid and the spot size. Artzner [17] has shown that an optimum ratio of subimage size to the grid spacing should be much larger than $\lambda/4\delta$, where δ is the sag of the microlens. The net result is that the sag of the microlenses will be small irrespective of the period and focal length of the microlenses.

A sample pattern taken from Ref. [16] is shown in Fig. 6.20 for a 22×22 array with a pitch of 0.185 mm and a lens focal length of 7 mm. Fig. 6.20a is that of the reference grid for an artificial point source (Fig. 6.19). Figure 6.20b is that of an actual sensing image from a 4.2-m telescope. Different sub-images have different shapes corresponding to different instantaneous atmospheric wavefront distortions.

This lens array has a sag of about 1.5 μm. The smallness of the sag represents one of the main challenges to the lens preparation for this application. One way to offset this is to reduce the power of the lens by near index matching the lens with a suitable liquid.

6.3.2. Parallel Processing (VCSELS)

With the advancement of GaAs technology, there has emerged a new area which is based on a vertical fabrication

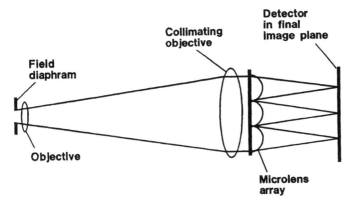

Figure 6.19 Optical diagram of a Shack-Hartman wavefront analyzer. The lens array images to an array of photodetectors.

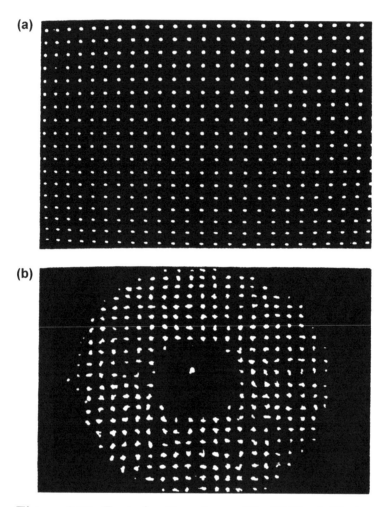

Figure 6.20 Typical pattern from a 22×22 Shack-Hartman array with a pitch of 0.185 mm and a focal length of 7 mm. The upper photo is the reference grid for an artificial point source, and the lower is the actual sensing image from a 4.2-m telescope. (From Ref. [16].)

architecture, that is, structures that are fabricated such that their output is normal to the face of the growth. An important example of this is vertical-channel surface-emitting laser, or VCSEL, for short [18–21]. A schematic representation of such a device is shown in Fig. 6.21, which is taken from Ref. [18].

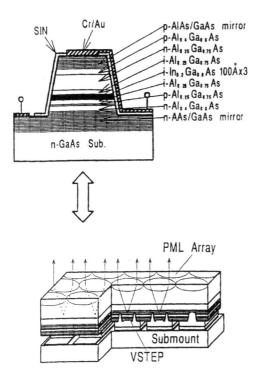

Figure 6.21 Representation of a VCSEL (vertical-cavity surface-emitting laser) and how a lens array would be positioned to couple out the light. (From Ref. [18].)

The ability to deposit precise thin layers of varying composition allows one to create not only the active lasing junction, but also the "mirrors" which in the case shown are distributed feedback gratings. This type of vertical architecture distributed over two dimensions is the ideal way to imagine a truly parallel system where each laser can be individually addressed. The vertical architecture is not limited to lasers. One can also construct amplifiers, detectors, and switches. Indeed, if optical computers ever become a reality, it may very well be based on this type of technology.

The ability to couple light in and out of these devices provides a real challenge because the lateral dimension of these devices is typically in the range of 5–50 μm. Clearly, the smaller the better in order to maximize the areal density of the device, but the lenses must also be smaller. In the lower portion of Fig. 6.21, we see a schematic representation of how the microlens array would be fitted to the VCSEL to couple light out. Many of the methods of making lens arrays will not work for lens diameters below 50 μm. What do work at the smaller diameters are the melted photoresist method and the diffractive element. The performance of diffractive lenses for this application has been reported down to a diameter of 43 μm.

6.3.3. Laser Eye Protection

One of the more interesting applications proposed for microlens arrays is where they would be utilized as an element of a broad band laser protection goggle. With lasers present everywhere on the battlefield, the need for eye protection became a real need. Moreover, the protection had to be more than for a fixed wavelength since many lasers are purposely randomly tuned to avoid jamming.

The optical construction is represented in Fig. 6.22. It is essentially a 1× telescope [22]. The parallel light from the

a)

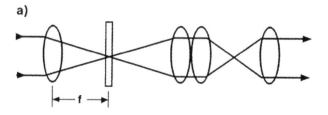

b)

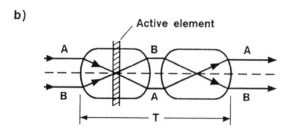

Figure 6.22 Concept of laser eye protection. Light is focused at a midplane containing the active switch material that would shut down at high intensity. Second array is to provide an erect image.

object is brought to a focus at the midplane of the lens and then emerges as a parallel beam. Since the intention is to use it as a goggle for human use, the images have to be erect. This is accomplished by using a second identical lens array. The active element is placed at the first focus and has the property that it blocks the light when the intensity reaches a desired threshold. Suffice it to say here that the nature of this limiter material is crucial in determining the usefulness of the device. The example we will use later on is a liquid crystal that changes from a transmitting to a scattering state when a certain intensity is incident. The purpose of the lens arrangement in Fig. 6.22 is to provide this high intensity at the focal spot. The impracticality of this concept in using conventional lenses is the total thickness of the device.

If f is the focal length of the lens corresponding to the required NA to produce the necessary spot size at the active plane to effect the switching, then the total thickness of the goggle could be no less than $4f$. For a conventional lens to cover the eye without restriction, the lens diameter should be at least 3 cm. Assuming that an NA of at least 0.25 is required, this would yield a total thickness of 24 cm.

The microlens approach is to replace the single element with an $N \times N$ array of microlenses operating in the same 1× mode, with a comparable NA. Since the lens diameter is smaller, the focal length will be correspondingly smaller, and consequently the overall thickness will be smaller. The expression for this condition is given by Eq. (5.6) with t being infinite, which yields the focal length as the lens thickness divided by twice the refractive index.

As an example, a lens array made up of 400-μm-diameter lenses on 480-μm centers was fabricated with an NA of 0.24 and an effective focal length of 0.83 mm. From the expression for the total thickness, this would yield something less than 5 mm. A ray trace of the operation of an array of such elements is shown in Fig. 6.23 .

There are some trade-offs with the microlens approach. For example, the active material can be switched by either peak fluence or intensity. In either case, the energy delivered to the active plane is per unit area. The individual microlens

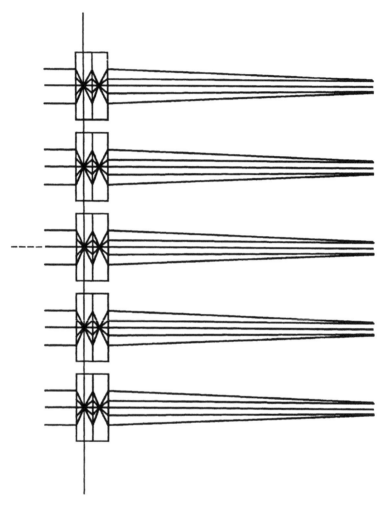

Figure 6.23 Computed ray trace of lens array acting as a one-to-one telelescope.

intercepts less light, yet the intensity must be sufficient to switch the active material. The device has been experimentally demonstrated using the above-mentioned smetic liquid crystal as the active medium (23). The output energy is plotted against the input energy (which would correspond to the ambient laser), as shown in Fig. 6.24. The lens used was that described above. The threshold is defined as the point

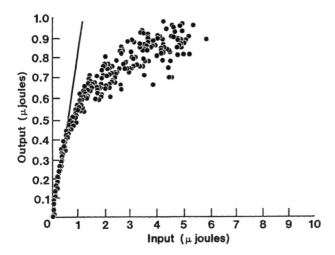

Figure 6.24 Actual optical clamping data obtained by using the optical arrangement shown in Figs. 6.25 and 6.26 using a two-dimensional array of 400-μm-diameter lenses imaging onto a nonlinear liquid crystal material. [23]

at which the slope deviates from unity, which in this case occurs at about 1 μJ. The maximum level, or so-called clamped value of the output, is the order of 1 μJ and occurs when the ambient energy is greater than 5 μJ. In general, eye safe levels are considered to be below 0.5 μJ.

A serious drawback to this device is the field of view. Unless the lenses could be fabricated with a curve, the vision of the wearer would be too restricted for practical eye use.

6.3.4. Projection LCD-TV

Projection LCD-TV is one approach to large-screen television. The active LCD matrix is contained in a 54 × 75-mm area. The light is projected through the panel onto the wall. One of the major problems with this projection approach is the optical efficiency. A number of attributes contribute to this problem, not the least of which is that it is a display based on polarized light, so one starts with a 50% efficiency. Color filters are another source of light loss. The feature that loses about 60% of the available light is the geometric obstruction on the panel itself produced by the conductors and the transistor gate. One can get an idea of this problem by looking at the structure of a typical panel, as shown in Fig. 6.25. The clear area, which is referred to as the aperture ratio, is only about 40%. A microoptic solution to this problem is to use a microlens array to focus the incident light through the open area, as shown in Fig. 6.26a, b. In (a), one can see the light lost by the nontransmitting area of the electronics. In (b), one sees how the microlens directs the light through the open "window" regions.

For the microlens approach to be viable for this application, it is necessary to produce an array with as close as a 100% fill factor as possible [23–25]. For certain of the fabrication techniques, this poses a problem because of the effect that the coming together of the boundaries has on the lens. Related to this is the problem associated with having to make lenses with noncircular boundaries. We shall see this for the GRIN and the SMILE lenses.

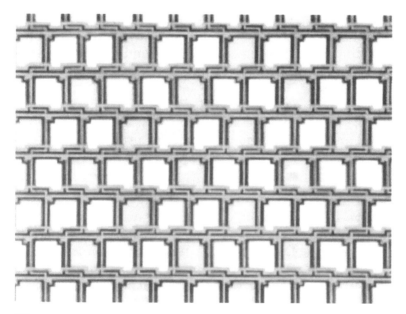

Figure 6.25 Photograph of a typical projection LCD TV panel. The pitch is approximately 100 μm.

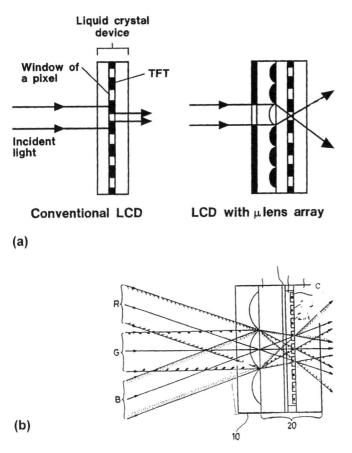

(a)

(b)

Figure 6.26 Two concepts to utilize lens arrays to improve brightness. The upper diagram (a) shows how the lens array would focus the light through the open areas of the LC panel shown in Fig. 6.25. The lower diagram (b) shows how the lens array can direct each color light to a separate pixel, obviating the need for a color dot array.

Referring to Fig. 6.26, the focal length of the lens should be close to the thickness of the cover glass. The standard cover glass thickness is 1.1 mm, but thickness as thin as 0.4 mm has been reported. Typical lens diameters are of the 100-μm order, so that the numerical aperture of the lenses is quite low. The problem this will produce is that this will limit the ultimate spot size. For the present pixel size shown in Fig. 6.25, the rough diameter of the clear opening is of the order of 60 μm. For an NA of 0.05, the diffraction limited spot size for the red beam would be 15 μm. The problem is that the projection lamp that is used is effectively an extended source, so that the light is not parallel and diffraction-limited performance cannot be achieved. The situation is indicated in Fig. 6.27. The beam spread comes from the off-axis points of the extended source. One can show that the spread angle θ is related to the object height through the simple relation

$$\tan\theta = \frac{z}{2f} \tag{6.3}$$

Here f is the focal length of the lens. The spot size that will form will simply be the product of this angle with the focal length of the lens. For a 5° beam spread, the minimum spot size would be 45 μm.

The spot-size issue is even further complicated by the physical layout of the pixels. The pixel separation in the x direction is not the same as in the y direction. In some designs, the rows are staggered, presenting an elongated

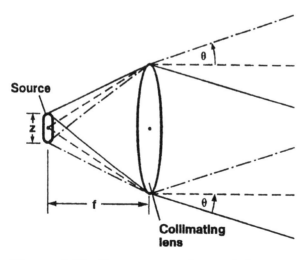

Figure 6.27 Consequence of extended source on the beam spread, and ultimately on the size of the spot that can be imaged.

hexagonal layout. What this means is that in order to fill space, one has to resort to lens shapes that are not symmetric. As an example of this, we will use the SMILE lens array [23]. The pixel pattern is shown in Fig. 6.28a. The x direction pitch is 160 μm and the y direction pitch is 190 μm. If one uses the pattern of 150-μm-diameter circular lenses, Fig. 6.28b, only 58% of the area would be covered, but the focus spot is well defined, as shown in Fig. 6.28c. If one resorts to oval-shaped lenses to effect better coverage, one suffers the consequence of having an anamorphic behavior, as shown in Fig. 6.29a.

(a)

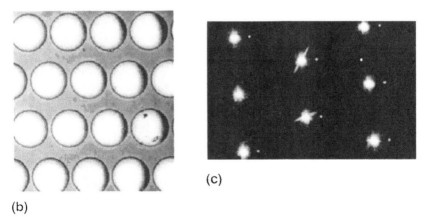

(c)

(b)

Figure 6.28 Array of circularly shaped lenses and the images formed to go through the LC pattern shown.

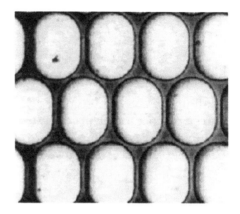

(a)

Figure 6.29 Array of oval-shaped lenses and hexagonal-shaped lenses and the images formed. Note the lenses are anamorphic. For the oval-shaped lenses, the image in each image plane is shown.

What this means is that the lens has a different focal length in the long and short directions of the lens. We show this in the images below that pattern. The best coverage design is that of lenses with a hexagonal shape, which is shown in Fig. 6.29b. There is still an anamorphic effect present which

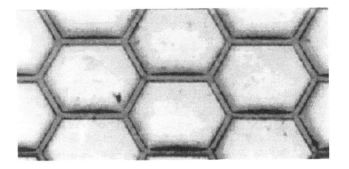

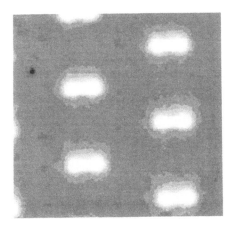

(b)

Figure 6.29 Continued

has an adverse effect on the spot size in that the image is elongated.

The planar GRIN approach [24,25] suffers from a similar problem. One would start with a patterned set of openings to allow the ion exchange to be accomplished. The pattern would be set by the x and y pitch. The length of time of the exchange would dictate the lens shape. This is schematically shown in Fig. 6.30. The planar GRIN lens suffers similar asymmetry problems, as demonstrated by the images formed (Fig. 6.31).

The ideal method of fabricating a lens array for this application is the diffractive approach (see the discussion in

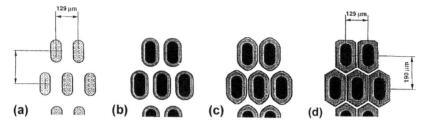

Figure 6.30 Formation of planar GRIN array for the same application. (a) Patterned substrate; (b) separated; (c) partially fused; and (d) closed packing structure (honeycomb shaped). The index region grows as the ion-exchange time through the aperture increases. Finally, when fronts meet, a hexagonal pattern results.

Chapter 4) because it decouples the shape of the lens from the performance. For example, one can almost seamlessly arrange the lenses in any configuration and still maintain the spherical-like lens imaging performance. Moreover, it is a low NA application so the resolution of the required fabrication is not an issue. A full 100% fill-factor layout is shown in Fig. 6.32a. The fact that the diffraction efficiency varies with wavelength is easily remedied by adjusting the focal length of each lens in the triad for the RGB wavelengths. The actual function of a diffractive array to increase the efficiency of a LCD projection panel is shown in Fig. 6.31b.

There have been reported efficiencies for the three lens fabrication methods, calculated as the percentage increase in the light through the array with the lens array in place, relative to none. They range from 50% to 90%.

6.3.5. 3D Images

People have always been fascinated by the ability to produce a three-dimensional image. Holography is by far the best-known example of how this can be accomplished. However, there is an older concept which does not rely on coherent optical properties at all. The method is called integral photography.

Integral photography consists of using a fly's eye array of lenses to record multiple images on a conventional

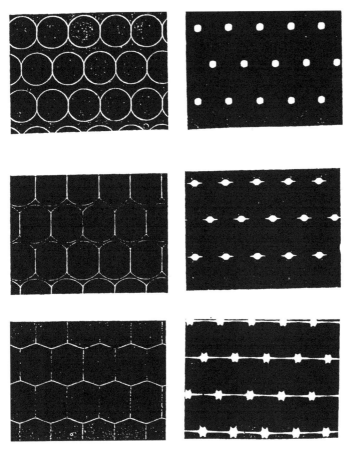

Figure 6.31 Lens shapes of GRIN lens arrays and corresponding images. (From Ref. [25].)

photographic plate, as shown in Fig. 6.33a [26–30]. Consider the recording from a point P on the object. Each lens in the array produces an image of point P at approximately the focal length of the microlenses. Putting it slightly differently, each lens forms an image with a different perspective. The photographic plate is developed and a positive is formed. The developed positive is registered behind the lens array and illuminated as shown in Fig. 6.33b. Each point on the image emits a spherical wave that is collimated by the fly's eye lens. All the beams reconverge to the original position of the object

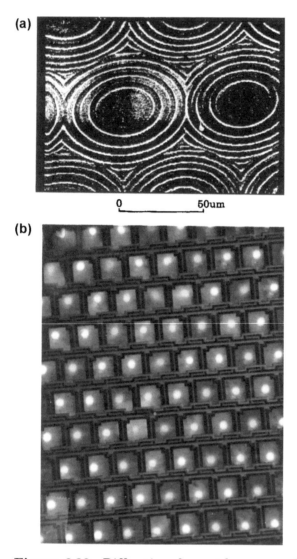

Figure 6.32 Diffractive element lens array for the same application, along with the images.

point. The image so formed is real and psuedoscopic, the same as would occur for a holographic image. To obtain a more practical concept where the image would be virtual and orthoscopic, a second recording has to be made. This can be done

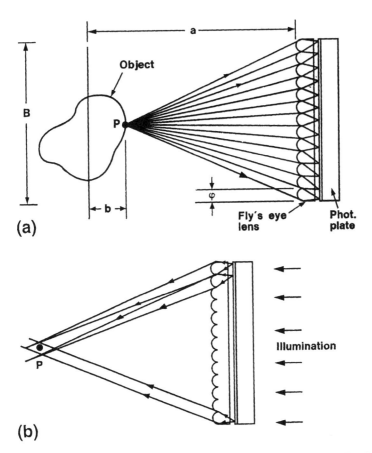

Figure 6.33 Concept of integral photography. Multiple images are formed on emulsion behind the lens array. Negative is developed and light is sent through the negative to the same lens array to form a real 3D image.

in one of two ways. The first is to make another integral photograph in the converging reconstructed beam as shown in Fig. 6.34a [31]. Then one makes a second exposure through the original as shown in Fig. 6.34b. Finally, the desired orthoscopic virtual image is obtained by illuminating the second exposure through the lens array as indicated in Fig. 6.34c. The disadvantage of this is in the reduction in the quality of the image.

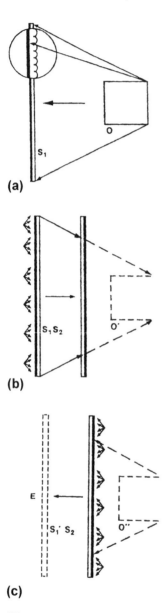

(a)

(b)

(c)

Figure 6.34 Scheme to produce desirable virtual orthoscopic image rather than real pseudoscopic image. The method is to make a second recording from the first as shown in (b). Illumination of this image from the right forms a virtual orthoscopic image when viewed from the left.

A second method is from a hologram of the imaging forming wavefront, as shown in Fig. 6.35 [28]. The problem here is that one has to introduce a different holographic process into mix.

Burckhardt [26] has analyzed the resolution of this method along with an estimation of the ideal microlens parameters to obtain this condition. His analysis shows that the optimum microlens diameter ϕ is given by the expression $1.24a \sqrt{\lambda/b}$, where, referring to Fig. 6.32a, a is the object distance, b is half of the object depth, and λ is the wavelength of light. For typical values of the parameters, $a = 50$ cm and $b = 5$ cm, and the estimate of the optimum lens diameter was 2 mm. This could easily be less than 1 mm for other choices of the parameters, or with somewhat less resolution at the values used. The expression for the maximum field angle was also estimated as ϕ/c, where c is distance between the lens array and the film plane, which is essentially the focal length of the microlens. For the above parameters, it is 10 mm.

The lens array requirements for this application are not particularly stringent, and would seem ideally suited for a molded plastic, or even glass, fabrication process. One would want a 100% fill factor which should be readily possible for a spherical lens figure on a square format.

6.3.6. 3D Optical Switch

The idea here is to build an $N \times N$ all-optical switch that is any element of the N inputs could be switched to any one of the N outputs. The input would be an $n \times n$ 2D array, that is $N = n \times n$. The light from each of the input fibers would be collimated by a 2D lens array and sent into the switching structure and then be refocused to the output array of fibers by an identical lens array. There are a number of microoptic challenges presented by this device concept. Some element must hold the array of fibers, a fiber array. (We actually discuss a method to fabricate such an element in Chapter 10.) This then is to be registered to a 2D lens array that directs the light into the switching structure. Then, of course, there are the switches, themselves.

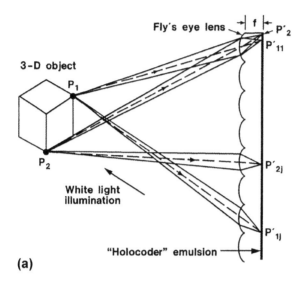

(a)

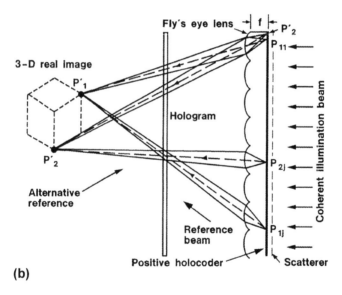

(b)

Figure 6.35 Another scheme to produce an orthoscopic image. In this case, a hologram is formed in the imaging beam as shown. The first image is formed in the normal way. Viewing the developed hologram will show a virtual orthoscopic image.

The initial attempt to construct such a device was using micromirrors, or MEMs to deflect the light. Schematically, the switching protocol is represented by a simplified ray diagram shown in Fig. 6.36. The motion of the two micromirrors controls the ultimate final output position of the beam. It should be clear that in switching from an arbitrary position within the input $n \times n$ array to another position in the output array, there is a worst case situation that is the one that corresponds to the longest distance of travel. This situation depicted in the figure.

The optical design must consider a number of interdependent factors. Among these are the overall switch size, the micromirror size, the mirror spacing, the angle of incidence to the mirror, and last but hardly least is the amount of mirror deflection. The total loss is also a critical factor in telecommunication applications. The most important relationship is how long a path must the beam travel, given the mirror deflections so that it corresponds to the separation distance of the mirrors. To complicate matters further, one must do this within the context of Gaussian beam propagation since the inputs are from single-mode fibers. There are a number of starting points for the optical design, the one that is chosen depends on what is the most critical design limitation. One might be presented by the physical size of the mirror, in not wanting to largely overfill the input mirror and consequently lose light. Depending on the overall path as shown in the figure, one might decide as the first iteration

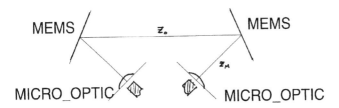

Figure 6.36 Schematic representation of MEMS-based 3D optical switch. Output from SMF is collimated by microlens onto a x–y deflectable micromirror, which then directs it to another deflectable micromirror and ultimately to the desired output port.

to design the optics so that this total path distance is equal to the confocal parameter of the microlens defined by $(2\pi/\lambda)w_0{}^2$ (see Appendix A). This would put the minimum beam waist at the midpoint of the path shown in the figure. One could write the following condition:

$$b = 2(z_0 + z_M) = \frac{2\pi}{\lambda}w_0^2 \qquad (6.4)$$

This yields the value of the minimum beam waist w_0 and then fixes the focal length of the microlens through the following relationship:

$$w_0 = \frac{\lambda}{\pi}\frac{f}{w_F}$$

Here w_F is the modal radius of the single mode fiber, usually about 10 μm at 1550 nm. Now one checks to see what the radius of the beam would be at the micromirror which is approximately $\sqrt{2}w_0$, and hopefully it is not too big. If it is, one goes through another iteration where now one chooses the total path to be smaller than "b".

In Fig. 6.37a, we show a picture of an assembled fiber array [32]. The fibers are aligned to the lens array. In (b) is shown a 30×30 array of micromirrors which would be aligned to the fiber array. Finally in (c) is shown a blow-up of the actual individual silicon-based micromirror. The mirror can deflect in both the x and y directions as one can see from the two sets of hinges. Foran introduction of the use of micromirrors for photonic device application, the reader is referred to Ref. [33]. An example of a 2D lens array used to collimate the fiber inputs into the micromirror is shown in Fig. 6.38 .

APPENDIX A. GAUSSIAN BEAM IMAGING

With the invention of the laser and subsequently, single-mode waveguides, one had to consider how beams emanating from such sources propagate in free space. What is different from the point of view of the simple plane wave solution of Maxwell's equations is that invariably the intensity

(a)

(b)

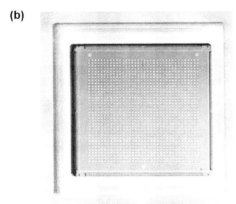

(c)

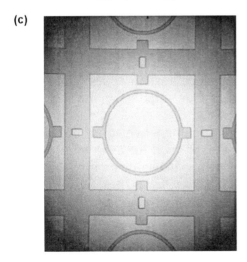

Figure 6.37 (a) Photograph of an assembled SMF fiber array. Each fiber is aligned to a microlens, (b) top view of the micromirror array (pitch 2.5 mm), and (c) blow up of an individual micromirror.

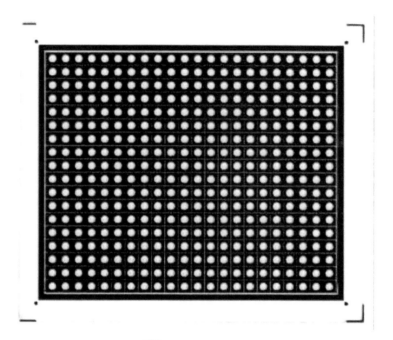

Figure 6.38 SMILE™ lens 16×20 collimating lens array, pitch 2.5 mm, clear aperture 1.26 mm focal length 4.2 mm.

diminishes in the transverse direction, and the wavefront is spherical, not flat.

Kogelnik and Li [34] were the first to look at this problem which we will essentially reproduce here in abbreviated form, since we really only want to provide some basis for Eqs. (6.2), (6.4), and (6.5).

One starts with Maxwell's scalar wave equation and the general solution, $\phi = A(x, y, z) \exp[-ikz]$ for a wave propagating in the z-direction. Substituting this solution into the wave equation and assuming that $A(x, y, z)$ is a slowly varying complex function, so one can ignore its second derivative with respect to z one gets the following equation:

$$\nabla_{\text{T}}^2 A - 2ik \frac{\partial A}{\partial z} = 0 \tag{A6.1}$$

Here the delta symbol refers to the transverse coordinates. Clearly, the complex function $A(x, y, z)$ represents the deviation from a plane wave produced by the transverse field profile and the curvature of the wavefront. A solution of this equation can be written in the following form:

$$A(x,y,z) = \exp\left[-i\frac{k}{2}\frac{(x^2+y^2)}{q(z)}\right]\exp[-\varphi(z)] \tag{A6.2}$$

This is an appropriate solution for the lowest order mode, or a gaussian intensity distribution. (Other solution would be more appropriate for higher order models.) The key function is $q(z)$ which will describe the gaussian field variation as well as the curvature of the wavefront. To make this more explicit, one can write the function in the two parts, the $R(z)$ representing the radius of curvature of the wavefront, and $w(z)$ measuring the decrease in field amplitude with z:

$$\frac{1}{q(z)} = \frac{1}{R(z)} + i\frac{2k_0}{w(z)^2} \tag{A6.3}$$

Substitution of Eq. (A6.3) into (A6.2) shows that $w(z)$ is just the gaussian width parameter $[r/w(z)]^2$, in other words measuring the $1/e$ points of the Gaussian beam as a function of z. The beam has its minimum waist when $R \to \infty$, and we designate this quantity as w_0. As a result of the solution given as Eq. (A6.2) to Eq. (A6.1), one can show that

$$q(z_2) = q(z_1) + z \tag{A6.4}$$

And combining this with Eq. (A6.3) yields the desired familiar results

$$w^2(z) = w_0^2\left[\left\{\frac{2z}{k_0 w_0^2}\right\}^2 + 1\right], \qquad R(z) = z\left[\frac{(k_0 w_0^2)^2}{(2z)^2} + 1\right] \tag{A6.5}$$

One can define an asymptotic spreading angle equivalent to the NA in geometric optics as shown in Fig. A6.1

$$\theta = \frac{2}{k_0 w_0} \tag{A6.6}$$

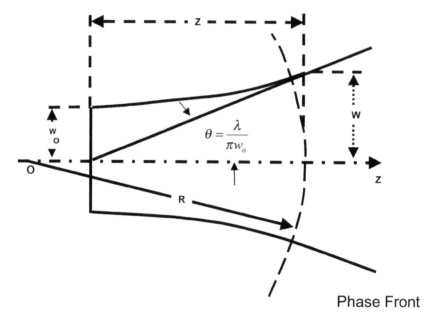

$$\theta = \frac{\lambda}{\pi w_o}$$

Phase Front

Figure A6.1 Definition of NA in gaussian optics.

The next issue is how a lens acts on such a beam. In the case of geometric optics, one can say that for a thin lens the lens of focal length f transforms an incoming spherical wave to an outgoing spherical wave which is simply written in the following way:

$$\frac{1}{f} = \frac{1}{R_1} - \frac{1}{R_2} \tag{A6.7}$$

For the case of the action of a spherical lens on the type of beam, we are discussing the same relationship holds except we generalize Eq. (A6.7) to include the Gaussian intensity as well. This means we replace the R with q and write Eq. (A6.7) as

$$\frac{1}{f} = \frac{1}{q_1} - \frac{1}{q_2} \tag{A6.8}$$

Combining this equation with Eqs. (A6.4) and (A6.5) yields Eqs. (6.2) in the text in one form or another.

REFERENCES

1. Iga, K.; Kokubun, Y.; Oikawa, M. *Fundamentals of Optics;* Academic Press: New York, 1984.

2. Nishizawa, K.; Oikawa, M. Microoptic research in Japan. In *Miniature and Microoptics,* SPIE; 1992; Vol. 175a, p. 54–66.

3. Oikawa, M.; Imanishi, H.; Kishimoto, T. High NA planar microlens for LD array. In *Miniature and Microoptics*, SPIE; 1992; Vol. 1751, p. 246–255.

4. Yariv, A. *Optical Electronics;* Holt-Sanders, 1985.

5. Borrelli, N. F. Private data.

6. McDonald, T. G.; Boysel, R. M.; Sampsell, J. R. Deformable mirror based 4×4 fiber optic cross-bar switch. In *OFC Paper* WM–22; July 22–26, 1990; San Francisco.

7. Hornbeck, L. J. Deformable mirror spatial light modulator. Proc. SPIE **1989**, *1150*, paper #6.

8. Boysel, R. M.; McDonald, T. G.; Sampsell, J. B. A linear torsion hinged deformable mirror device for optical switching. In *Tech. Digest, Topical Mtg. Photonic Switch*; Optical Society of America: Washington, DC1989; 183–185.

9. Florence, J. M. Joint-transform correlator systems using deformable mirror. Opt. Lett. **1989**, *14*, 341–343.

10. Collins, D. R.; Sampsell, J. B.; Hornbeck, L. J.; Penz, P. A.; Gately, M. T. Deformable mirror device spatial modulator. Appl. Opt. **1989**, *28*, 4900–4907.

11. Borrelli, N. F.; Morse, D. L. Microlens array produced by a photolytic technique. Appl. Opt. **1988**, *27*, 476–479.

12. Borrelli, N. F. Private data.

13. Borrelli, N. F. Private data.

14. Stauffer, N.; Wilwerding, D. Sci. Honeyweller **1982**, *3*, 1–10.

15. *Handbook of Plastic Optics*; US Precision Lens Inc: Cincinnati, OH, 1983.

16. Borrelli, N. F. Optoelectronic interconnects using microlens arrays formed in glass. In Proceedings of 43rd Electronics & Components Conference, Orlando, FL, 1993.

17. Montgomery, P. C.; Fillard, J. P. In *Miniature and Microoptics*, Proc. SPIE; 1992; Vol. 1751, p. 76–85.

18. Artzner, G. Microlens arrays for Schack–Hartmann wavefront sensor. Opt. Eng. **1992**, *31* (6), 1311.

19. Kawai, S.; Araki, S.; Kasahara, K.; Kubota, K. Optical interconnections using microlens arrays for parallel processing. In *Miniature and Microoptics*, Proc. SPIE; 1992; Vol. 1751, p. 255–268.

20. Oikawa, M.; Okuda, E.; Hamanaka, K.; Nemoto, H. Integrated planar microlens and its applications. In Proc. SPIE; 1988; Vol. 1898, p. 3.

21. Kasahara, K.; Tashiro, Y.; Hamao, H.; Sugimoto, M.; Yanase, T. Double hetero-structure optoelectronic switch. Appl. Phys. Lett. **1988**, *52*, 679–681.

22. Numai, T.; Sugimoto, M.; Ogura, I.; Kosaka, H. Laser operation in VSTEPs. Appl. Phys. Lett. **1991**, *31*, 1250–1252.

23. Optical Shields, Inc./Corning, Inc. Final report to DARPA, 1990.

24. Borrelli, N. F. Efficiency of microlens arrays for projection LCD. In Proceedings of 44th Electronic & Components Conference, Washington, DC, May 1–4, 1994.

25. Hijikigawa, M. Society for Information and Display. SID Digest **1992**, 265–267.

26. Hamada, H. et al., SID Digest, **1992**, 269 paper 239.

27. Burckhardt, C. B. Optimum parameters and resolution of integral photograph. Opt. Soc. Am. **1968**, *58* (1), 71–76.

28. Lippmann, G. J. Phys. Theorique et Appliquee **1908**, 7, 446–450.

29. Pole, R. V. 3-D imagery and holograms of objects illuminated in white light. Appl. Phys. Lett. **1967**, *10* (1), 20–22.

30. deMontebello, R. L. Wide angle integral photography. Proc. SPIE **1977**, *120*, 73–91.

31. Dudley, L. P. A new development in autosterecoscopic photography. J. SMPTE **1970**, *79*, 687–693.

32. Burckhardt, C. B. Formation and inversion of pseudoscopic images. Appl. Opt. **1968**, *7* (3), 627–631.

33. Corning Incorporated, OND Division. Private data.

34. Proceedings of IEEE International Workshop on Micro Electro-Mechanical Systems (MEMS 98'), 1998.

35. Kogelnik, H.; Li, T. Appl. Opt. **1966**, *5* (10), 1550–1567.

7

Gratings

7.1. INTRODUCTION

In Chapter 1, we mentioned that the area of *microoptics* constitutes a much larger topic that includes devices and applications beyond the intended scope of this book. Our main objective was to concentrate on fabrication and application of the important subset of microoptics; namely, microlenses. A nonexhaustive listing of microoptical elements of importance is given in Table 7.1 along with an important application area. However, many of the methods of lens fabrication described in the previous chapters are equally applicable to making some of these microoptical elements. Two important structures for which this is true are gratings and planar waveguides. The implementation of the former leads to devices that can be used to redirect light, select out specific wavelengths, or both. The fabrication of the latter in various configurations leads to optical devices such as demultiplexers, directional couplers, and splitters [1,2].

The fabrication of planar optical waveguides is a topic that is amply and well covered in the literature and other

Table 7.1 Listing of Microoptical Elements

Element	Example	Application
Polarizers	Dielectric/metal stack (Lamipol[a])	Optical isolator[b]
Polarization separator[c]	Wedge of birefringent crystal	Polarization-independent isolator
Waveplates	Slab of birefringent material	Isolators
Magnetooptic elements[d]	Slab or thin film-of Fe–garnet crystal	Isolators
Filters	Dielectric stack	Demultiplexers
Gratings	Fiber Bragg	Filters/Demux
Waveguides	Etched planar in SiO_2	Planar integrated optics

[a] Tradename of Sumitomo Osaka Cement Co. It is a way to make a wire-grid type polarizer for the NIR by alternating very thin layers of metal and dielectric plates, and then slicing in cross-section.
[b] Optical isolators are devices that allow light to pass in the forward direction, but not in the reverse direction. They are used in such equipment as solid-state lasers and optical amplifiers. They are made up of a series of components.
[c] Element that separates the two polarizations into two paths to provide polarization-independent isolation.
[d] Faraday rotators: These are the essential elements in the optical isolation because they provide the nonreciprocity property.

texts and monographs [1,2]. The common elements of the fabrication methods covered in the previous chapters are the photolithographic patterning followed by etching, and the use of ion exchange. Although one of the main driving forces for microoptic elements and devices is integrated optics, it is not possible to do justice to this topic here. Rather, we will deal with two of the other important microoptic elements: gratings and the elements that make up optical isolators. These two topics are not often discussed in detail in the microoptic context. Gratings fit naturally into the context of this book because many of the methods used to make gratings are similar to those described previously in the fabrication of microlenses. The elements that go into optical isolators do not share the commonality of fabrication, but they do share the same overall area of application and are used in conjunction with other microoptical elements, such as microlenses. Because there have been few, if any, reviews of the elements that go into optical isolators, it was deemed worthwhile to include some discussion in this book.

The descriptions given in the following Chapters 8 and 9 will be somewhat brief, compared with those given in the previous chapters, because much of the material properties and processes have already been discussed in the previous chapters. The layout of the chapters will be the following: First we will give a very brief review of the physics of the optical element, to give some context to the subsequent descriptions (diffraction grating and optical isolators). Then we will list the ways these elements can be fabricated within the microoptics format, and finally, we will discuss some of the most important devices in which these elements are used. We will deal with gratings in this chapter and optical isolators in Chapter 8.

7.2. TYPES OF DIFFRACTION GRATINGS

In Sec. 4.1.3, we discussed the physical basis of diffraction gratings from the viewpoint of a surface relief pattern providing the periodic phase difference. This is certainly one primary way that diffraction gratings can be made. However, there is a much more general description of diffraction gratings of which the surface relief type is a special case. Actually, the underlying parameter of significance in a grating is the phase difference. This pattern can be produced in several ways, which we express as follows:

$$\Delta\phi(x) = \sum_m \Delta\phi_{0m} \cos\left(\frac{2\pi m x}{\Lambda}\right) \tag{7.1}$$

Here, Λ represents the grating period, the phase ϕ is expressed in waves, $\phi = nZ/\lambda$ where n is the refractive index and Z is the thickness. One can see the phase difference can be created either by letting Z vary with x in a periodic way, which would be the surface relief type, or by letting the refractive index vary in a similar way. The way we have written Eq. (7.1) amounts to a Fourier series; thus we are allowing a variety of functional forms of the way the phase can vary through a given period.

Another useful way to classify gratings is by the characterization: *thick* and *thin*. The methods of making gratings (discussed later on) will clearly depend on what type of grating one wants to make; hence, this classification applies equally to the fabrication [3,4]. The distinction originates because the phase change of Eq. (7.1) can be accomplished with a small refractive index over a very long path. From a light-propagation standpoint, this represents a much different case and, as a matter of fact, is closer to the way one would deal with a waveguiding structure (see Ref. [2] in which this is referred to as the *coupled mode treatment* of gratings). On the other hand, if the grating were very thin, then small changes in the path length have little effect on the diffraction efficiency. Such a change might occur by a differing angle of incidence. This is in contrast to the *thick* example, in which very small changes in the path length have a very large effect on the diffraction efficiency. So much so, that in the thick limit the efficient diffraction occurs at only one angle of incidence. This angle is called the *Bragg angle*. A number of gratings are schematically shown in Fig. 7.1.

Magnusson and Gaylord [3] have developed a mathematical description of this thick and thin classification that is of considerable help in determining the ultimate behavior of the grating. The parameter of consequence is called the *Q parameter* and is defined as

$$Q = \frac{2\pi Z \lambda}{n_0 \Lambda^2 \cos \Theta} \tag{7.2}$$

where λ is the wavelength of light, Z is the thickness of the grating, Λ is the grating period, and Θ is the angle that the light beam makes with the grating normal. The reader is referred to Fig. 7.2 for the definition of the terms.

7.2.1. Thin Gratings (*Q*, Small)

From the Magnusson and Gaylord [3,5] analyses, in the limit of small Q, that is less than 1, as defined in Eq. (7.2), one can

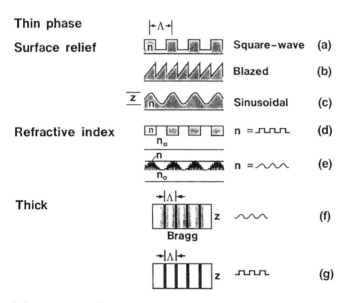

Figure 7.1 Grating types: examples of thin gratings: (a–c) surface relief gratings, with the geometric shape as shown; (d, e) thin phase gratings where the variation in phase derives from the index change rather than surface relief; (f, g) thick gratings stemming from the periodic refractive index throughout the depth.

express the behavior of various thin phase gratings in terms of their diffraction efficiency. The expression for the three most common thin gratings is listed in Table 7.2. Generally, these gratings produce multiordered diffraction patterns. For example, for the sinusoidal grating, such as would be produced by the interference of two monochromatic beams (holographic-grating pattern), the expression for the diffraction efficiency is

$$\eta(i\text{th order}) = J_i^2 \left(\frac{2\pi n_1 Z}{\lambda \cos \Theta} \right) \qquad (7.3)$$

where J_i is the ith order Bessel function, and the other terms are as defined in the foregoing. An exception to the multiordered behavior can be achieved with the use of a blazed grating, which is a special case of the sawtooth index pattern

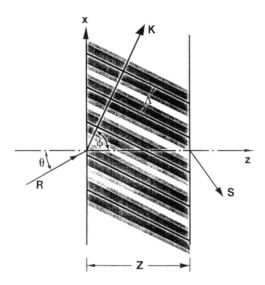

Figure 7.2 Model of thick hologram with slanted fringes: the spatial modulation of the refractive index is indicated by the shaded regions. The grating has period Λ, and grating vector $\mathbf{K}$, as shown. The incidence angles θ and the grating angle ϕ are defined.

shown in Table 7.1. In this example, as shown in Eq. (4.16) in different notation, the diffraction efficiency is given by the expression

$$\eta(i\text{th order}) = \frac{\sin^2[\pi(\gamma - 1)]}{[\pi(\gamma - 1)]^2} \tag{7.4}$$

Table 7.2 Diffraction Efficiency of Thin Phase Gratings

Type	Efficiency	Comment
Sinusoidal	$J_j^2(2\gamma)$	$j = $ order
Square	$\cos^2(\pi\gamma/2)$	0th order
	0	Even order
	$(2/j\pi)^2 \sin^2(\pi\gamma/2)$	Odd order
Sawtooth	$\sin^2(\pi\gamma)/[\pi(\gamma + j)]^2$	$j = $ order

$\gamma = \pi n_1 d / \lambda.$

where $\gamma = \pi n_1 Z/\lambda$. The difference in the notation between Eqs. (4.16) and (7.4) is because for a surface relief grating (see Fig. 7.1a,b,c), the phase difference is fixed by the depth of the groove. In other words, the phase shift is expressed by the term $2\pi(n_0 - 1)Z/\lambda$. Whereas, for the thin phase grating (see Fig. 7.1d,e), the phase shift is determined by the refractive index difference, $2\pi n_1 Z/\lambda$. For the surface relief grating, one ensures the optimum efficiency by making the depth correct for the argument of Eq. (4.16) to be 0.

An important point in the fabrication of these thin gratings is that quite different methods are required. For the surface relief type, methods such as etching or molding of the surface of the substrate are appropriate. On the other hand, for the thin phase gratings, the photosensitive process whereby the refractive index is locally ordered is the method of choice. We will go into much more detail about the methods when we describe the fabrication methods.

7.2.2. Thick Gratings

For the parameter Q, as defined in Eq. (7.2), when its value is large, $Q > 10$, the diffraction behavior changes from a multi-ordered phenomenon to that of a single-order one when the Bragg condition is met; namely,

$$\cos(\phi - \Theta) = \frac{(\lambda/n_0)}{2\Lambda} \tag{7.5}$$

where Θ is the angle of incidence, ϕ is the angle that the grating vector makes with the surface normal, and n_0 is the refractive index of the grating medium (see Fig. 7.2).

For a holographic grating (sinusoidal index profile), the diffraction efficiency is given by the expression

$$\eta = \sin^2\left(\frac{\pi n_1 z}{\lambda \cos \Theta_0}\right) \tag{7.6}$$

Clearly, from the equation, this type of grating can reach 100% efficiency when the argument of the sine reaches 90°. Kogelnik [4] has shown by coupled mode analysis of the behavior of thick gratings that the Bragg condition corresponds to a momentum-conserving relation between the wave vectors of the input beam, the diffracted beam, and the grating vector. Thick gratings have various properties that make them particularly interesting from an optical applications standpoint. In particular, the sensitivity of the thick grating to small changes in wavelength and angle provides unique applications that will be discussed later. Their behavior is shown in Fig. 7.3. Although the ν and ξ parameters are somewhat obscure, one can think of ν as the measure of the index modulation n_1, and ξ as a measure of the deviation from the Bragg condition of either the incident angle or the wavelength. Here, the diffraction efficiency is normalized to the value at $\xi = 0$, which corresponds to $\sin^2 \nu$.

These types of gratings are invariably formed in solids by a photosensitive technique, usually interfering coherent beams provide the required exposure. This will be covered in some detail when we discuss the actual fabrication process.

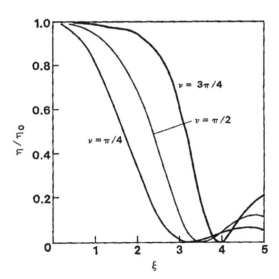

Figure 7.3 Representative behavior of the falloff in diffraction efficiency with deviation from the Bragg condition. The x-axis can be thought of as the deviation in the angle or the wavelength from the optimum. The curves are given for gratings of three strengths, as indicated by the phase angle. (From Ref. [4].)

There is sufficient literature on all aspects of holography [6,7] that it is not considered necessary to deal with it here other than how it relates the actual fabrication. Our main concern is its usefulness as a grating in the field of microoptics and with the various materials and methods by which they can be made. In a subsequent section, we will list and discuss the types of devices that use microgratings.

7.3. FABRICATION OF MICROGRATINGS

In the following sections, we will describe the various methods employed for making gratings on the microoptic scale. The dimension here is roughly 1 mm, or less, in spatial extent. From the foregoing discussion, the methods will be broken down into those applicable for the fabrication of thin surface relief gratings and those appropriate for thick holographic-type gratings.

7.3.1. Surface Relief Gratings

The methods to provide finely spaced patterns in surface relief of a material are similar to, if not substantially identical, with those for making microlenses, described in the previous chapters. Nonetheless, we will review the methods with specific attention to the aspects that affect the grating performance.

7.3.1.1. Photolithography

Photolithography is the most straightforward method used to make surface relief gratings, although it is not necessarily the best one. This will depend on such factors as determined by its ultimate use. The photolithographic method is used when high-resolution gratings are required. We have discussed the basic method in Sec. 4.2 for the fabrication of diffractive lenses, and for convenience it is outlined again in Fig. 7.4. The photoresist is spun on a suitable substrate and exposed through a mask that contains the appropriate spatial pattern. After development, the grating pattern is in the resist layer.

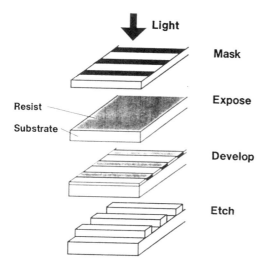

Figure 7.4 Representation of the classic photolithographic production of a surface relief grating.

The grating pattern is then produced in the substrate by a suitable etching technique, such as reactive ion etching (RIE).

In general, the gratings require high resolution—spacing of the order of 0.5 µm—thus, the mask must be fabricated with the highest-resolution technique that is available. There are two methods that can be used to prepare such masks. The first uses standard e-beam lithography writing onto resist that covers a metal film. After development of the resist, followed by dissolution of the exposed metal, a pattern of metal on glass is achieved. The type of relief pattern that will be produced by the mask can be varied to some extent by the e-beam-writing methods. For example, as mentioned in Sec. 4.2.1, one can vary the transmittance through the pitch of the pattern, by half-tone techniques that approximate transmittance functions that are sawtooth, or sinusoidal. The former pattern would allow a blazed grating to be made, and the latter a pattern approximating a holographic grating.

The second method is the holographic method for which the interference of two mutually coherent beams is used to expose the resist (Fig. 7.5). This method produces a sinusoidal

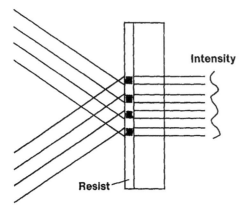

Figure 7.5 Representation of the interference method of exposure to form a sinusoidal grating in a photosensitive film.

pattern that may be desirable. Another advantage of the holographic exposure method is that it is capable of a much larger area of exposure. The e-beam system operates in a step and repeat mode; hence, stitching errors are possible.

Recently, a third type of mask called a *phase mask* has found its niche in the area of exposing fiber Bragg gratings. We will discuss this important application later in this chapter. A phase mask is itself a diffraction grating that, in turn, is used as a mask to produce gratings in photosensitive materials. The grating is made in a way, usually using the holographic exposure method [8], such that it maximizes the +1 and −1 orders and suppresses the 0th order. It is then used as shown in Fig. 7.6. The advantage of this type of mask is that it is essentially equivalent to a holographic-type exposure, but requires very little coherence length of the laser. In the conventional holographic exposure, the coherence length of the laser sets the allowable path length difference. The excimer lasers are not very satisfactory coherent sources, with coherence lengths of less than 100 μm. With the exposure geometry shown in Fig. 7.6, this is not a problem because the beam is essentially split at the grating – sample interface.

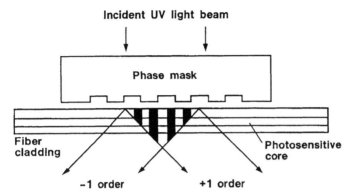

Figure 7.6   Diagram showing the use of a phase-mask to expose a sinusoidal grating.

7.3.1.2. Molding

The molding of diffraction gratings strongly depends on the nature of the material requirements. For most applications, the requirement is for a robust material, such as silica. To mold into glass requires the use of a master. The master would contain the opposite relief pattern, as shown in Fig. 7.7. In Chapter 2, we discussed the molding of aspheric lenses. This same method can be applied to the molding of diffraction gratings. Here, a special low-temperature glass was developed that was compatible with the molds [9]. For the grating,

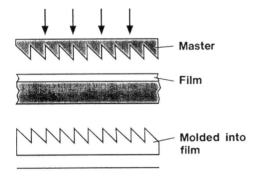

Figure 7.7   Schematic representation of the molding of a surface relief pattern.

one can actually use a silica master. In Fig. 7.8A and B, we show an AFM surface profile of the silica master and the resulting pattern in the glass.

Another similar approach is to mold into a soft layer deposited, or otherwise produced, on the surface of a hard material. This layer could be a polymer, or a sol–gel-derived layer. Figure 7.9 shows the pattern produced in a sol–gel layer (mixture of polymer and TEOS-derived glass) produced by the same master as used to make the grating shown in Fig. 7.8.

7.3.2. Photosensitive Films

The type of grating we have described is a phase grating, of which the surface relief type is the important, but nonetheless, special example. The more general approach is to apply a thin photosensitive film to a hard substrate. *Photosensitive* means that on exposure to light the film will undergo some change, structural or otherwise, that will result in a change in refractive index. There are examples of this type of material, the most common being photographic emulsion. For example, in Kodak 649-F, the conventional use of a silver-based emulsion is to produce an image through a development process that reduces silver halide to silver. However, by special development conditions, one can bleach the silver and end up with an almost pure refractive index change [10]. This is how holography was performed in the early days. The problem is the durability of the resulting material. A related process makes use of what is called *dichromated gelatin* [11]. This material can be prepared on glass, and when exposed to light, cross-linking of the gelatin molecules occurs. Much like photoresist, this causes changes in solubility or volume, which lead to phase changes in proportion to the exposure. More recently, families of photopolymers have been developed that undergo structural changes, such as polymerization on exposure to light, which lead to refractive index changes.

The thickness of these layers can vary from a few tenths of a micrometer to tens of micrometers. The thickness can

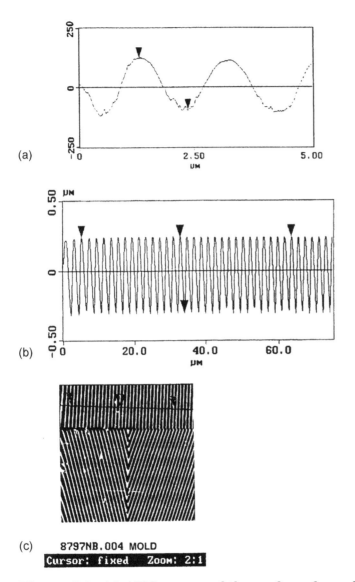

Figure 7.8 (a) AFM pattern of the surface of a molded grating embossed into a special low-temperature glass; (b) the actual mold pattern is shown. The important point to note is the depth of the pattern. For the mold, the maximum depth was 556 nm and the molded part yielded 543 nm.

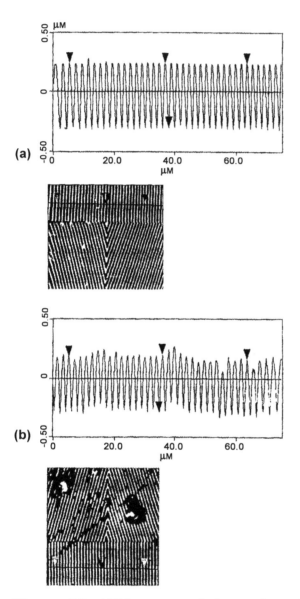

Figure 7.9 AFM pattern of the surface of a molded grating embossed into a polymer/sol–gel composite: (a) the silica mold and (b) the sample. Note that the depth here is 493 nm, compared with the mold depth of 556 nm.

bridge the gap between the thick and thin regimen discussed earlier.

7.3.3. Thick Gratings

The fabrication of thick gratings follows the method by which one would make a plane-wave hologram. This was schematically shown in Fig. 7.5 and we expand on it in Fig. 7.10. The coherent phase difference between the two beams **U** and **R** produces an interference pattern, such that the intensity is given by the following expression:

$$I = (\mathbf{U} + \mathbf{R})(\mathbf{U} + \mathbf{R})^*$$

(7.7)

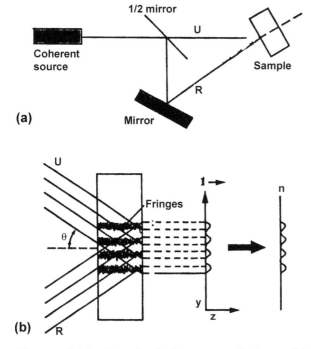

Figure 7.10   Standard plane wave holographic exposure process: (a) the coherent beam is split into two and is made incident onto the photosensitive body at equal angles to produce (b) an unslanted fringe pattern. The interference of the two coherent beams produces a sinusoidal intensity pattern.

where the asterisk stands for the complex conjugate. In the simple plane wave, **U** and **R** differ in phase only by a value of δ. This means one can write Eq. (7.7) as

$$I = E_0^2\{1 + \exp[i\delta(y)][1 + [-i\delta(y)]\} = I_0 \cos^2\left[\frac{\delta(y)}{2}\right] \quad (7.8)$$

This sinusoidal intensity pattern is incident on and through a photosensitive material. The material responds to the exposure in a way such that a permanent refractive index variation is subsequently established. (This property of permanence is important because the applications in this book are to make robust optical elements. This is not to say that there is a whole other technology that is built on being able to make and erase gratings with light or heat.) This need not be a direct response, but may require a further step, as we shall see, such as heating. It is important to make a distinction between the nature of the initial change that occurs in the body and what ultimately happens. For example, the initial state of the material may absorb the light, promoting a structural change that will eventually be a pure refractive index grating. To the extent that absorption persists, the diffraction efficiency decreases.

As a result of the photoprocess induced by the intensity variation, a phase function is produced

$$t(y) = \exp[2\pi i\phi(y)] \quad (7.9)$$

where the phase function $\phi(y)$ is given by the product of Eq. (7.8) with the function that represents the manner in which the refractive index develops with intensity. For the simple plane case, this just means that a sinusoidal phase grating has been produced, the depth of which will depend on the specific material response.

7.3.3.1. Photosensitive Materials

There are several examples of *photorefractive* materials; that is, materials that when exposed to light can undergo some degree of *permanent* change leading to a refractive index change. We have made a fairly representative list of the

Table 7.3 List of Photorefractive Materials

Ferroelectric single crystals
Polymer guest/host
Photosensitive polymers
Photosensitive glasses
Porous glass
UV-photosensitivity (fiber Bragg grating)
As_2S_3

materials that have been reported to have such an effect in Table 7.3. The mechanism for the production of the refractive index change is quite distinct for the various materials. We shall very briefly review each of the materials in Table 7.3.

Ferroelectric single crystals: $LiNbO_3$ and $BaTiO_3$ can be considered the "original" photorefractive materials, and even today to many, the photorefractive effect refers exclusively to the effect observed in these materials. There is much literature on this and related materials [12–14]; thus, we will touch only briefly on the phenomenon. $LiNbO_3$ is a pseudocubic single crystal of the perovskite family. The important properties of this class of materials are that it is ferroelectric, and it is photoconductive. The ferroelectric materials are highly polar, and undergo a spontaneous polarization at a temperature termed the *Curie temperature*. Because of this highly polar structure, these materials possess a large electrooptic effect, which is expressible as

$$n(\text{par}) - n(\text{per}) = \Delta n = rE \qquad (7.10)$$

where $n(\text{par})$ is the refractive index parallel to the applied electric field, and $n(\text{per})$ is for the perpendicular. The photoconductive property derives from defect centers and impurities from which electrons can be promoted into the conduction by exposure to light with sufficient energy. The level of defects required to produce the photorefractive effect is roughly parts per million (ppm).

The induced photorefractive effect occurs in the following sequence of steps [15,16]. On exposure, the mobile electrons are promoted from the defect into the conduction band, leaving behind a positively charged site. Eventually, the charge drifts into the unexposed region where it becomes permanently trapped. An electric field is developed between the region of negative charge (dark) and the positive charge (bright). This strong internal field, estimated to be $10^4\,\mathrm{V/cm}$, induces a refractive index change through the electrooptic effect (see Eq. (7.10)).

The production of the diffraction grating follows the standard process indicated in Fig. 7.10. The light source is in a spectral range sufficient to provide the excitation of the electrons out of the defects, and this depends on the specific crystal and the nature of the defects. The band gap of most of these crystals is in the region of 3–4 eV; however, the defects' absorption that leads to the photoelectrons extends through the visible portion of the spectrum. The dependence of the index change on the exposure is shown in Fig. 7.11 [17]. It is characteristic that this effect requires only small intensities to produce significant index changes. This means that the production of photocarriers is a very efficient process which, when coupled with a large electrooptic effect, produces the large index change.

Because electrons are trapped in the dark regions, it is possible to thermally stimulate them out of the traps.

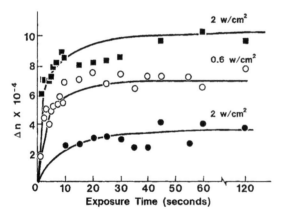

Figure 7.11 Induced refractive index change in LiNbO$_3$ as a function of time for three exposure intensities. (From Ref. [17].)

Subsequently, they can recombine with the trapped holes. The energy depth of the trapped electrons determines the thermal stability of the photorefractive effect. In other words, one can thermally erase gratings.

As we will see in the applications section, LiNbO$_3$ is a versatile substrate in the planar-guided wave application. It provides a platform for much functionality; the ability to form gratings is one important example.

Polymer guest–host: In the past few years, an all-organic version of the photorefractive effect has evolved [18,19]. The basic idea is to mix into a polymer host, various organic molecules that will function as the required agents to produce the refractive index change. An example is as follows: One dissolves into the polymer matrix a molecule that is photoreducible; that is it can promote a hole into the polymer. Examples of good charge generators are donor–acceptor charge transfer complexes, examples of which are shown in Fig. 7.12. The required next element is a molecule to provide charge transport. The charge-transporting function is usually provided by a charge transfer agent that is in sufficient concentration to permit hopping of the hole or electrons. These are chosen from the class of molecules such as carbazoles,

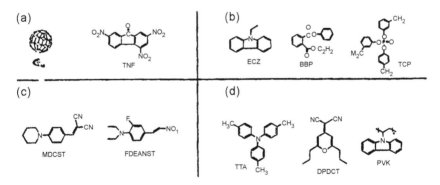

Figure 7.12 Listing of some of the molecules used for the organic photorefractive effect: (a) photosensitizers to produce the carriers; (b) plasticizers to control the stiffness of the polymer; (c) NLO chromophores to produce the electrooptic effect; and (d) transporting molecules and polymers to produce the conduction. (From Ref. [18].)

hydrozones, and aryl amines. The next molecule needed is one that provides relatively deep traps for the photogenerated mobile species. These trapping sites are intrinsic to the polymer configuration.

With these three elements, one can envision the evolutionary formation of a space charge region, as shown in Fig. 7.13. Unfortunately, the natural fast diffusion or large drift mobility that occurs in the inorganic version is not present in the organic version, so that a bias voltage is maintained both during exposure and after. This bias field is close to $90\,V/\mu m$ [19].

The final ingredient in the mix is the molecule that will provide the electrooptic effect as a result of the internal space charge field. Here one chooses to include a high concentration of a molecule with a large optical nonlinearity. These nonlinear optical chromophores (NLO) are molecules that

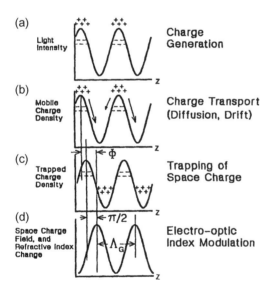

Figure 7.13 Schematic of the photorefractive process: (a) carriers are formed in the illuminated area; (b) charge is separated either by drift or the bias field; (c) mobile carriers are trapped in the dark regions; and (d) the space charge field provides the electrooptic effect. (From Ref. [18].)

have structures that involve a large dipole moment change with applied field. The symmetry-breaking step that is required here is that the sample be *poled*. This means that all of the mix is now complete and the resulting refractive index change is achievable.

The exposure for grating production in these materials is the standard plane-wave holographic exposure (see Fig. 7.10). One significant difference between the organic version and the inorganic one is that a larger Δn is achievable in the organic material [19].

Photosensitive polymers: We include here all of the polymeric materials that undergo structural changes when exposured to light, usually followed by some thermal treatment. Conventional photoresist is a good example of this class. The dichromated gelatin mentioned in the foregoing section is another example. The reason they are included here is that these layers can be thick enough that the grating can exhibit thick grating behavior. The gratings are made as shown in Fig. 7.10, with the specific exposure conditions determined by the particular photosensitive material.

Photosensitive glass: In Sec. 3.4.3, we described the composition of certain special glasses, and how the photosensitive development of a second microphase could be used to alter the refractive index. Reviewing the mechanism that was described in the previous section, the UV light promotes an electron from a donor, such as Ce^{+3}, that ultimately becomes trapped. The thermal treatment that follows the exposure frees the electron from the trap and reacts with the mobile Ag^{+1} ions that are contained in the glass. The reduced silver agglomerates to such a size to nucleate a phase separation of an NaF phase. If one "overexposes" the sample, thereby creating many nuclei, then the growth of the individual particles is limited to a size sufficiently small to keep the glass transparent. The refractive index increases where the crystal is produced because the fluoride is removed from the surrounding glass.

Efimov et al. [20a] have done a detailed study of the induced refractive index phenomenon as related to hologram writing anfd diffraction gratings in general in the NaF-based glasses. They coined the word photo-thermal glasses, or PTG to distinguish them. Their exposure utilized a 1 mW He–Cd laser passing through a 1-10X telescope and a spatial filter. The exposure times could be inferred from the reporting of the exposure fluence which was 400 mJ/cm^2 if the spot size were known. After the exposure, they studied the effect of time at temperature 520°C on the ultimate refractive index change. This is shown in Fig. 7.14a. The diffraction efficiency of a grating frequency of 600 mm^{-1} along with the calculated induced index as a function of dose is given in a related paper [20b] and is shown in Fig. 7.14b and c.

In this material, and for any photosensitive material for which the photorefractive process is initiated by an absorption process, the depth of penetration of the excitation and, hence, the value of the induced refractive index change, are limited. They report an absorption coefficient of 0.5 cm^{-1} at the excitation wavelength; this is only a 10% variation in exposure and is experienced through a 1 mm sample. Consequently, the thickness of the grating can be adjusted by the choice of the excitation wavelength. For photosensitive glasses, the peak of the excitation is at 307 nm. However, by using an exposure wavelength of 337 nm, which is on the tail of the excitation peak, one can obtain a deeper grating, at the expense of a corresponding longer exposure time.

Doped porous glass: As with the photosensitive glasses (see Sec. 3.4.3), permanent refractive index changes can be produced in porous glass that has been impregnated with photosensitive materials, we make a distinction between the two different ways the porous medium can be used to this end.

In the first method, a photosensitive organometallic compound is loaded into the porous glass through vapor loading or through the use of a solvent that is subsequently evaporated off. Here, the action of the light is to alter the physiosorbed molecule in such a way that it bonds more strongly to the surface of the porous glass. Subsequent heating oxidizes the metal, and the result is an oxide phase, contained within the pores, that results in an index

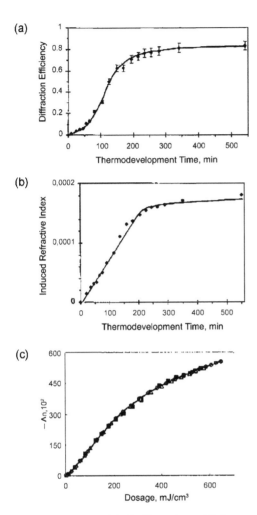

Figure 7.14   (a) Diffraction efficiency for photothermal glass as a function of thermal development time for a fixed temperature, (b) calculated induced refractive index for the same treatment, (c) calculated induced refractive index change as a function of exposure dose. (Taken from Refs. [20a] and [20b].)

of the refractive change [21,22]. An example of holographic formation is shown in Fig. 7.15. Multiple gratings were formed in a porous glass sample impregnated with an orga- nometallic titanium compound. The exposure arrangement

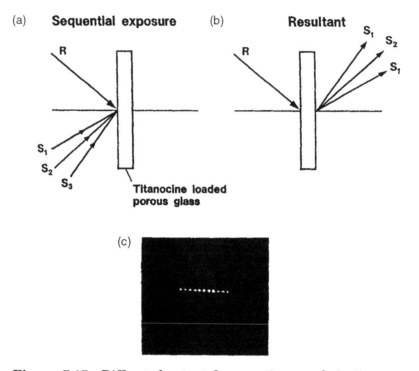

Figure 7.15 Diffracted output from gratings made in titanocene-loaded porous glass: the exposure was $1\,\text{W/cm}^2$ for 5 min. Glass was heated to 600°C for 1 hr after exposure. The exposure arrangement is shown. Eleven gratings were exposed about 2° apart. The reconstruction was with the single reference beam, as shown.

is also shown. The exposure was accomplished with a 10-mW He–Ne laser with exposure times of minutes. The sample after exposure was heated to 600°C. The diffraction efficiencies of gratings prepared in this way were over 80%. A wide range of organometallic compounds can be used for this purpose [20].

The second method [23,24] involves the infiltration of the porous structure with a photosensitive material such as a photoresist. The photoinduced refractive index change derives from the resist in a manner in keeping with the composite nature of the structure [see Chapter 3, Eq. (3.50)]. There are several important steps that can be used to modulate the index in the resist-impregnated porous glass. These involve the various changes that the resist undergoes. The inherent effect is from the structural changes in the resist, such as the density and volume. However, one can proceed further and use the solubility difference between the unexposed and unexposed resist to enhance the contrast. Moreover, one can use the retained resist as a protection when the glass is reexposed to a leaching bath. After the resist is removed, the index pattern is purely structural in that the exposure pattern is embedded in the pore structure. In Table 7.4, we reproduce the refractive index (RI) changes achieved after the various stages [23].

UV photosensitivity in SiO_2–GeO_2 waveguides: One of the more interesting phenomena that has been uncovered over the last few years is what has come to be called *fiber Bragg gratings*. This name has arisen because of its wide application as a Bragg grating in optical fiber devices as we will see in the later section.

What the term really means is that in certain binary silica systems, a UV photosensitivity exists. In particular, the binary system SiO_2–GeO_2 is of most interest because it has the largest effect and is the basis for most of the optical fiber in use today.

The amount of work on this effect could fill a chapter in itself, perhaps even a book. Consequently, here we can only briefly describe the nature of the effect and its proposed origin. The original report of an induced index change of the order of 10^{-5} in a single-mode GeO_2–SiO_2 optical fiber is attributed to Hill et al. [25]. The initial experiment used green light into the core of a single-mode fiber. After prolonged exposure, there appeared a weak reflection indicating that some sort of grating was forming as a consequence of the

Table 7.4 Induced Index Change After Stages of Porous Glass Composite Treatment

Stage	Exposed resist	n Resist removal	Releaching
R1	0.01	0.03	0.15

green exposure. However, the real interest began when it was reported that one could induce the change much more efficiently using deep UV radiation. Moreover, the exposure could be made from the side of the fiber passing through the transparent silica cladding [26]. This is the so-called side-writing technique (Fig. 7.16). With reference to the Bragg condition given by Eq. (7.5), the condition corresponds to $\Theta = \phi = 90°$. Measured reflection efficiencies obtained with the side-written exposure geometry were now consistent with that predicted from a Bragg grating with an induced refractive index change of greater than 10^{-4}

Although there is no total agreement on the exact mechanism for the induced index change, there is a consensus on certain aspects. First, there is a Ge defect that has a distinct optical absorption at 240 nm. This absorption band is bleached when exposed to light within its absorption spectrum. Finally, new absorption bands, extending to the higher energies, are created as a consequence of the disappearance of the 240-nm absorption. This is shown in Fig. 7.17, measured on a 0.5-mm-thick slice from a preform. (The single-mode waveguide is made by redrawing the preform into fiber.)

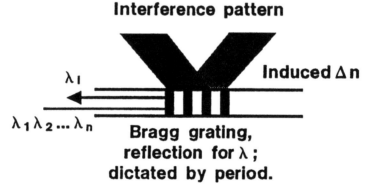

Figure 7.16 Schematic drawing of the exposure arrangement to develop a fiber Bragg grating. Two interfering beams are exposing the core of a single-mode fiber through the cladding from the side. Typical exposure of SiO_2–GeO_2 fiber is 100 mJ/cm^2 for 10–20 min at 10 Hz with 15 nsec pulses from a 248 nm excimer laser.

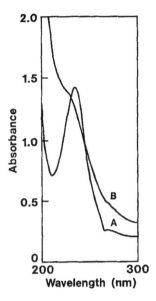

Figure 7.17   A UV absorption spectrum of 0.5-mm-thick section of
93% SiO_2–7% GeO_2 waveguide core-cane blank showing the characteristic defect absorption feature at 240 nm (A) before and (B) after
exposure to 248 nm excimer radiation.

There is a correlation between the existence of this absorption
center and the subsequent ability to produce a grating in the
fiber. The defect center has been attributed to an oxygen defi-
ciency [27]. In support of this are the data that show the
strength of the 240-nm-absorption band as a function of the
oxygen partial pressure used in consolidation of the soot preform. This is the so-called outside process (Fig. 7.18). There
are other processes used to prepare waveguide blanks for
which it is harder to pinpoint the conditions that favor the
formation of this oxygen-deficient center.

The mechanism for the photoinduced refractive index
change, and hence, the grating, is given by the following
explanation. The 240-nm-absorption band is bleached, which
corresponds to the destruction of the Ge-oxygen-deficient
center. The electrons and holes produced in the process
become trapped and produce new strong absorption features

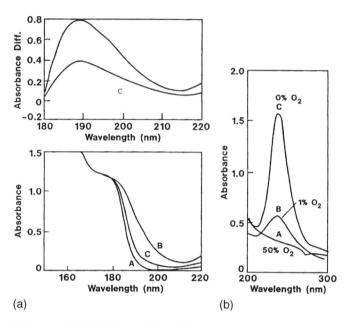

(a) (b)

Figure 7.18 Ultraviolet absorption spectra showing the dependence of the 240-nm-absorption band on the oxygen partial pressure during consolidation of the porous preform. Samples are 0.5-mm slices of the consolidated blank: curve A represents sample from consolidation in 50% O_2/50% He, whereas that of curve B is of 1% O_2, and C that of <1 ppm O_2. Both VUV and near UV regions are shown, as well as the difference curves for the VUV.

in the vacuum UV portion of the spectrum. An example of this phenomenon is shown in the absorption spectrum, before and after exposure in Fig. 7.19. This is the result of the exposure at 248 nm (100 mJ/cm^2, 20 Hz for 10 min) of a slice of a waveguide blank containing 40% GeO_2. One can clearly see the disappearance of the 240-nm-absorption band as a consequence of the 248-nm exposure, and the increase in the absorption at shorter wavelengths. One can measure a significant number of paramagnetic species after exposure, corresponding to trapped holes and electrons. The most significant is the Ge E' center, which is a hole trapped on a three-bonded Ge (Fig. 7.20). Also shown is a possible origin of this center; namely, the Ge–Ge bond, which when broken

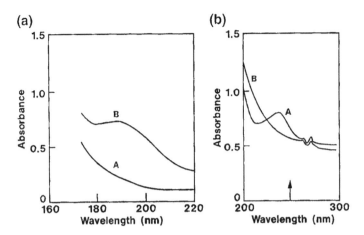

Figure 7.19 Typical absorption changes optically induced in 0.5-mm slice of core-cane 40% GeO_2 sample by 248 nm exposure ($100 \, mJ/cm^2$, 10 Hz, 10 min): (a) before exposure; (b) after exposure.

by the action of the light, produces the E' center. The electron promoted out of this bond is also trapped, and corresponding electron spin resonance (ESR) signals are also observed [28].

Whatever the specific details of the reaction are, the refractive index change at wavelength λ is proposed to come about through the Kramers–Kronig relation

$$\Delta n(\omega, I) = \frac{c}{\pi} \int\limits_0^\infty \frac{\Delta\alpha(\varpi, I)}{\omega^2 - \varpi^2} \, d\varpi \qquad (7.11)$$

Here, the light induced refractive index is related to the induced change in the absorption. (This relation is derived in Appendix A.)

This is just the mathematical way of saying that if significant absorption changes are produced in the deep UV portion of the spectrum, then the refractive index will be influenced at all longer wavelengths.

A much larger photoinduced refractive index change could be produced in GeO_2–SiO_2 compositions by

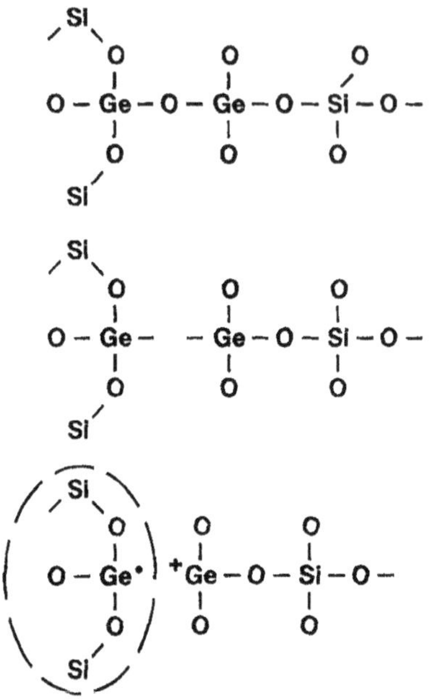

Figure 7.20 Schematic representation of the possible defect structure in SiO_2–GeO_2: (a) A 2D representation of the Si–Ge network; (b) the oxygen vacancy; and (c) the way the defect would interact with light.

impregnating the fiber with molecular hydrogen before the exposure [29,30]. This impregnation is carried at room temperature, or slightly above to ensure that no reaction occurs. The pressure of the hydrogen is high to allow the order of 10^{20} molecules per cubic centimeter to be dissolved. The mechanism is thought to involve the direct attack of the Si–O–Ge bond with molecular hydrogen in the presence of the radiation [28], leading to GeH and SiOH species. Both of these are observed after exposure, as shown in Fig. 7.21. Another possibility is the formation of SiOH and a $\equiv Ge^{+2}$ center [27]. The increase in the absorption in the deep UV as a consequence of the exposure is large, as shown in Fig. 7.22. Note that the magnitude of the photoinduced absorption in the hydrogen-loaded preform slices is independent of the initial defect absorption at 240 nm. As a matter

(a)   (b)

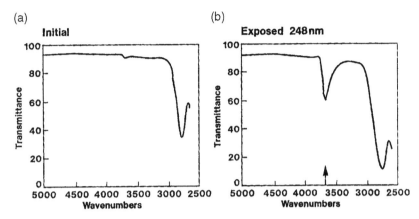

Figure 7.21 Near infrared transmission spectra of hydrogen-loaded core-cane blanks, before and after 248-nm exposure. Note the increase in the absorption at $3670\,\mathrm{cm}^{-1}$, which is attributed to the OH stretching mode.

of fact, in the hydrogen-loaded example, the excitation is equally effective at 193 nm. The explanation for this resides in the suggestion that the process here is dominated by a two-photon-initiated exciton formation that, on being trapped, reacts with H_2. We give a brief summary of the phenomenon in Table 7.5 in its various manifestations.

The exposure method is generally to place a phase mask, corresponding to the period that will yield the desired peak wavelength reflectivity, in contact with the single-mode fiber over the region where the fiber is to be formed. The exposure conditions are somewhat different, but in general, a pulsed UV source is used. This is typically an excimer laser, KrF or ArF. There is some indication that 193-nm exposure is better than 248 nm [31]. The reason is unclear, although the higher energy would favor processes dependent on a two-photon process. The use of the phase mask relaxes the requirement of long coherence length that would be needed for the conventional holographic exposure. Nonetheless, one could also use the more coherent excimer-pumped tunable dye lasers as the source. The exposure fluence is in the vicinity of 100–$200\,\mathrm{mJ/cm}^2$, operating at 10 Hz for tens of minutes. It is also

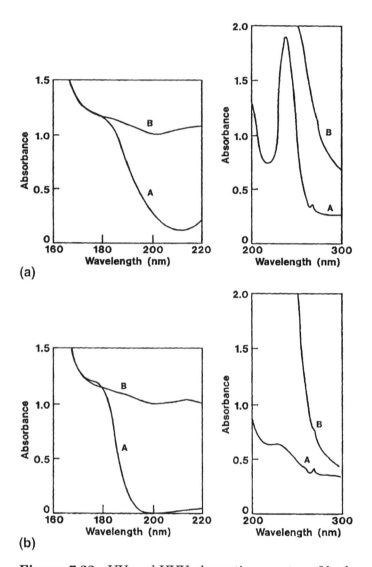

Figure 7.22 UV and VUV absorption spectra of hydrogen-loaded 7% GeO_2–93% SiO_2 core-cane samples, curve A before and curve B after 248-nm exposure. (a) Curves for a sample with a high concentration of oxygen vacancy defects, as indicated by the strength of the 240-nm-absorption band; (b) spectra for a sample with little, or no oxygen defects. In both cases, the amount of induced absorption is equivalent, irrespective of the initial difference in the oxygen vacancy concentration.

Table 7.5   Relative Comparison of Mechanisms

Initial state	Defect	Reaction	Products	$\Delta n \times 10^{-4}$
As made[a]	=Ge:, =Ge– Ge=	Bleaching of Ge–Ge	Ge', Ge-1,2	5
H_2 low temperature[b]	=Ge–O–Si =Ge–O–Si	Bond scission Bond scission	GeH, SiOH =Ge^{+2}	40

[a] Number of defects and mix varies considerably with method of preparation and composition.
[b] Samples are usually impregnated at <150°C at pressures >100 atm.

possible to use a doubled cw Ar ion laser. Here, although the peak power is low, the average power is comparable with the pulsed sources for which duty cycles are only 10^{-9} [32]. There are also reports of single-shot production of gratings aimed at producing gratings "off the draw" [33]. Here the peak fluence is much higher, and it is likely that the gratings are forming from a contribution other than the one just described. The oxygen defect that produces the optical absorption at 240 nm is necessary to couple the light. The major effect may be thermal. Successive pulses produce large swings in the measured efficiency, indicating a different mechanism than pure photochemistry [33]. Nonetheless, the grating is stable, once made.

One last point of practical significance is the thermal stability of these gratings. The thermal stability is characteristic of a material with a distribution of trap depths. Erdogan et al. [34] have analyzed the dependence on temperature. To obtain long-term stability, the gratings are heated to somewhere near 300°C to remove the unstable portion, leaving the long, slowly decaying part of the distribution.

As$_2$S$_3$: Refractive index changes can be readily induced in As$_2$S$_3$ glasses and films [35,36] by exposure to light near the bandgap. More appropriately, one might include the family of glasses As–Ge–S–Se [36]. The origin of the optically induced changes is not fully understood, although it has been the subject of much study [34]. Two proposed explanations are based on light-induced structural changes, sketched in Fig. 7.23. The one mechanism is a double-well

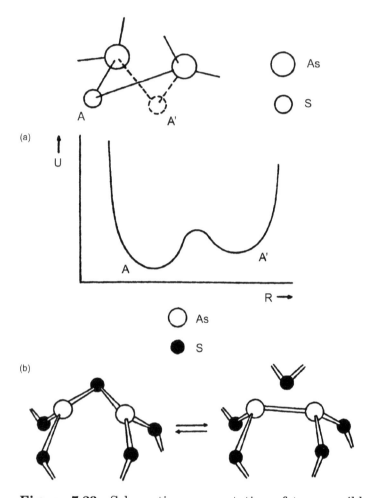

Figure 7.23 Schematic representation of two possible mechanisms leading to light-induced refractive index changes in As_2S_3. (a) A double-well mechanism is proposed; (b) a bond breakage and re-formation is considered. Each of these would lead to a structural change that would cause the index to be altered. (From Ref. [36].)

configuration in which the sulfur ion can be optically excited to the other site, or in the mechanism the proposed optically induced change is where a As–As bond is formed. The exposure also produces optical absorption on the edge, referred to as *photodarkening* (Fig. 7.24). Here the absorption on the

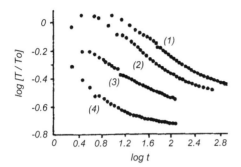

Figure 7.24 Log of measured transmission change in As_2S_3 film at 514 nm as a function of time for the powers of (1), 0.4; (2), 0.9; (3), 3.5; (4), 20 W/cm^2. (From Ref. [37].)

absorption edge is monitored in time as a function of exposure [37].

In addition to uniform refractive index changes, there are also reports of photoinduced optical anisotropy [36]. Here, linearly polarized light is used to produce some kind of defect alignment rendering the sample birefringent. An example of this effect is shown in Fig. 7.25 where a linearly polarized He–Ne laser was used to produce the birefringent pattern viewed under crossed polarizers.

Holograms have been written in As_2S_3 films. First-order diffraction efficiencies of over 80% have been reported in a 10-μm-thick film using a He–Ne writing beam [38]. The exposure was 15-m W He–Ne laser onto a 1-mm^2 spot for 10 sec.

Photosensitive Ge-doped aluminoborosilicate glasses: Gratings have been written in photosensitive glasses with a composition based on a Ge-doped alumino-borosilicate. The photorefractive effect in this conventionally melted glass is modeled after that in Ge-doped silica fiber as discussed in detail in the previous section. Specifically, the use of molecular hydrogen is required to produce a large photorefractive effect. One observes experimentally all of the same attributes of the photoreaction in these glasses as one sees in the Ge-doped fiber indicating the same H_2 mediated mechanism is operative [34a] in Fig. 7.26 is shown the induced refractive index, measured in 633 nm waves, as a function of time of

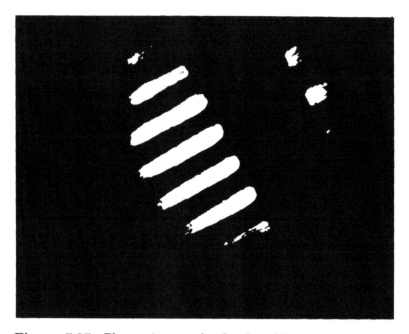

Figure 7.25 Photomicrograph of induced birefringence in As_2S_3 glass: the exposure was polarized with an He–Ne laser at 633 nm, 15 mW, 1.5 mm beam, for 15 min. Picture was taken under crossed polarizers with the exposure polarization direction oriented 45° relative to the polarizer direction.

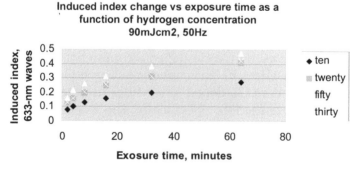

Figure 7.26 Induced refractive index change as a function of exposure time at different molecular hydrogen content for Ge-doped borosilicate glass 248 nm exposure was $90\,mJ/cm^2$ at 50 Hz.

248 nm excimer laser exposure for three different hydrogen loading concentrations. The induced index is just the number of waves per unit thickness (0.8 mm in this case) multiplied by the wavelength. In Fig. 7.27a is shown 1550 nm reflectively of a apodized grating made in the glass by the schematic exposure shown in Fig. 7.27b. We will see in a later section the application of this material as a basis for a tunable filter for optical measurements.

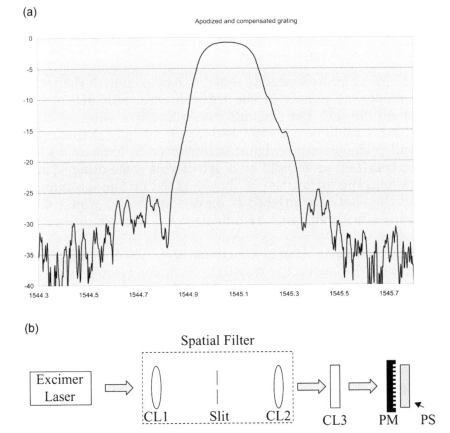

(a)

(b)

Figure 7.27 (a) 1500-nm reflection spectrum of grating made in Ge-doped borosilicate glass by 248-nm exposure through a phase mask, and (b) schematic of experimental exposure set-up.

7.4. APPLICATIONS

The applications of diffraction gratings of small dimension can be broadly classified into two areas: optical interconnects and optical waveguide components and devices. The grating utilization for the two areas is quite distinct. In the optical interconnect application, the grating is primarily used to redirect the light. In the optical waveguide application, the grating is primarily used to separate different wavelengths and to modify propagation characteristics. We will cover a few of the representative applications in the following sections.

7.4.1. Optical Interconnects

The optical interconnect application has to do with the transfer of optical signals in the three-dimensional (3D) spatial domain [39–41]. For example, the connection between chips or boards could be accomplished faster optically than electronically. The electrical signal is converted to light by a LED, and transmitted as light to a detector at some other spatial location. The advantage of this is based on the assumption that the electronic method is limited by the RC time, which becomes more an issue as dimensions of the circuit become smaller.

A couple of simple functions are shown in Fig. 7.28. Figure 7.28A shows the way light from one place on the chip might be communicated to another position on the same chip. The diffraction grating would be a thick or phase holographic type for high efficiency. Figure 7.28B shows the communication between boards. In this case, there are both fan-out type gratings, one beam to N, and fan-in gratings, n beams to one.

In this function, the grating element is used in combination with other microoptic elements. For example, the output from the source is collimated for the diffraction grating to be effective, similarly a lens would be required to focus the light onto the detector.

A wonderful example of the power of diffractive optics is in the ability to combine optical functionality. For example, in

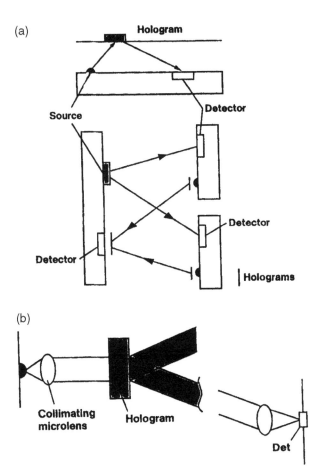

(a)

Hologram

Source

Detector

Detector

Detector

Holograms

(b)

Collimating
microlens

Hologram

Det

Figure 7.28 Schematic drawings of the way gratings can be used to optically connect places on (a) the same chip, and (b) on separate chips or boards.

the foregoing application, there are two elements required, one to collimate the beam from the source and the other to diffract it with the grating to form a fan-out pattern. Kostuk et al. [39] have designed a diffractive element that combines both functions. We schematically show the combined function in Fig. 7.29a and a picture of the eight-level diffractive element in Fig. 7.29b.

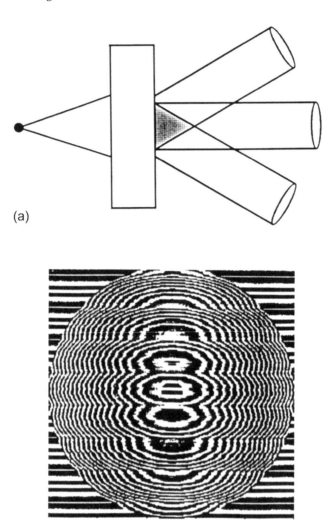

(a)

(b)

Figure 7.29 Schematic of a diffractive element that combines (a) collimating lens and fan-out function; (b) a photograph of the actual diffractive element. (From Ref. [39].)

A somewhat different, but related, grating function has to do with the beam tracking of a compact disk (CD) player [42]. The optical device is shown in Fig. 7.30. There are actually two gratings that are used. The first is a simple thin

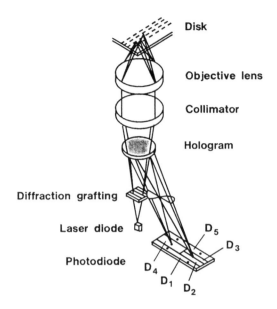

Figure 7.30 Schematic of the read element of a CD player showing the position of the two gratings. The function of the holographic grating is to diffract the return beam onto a detector array to provide tracking information. (From Ref. [40].)

phase diffraction grating placed right after the laser diode as shown. Its function is to split the beam into three, a strong zeroth order, and two weaker first-order beams. The three beams then pass through a segmented holographic-type grating that focuses them onto the disk. After reflection, the beams are directed differentially by the different segments onto the five-element photodetector array. Comparison of the signal from the detector array leads to a discernment of the tracking error.

7.4.2. Optical Waveguide Components

The fiber Bragg grating is by far the most important example of the use of a grating structure in the optical fiber area, in particular in wavelength division multiplexing (WDM). Its function is to select a wavelength from a number being carried in the fiber and to reflect it backward where it can be coupled out. Before we address the specific applications, it will be useful to present the phenomenon in terms of mode coupling. With this background, one will be able to see how gratings can be used for a wide range of applications.

The general formulation is written in terms of the perturbing influence of a periodic refractive index pattern (grating) on the mode propagation in a given waveguiding structure. It can be shown (1) that the coupling between two modes, represented by form of Eq. (7.12) will occur if the condition expressed by Eq. (7.13) exists between their propagation constants:

$$E_k = A_{0k} \exp[2\pi j \, \beta_k(z)] \tag{7.12}$$

$$\beta_k - \beta_m = p\left(\frac{2\pi}{\Lambda}\right) \tag{7.13}$$

Here, p is an integer and Λ is the grating period. One represents the grating as variation in the dielectric constant of light guiding medium

$$\epsilon = \epsilon_0 + \Delta\epsilon \tag{7.14}$$

For the cases with which we will be dealing, the variation in ϵ will be in the propagation direction; that is, $\epsilon = \epsilon_0 + \Delta\epsilon(z)$. Starting with Maxwell's equations and representations in the form of Eq. (7.12), together with making all the standard assumptions [1], one obtains the standard coupled mode equations

$$\frac{dA_1}{dz} = \sum_j \kappa_{1m} A_m \exp(-j\Delta\beta_{1m}z)$$

$$\frac{dA_2}{dz} = \sum_j \kappa_{2m} A_m \exp(-j\Delta\beta_{2m}z) \tag{7.15}$$

$$\frac{dA_l}{dz} = \sum \kappa_{lm} A_m \exp(-j\Delta\beta_{lm}z)$$

where the κ_{lm} are the coupling constants that are defined by the overlap of the lth and mth mode, in the following way:

$$\kappa_{lm} = \iint E_l^*(x,y)\Delta\epsilon(x,y)E_m(x,y)\,dx\,dy \qquad (7.16)$$

and $\Delta\beta$ is defined by

$$\Delta\beta_{lm} = \beta_l - \beta_m + p\left(\frac{2\pi}{\Lambda}\right) \qquad (7.17)$$

There are two applications that can now be discussed in terms of this development.

The first is the use of the fiber Bragg grating as a highly selective filter in WDM systems [26]. In this example, the fiber is carrying many wavelengths and the grating is fabricated at the appropriate spacing to reflect one. This is schematically shown in Fig. 7.31. The performance of an actual fiber grating made in the manner described in the foregoing is shown in Fig. 7.32, both in reflection and transmission. This represents a rather simple example in that one has to concern

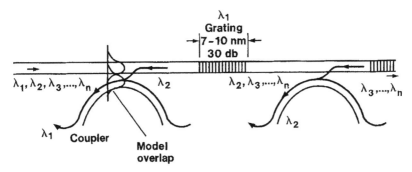

Figure 7.31 Schematic of the way fiber Bragg gratings would be used to remove selected wavelengths traveling in a WDM system. The trunk carries all of the wavelengths. After encountering a grating made to highly reflect one of the wavelengths, it is picked up by the directional coupler and carried out.

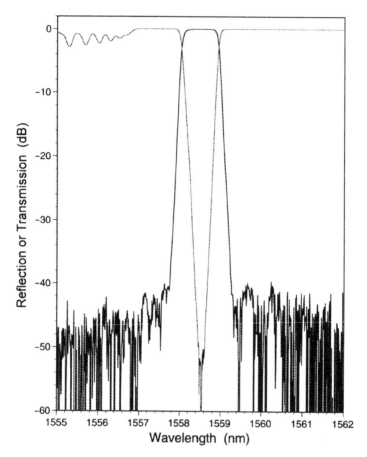

Figure 7.32 Actual transmission (red) and reflection (blue) curves vs. wavelength of a fiber Bragg grating. In this example, the GeO$_2$content was approximately 10 mol%, and the fiber was hydrogen loaded before exposure to 240-nm radiation.

oneself with only a single forward and backward mode, which reduces Eq. (7.15) to just two terms $A_1(\beta)$ and $A_2(-\beta)$. One can make $\Delta\beta = 0$ by choosing $\beta_1 = \pi/\Lambda = -\beta_m$, according to Eq. (7.17). The solution for the reflectivity—that is, $A_2(z)/A_1(0)$—is simply given by

$$R = \tan h^2(\kappa z) \tag{7.18}$$

If one assumes a sinusoidal form for the axial refractive index, that is, Eq. (7.14) is written as

$$n(z) - n_0 = \Delta n \cos(2\pi z / \Lambda) \tag{7.19}$$

then κ in Eq. (7.18) is simply given by the quantity $(\beta/2)\Delta n$. For this geometry, the Bragg condition (see Eq. (7.5)) gives $\Lambda = (\lambda_0 / 2n_0)$, thus one can see that $\kappa = \pi \Delta n / \lambda_0$. Theoretically, a peak-to-valley index change of 2×10^{-4} (corresponds to a $\Delta n = 1 \times 10^{-4}$) would produce a 96% reflection (14 dB) for a 1-cm-long grating. The actual specifications for this application are quite stringent not only in terms of the magnitude of the reflectivity, but also the flatness of the reflectivity vs. wavelength. To achieve this flatness, very deep gratings are needed (large Δn). A problem that is encountered in the grating formation stems from the fact that one is using Gaussian-shaped beams, so that Eq. (7.19) has a Gaussian envelope. To compensate for this, there are several tricks that could be termed *apodization*. One way is to pull the sample away from the mask. In this way, the diverging beams provide a uniform exposure on the ends to raise the average refractive index. There are also such things as apodized masks [8].

There is another way one can obtain an estimate of the induced refractive index change. This comes from the measured shift in the peak reflection wavelength as the grating develops. From the Bragg condition, one sees that the peak wavelength is just the product of the grating spacing and the refractive index of the medium in which the grating resides. Here, in which the refractive index change is induced by the laser exposure, the average index is also changing. One can see from Eq. (7.19) that the average index change is Δn, so that the peak reflection wavelength of the grating would shift to longer wavelengths by the amount $\lambda \Delta n$. For the foregoing example, one should see a shift of 0.15 nm for a grating made for 1500 nm. In general, the measured wavelength shift indicates a refractive index change in excess of what is determined from the reflectivity. This occurs because the actual contrast in the material produced by the interfering beams is coupled with the explicit dependence of the refractive index with the exposure intensity. It should be clear that, to the extent that the intensity pattern produced by the interfering beams does not produce high contrast, for whatever reason, then the average index will increase and the Δn will decrease. This will produce a larger wavelength shift, larger n_{avg}, yet a lower reflectivity, smaller Δn. Generally speaking, the hydrogen-impregnated samples yield a better agreement between the two index change numbers.

The next example for the use of the fiber Bragg gratings is for selective wavelength filtering [43,44]. For example, in a Er-doped fiber amplifier, the gain may not be flat over some wavelength interval. Thus, one designs a grating such that it couples light from the core mode to lossy cladding modes over the appropriate wavelength interval.

The loss should exhibit a wavelength dependence over the interval to compensate for the variation in gain. To accomplish this, one generally uses forward coupling; that is, one couples the core mode to lossy cladding modes traveling in the same direction. This means that the βs are all positive. From Eq. (7.17), one can see that the grating spacing will now be large. Because of this, these gratings are often called long-period gratings. Typically the grating period is 0.4 mm to effectively couple the core mode to lossy cladding modes. The modes one is trying to couple are those that have some overlap with the core mode, consistent with having a nonzero integral of Eq. (7.17) and, at the same time, extend out into the cladding sufficiently to access the surface so that they can escape. A schematic of this application is shown in Fig. 7.33. The coupling constants are smaller here because the modal overlap is not large. This requires a larger value of Δn to make κ a reasonable number, as indicated by Eq. (7.17).

Another possible application is the use of fiber gratings for dispersion compensation [45]. The width of a pulse propagating down a fiber broadens in time as a consequence of wavelength dispersion. This just means that the band of wavelengths that are contained within the time pulse do not travel at the same speed and, as a result, spread apart. To compensate for this, one could design a series of gratings that

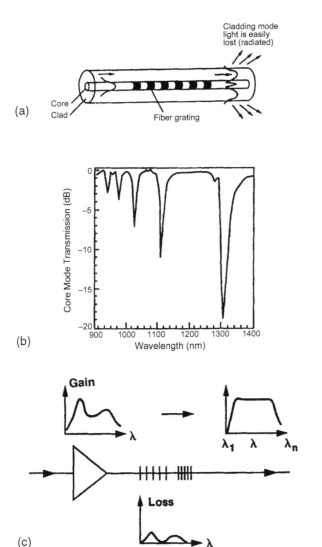

Figure 7.33 Representation of a long-period grating used for passive gain equalization: (a) the modal profile of the core mode and a typical lossy cladding mode with sufficient overlap; (b) the transmittance vs. wavelength, indicating the wavelength of the various modes that are being coupled out; (c) an idealized representation of the gain-flattening function. (From Ref. [44].)

couple wavelengths into the cladding for some distance where their speed is different to compensate for the speed difference in the core mode. This is most efficiently done with a long-period *chirped grating*. A chirped grating is one for which the pitch is a function of the distance. With appropriate design of the chirp, each wavelength can be coupled to a cladding mode accordingly to provide compensation.

Planar-based: A more conventional grating, although hardly in a conventional way, is one that uses demultiplex light in the manner shown in Fig. 7.34a. This is a planar device by which the output of a waveguide-carrying multiple wavelengths is directed at a grating. The function of the grating is to separate the wavelengths into separate paths,

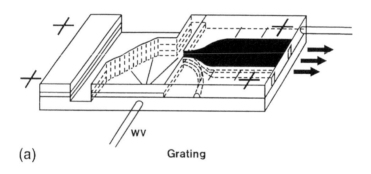

(a)   Grating

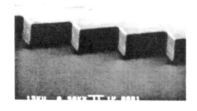

(b)

Figure 7.34 (a) Schematic of planar grating-based wavelength demultiplexer made by photolithography and reactive ion etching; (b) SEM photomicrograph of the grating pattern in the SiO_2.

ultimately coupled back out into separate waveguides. The grating is etched into a planar material from the top using photolithography and RIE. A scanning electron micrograph (SEM) of a portion of the etched grating made in a CVD-prepared silica film in shown in Fig. 7.34b [46]. The reported performance here was for an individual channel bandwidth of 0.45 nm, the crosstalk (isolation between channels) was 15 dB. Issues concerning wall verticality and edge feature sharpness contribute to the overall loss of the device, which is polarization-sensitive and more than 20 dB.

One can also use the same idea using planar photorefractive Bragg gratings, as shown in Fig. 7.35. Here the entire structure is in the planar waveguide framework. The multi-wavelengths guided in encounter the slanted chirped Bragg grating, sending each wavelength out at a different angle. The reported performance of this grating is for a 50 GHz spacing, the isolation is 9 dB [46].

There is a significant body of work devoted to the so-called "planar waveguide" technology. Often analogy is made to electronic circuits where essentially the entire optical function is confined to a plane. The grating demultiplexer discussed above is an example of such a device. We included this because it was another way to make a grating device not because it was a planar-like structure.

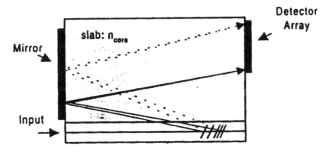

Figure 7.35 Schematic drawing of the demultiplexer based on a photorefractive Bragg grating made in a planar waveguide format. Multiple wavelengths enter the planar guide, then encounter a chirped Bragg grating. The diffraction angle is a function of the wavelength, as shown. (From Ref. [46].)

However, there is an architecture which utilizes a 3D grating, but also might be viewed as a planar structure. An example of this architecture is a device termed a "tunable filter". The function here is to have a filter with 50 GHz bandwidth tunable over maybe 30–50 nm of the communication band. The scheme is to write 50–100 50 GHz-spaced gratings onto a flat piece of photosensitive glass (or any other suitable solid medium). The gratings are fabricated from the top face using a 248-nm excimer laser through a phase mask. However, they are utilized in the direction normal to this as shown. In other words, the grating is addressed from the edge face. The schematic of this is shown in Fig. 7.36. The input/output fibers are shown and the input fiber makes a slight angle with the normal to the edge face so that the reflected beam is directed to the output fiber.

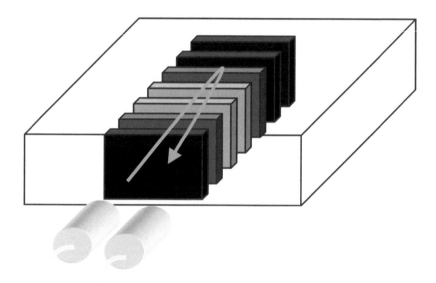

Figure 7.36 Schematic representation of a tunable filter device based on an array of gratings made in a photosensitive planar medium. Input and output fibers are slightly tilted to allow diffraction beam to picked up by adjacent fiber as shown.

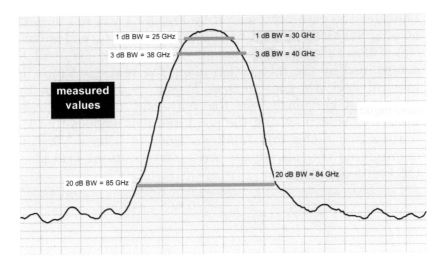

Figure 7.37 Measured values of a nominal 50 GHz grating made n the Ge-doped borosilicate glass. Grating was exposed from the top as shown in Fig. 7.25b and read in the manner shown in Fig. 7.36.

The tunability comes from moving the collimated output of a single mode fiber to the desired grating. A grating written in such a fashion as described is shown in Fig. 7.37. This device concept utilizing the bulk photorefractive material is able to produce a 25 GHz filter.

APPENDIX A. "DERIVATION OF KRAMERS–KRONIG RELATION"

This technique to calculate the induced refractive index is so commonly referred to in the literature and used in estimating the changes in refractive index from absorption data that it is worthwhile to understand its physical basis.

Its formal basis is the well-known Cauchy integral formula from the theory of complex variables

$$f(z) = \frac{1}{2\pi i} \int_C \frac{f(z')}{z - z'} \, dz' \tag{A7.1}$$

Consider the case represented by Fig. A7.1 where there are no singularities enclosed by the closed path, C, and further $f(z) \to 0$ as $|z| \to \infty$. This means the value of $f(z)$ will be determined by the integration along the real axis since the integral over arc goes to 0 as $R \to \infty$ as assumed. In the event of a singularity on the real axis, one then draws the arc as shown around the point and it is obvious that the contribution along this small arc goes to 0 as $\varepsilon \to 0$, thus maintaining the fact we can write $f(z)$ as an integral along the real axis

$$f(z) = \frac{1}{2\pi i} P \int\limits_{-\infty}^{\infty} \frac{f(z')}{z - z'} dz \qquad (A7.2)$$

The symbol P represents the condition of principal value which allows for the possibility of singularities on the real axis. What is amazing about this formulation is that by separating $f(z)$ into real and imaginary parts, one can now write an integral of the $\mathrm{Re}[f(z)]$ in terms of $\mathrm{Im}[f(z)]$ and vice

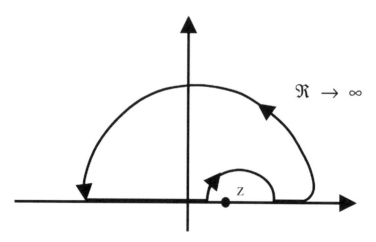

Figure A7.1 Contour of path defining the integral described as Eq. (A7.1).

versa

$$\mathrm{Re}[f(z)] = \frac{1}{\pi} P \int\limits_{-\infty}^{\infty} \frac{\mathrm{Im}[f(z')]}{z - z'} \mathrm{d}z',$$

$$\mathrm{Im}[f(z)] = \frac{-1}{\pi} P \int\limits_{-\infty}^{\infty} \frac{\mathrm{Re}[f(z')]}{z - z'} \mathrm{d}z'$$

(A7.3)

One now needs to show that the dielectric function $\varepsilon(\omega)$ of a dielectric, or the equivalent linear electric susceptibility, $\chi(\omega)$, satisfies the condition of $f(z)$ stated above. We need to establish that $\varepsilon(\omega)$ must have no singularities in the upper half of the complex plane, $\mathrm{Im}\ z > 0$, and that $\varepsilon(\omega) \to 0$ as $|\omega| \to \infty$. One defines the dielectric or susceptibility in terms of the response of the medium, the polarization, to the applied electric field, where $R(t-t')$ is the linear response function

$$P(t) = \int\limits_{-\infty}^{\infty} R(t')E(t - t')\mathrm{d}t'$$

(A7.4)

Because of causality, one must demand that $R(t') = 0$ for $t' < 0$. The dielectric function, or more appropriately the linear susceptibility, is the Fourier transform of this response function. In other words, it represents what the response of the dielectric medium is in the frequency domain. The dielectric constant and the susceptibility are simply related by the expression $\varepsilon = 1 + 4\pi\chi$, and it makes no difference which one we adopt

$$\chi(\omega) = \frac{(\varepsilon - 1)}{4\pi} = \frac{1}{\sqrt{2\pi}} \int\limits_{-\infty}^{\infty} R(t') \exp(i\omega t')\, \mathrm{d}t'$$

(A7.5)

One can show that if $R(t') = 0$ for $t' < 0$, then this implies that $\chi(\omega)$ has no singularities for $\mathrm{Im}\ \omega > 0$, and further that $\chi(\omega) \to 0$ as $|\omega|$ goes to infinity [48].

We have established that the dielectric function satisfies the condition allowing it to be represented by Eq. (A7.2) and its consequences in Eq. (A7.3). Equation (A7.3) requires a bit more algebraic manipulation to turn into Eq. (7.11) in the text. One assumes that $f(z)$ is either even or odd so as to be able to break the integral into a part form 0 to ∞, and $-\infty$ to 0. Further, we cast it into the optical representation of $f(z)$, namely, the complex dielectric constant and write it in the following form:

$$\mathrm{Re}[f(z)] = \varepsilon_1(\omega) = \frac{2}{\pi} P \int_0^\infty \frac{\omega' \varepsilon_2(\omega')}{\varpi^2 - \omega'^2} \, d\omega' \qquad (A7.6)$$

Further, one could use the relationships of the real and imaginary parts of the dielectric constant to the complex refractive index $(n + ik)$, $\varepsilon_1 = n^2 - k^2$, and $\varepsilon_2 = 2nk$, with the absorption coefficient relation to k through $\alpha = (4\pi/\lambda)k$. An alternative way of writing Eq. (A7.6) is the following:

$$n(\omega) - 1 = \frac{1}{2\pi} P \int_0^\infty \frac{\omega' k(\omega')}{\omega^2 - \omega'^2} \, d\omega' \qquad (A7.7)$$

Even a simpler approach is to deal with just the differences, in other words $\Delta n(I)$, and $\Delta \alpha(I)$, since that is the context it is used in the text. This is the form that is written in Eq. (7.11).

REFERENCES

1. Yariv, A.; Yeh, P. *Optical Waves in Crystals*; Wiley-Interscience: New York, 1984.

2. Hutcheson L.D., Ed. *Integrated Optical Circuits and Components*; Marcel Dekker: New York, 1987.

3. Magnusson, R.; Gaylord, T.K. Diffraction regimes of transmission gratings. J. Opt. Soc. Am. **1978**, *68* (6), 809–814.

4. Kogelnik, H. Coupled wave theory for thick hologram gratings. Bell Tech. J. **1969**, *48* (9), 02909–02947.

5. Magnusson, R.; Gaylord, T.K. Diffraction efficiencies of thin phase gratings with arbitrary grating shape. J. Opt. Soc. Am. **1978**, *68* (6), 806–809.

6. Smith, H.M. *Principles of Holography*; John Wiley & Sons: New York, 1975.

7. Collier, R.J.; Burckhardt, C.B.; Lin, L.H. *Optical Holography*; Academic Press: New York, 1971.

8. Lasiris Inc. Quebec Canada, H4R 2K3, product brochure.

9. Dahmani, B. Fountainebleau Research Center, Corning Inc. Private data.

10. Smith, H.M., Ed., *Topics in Applied Physics, Holographic Recording Materials*; Springer-Verlag: Berlin, 1977.

11. Brandes, R.G.; Francois, E.E.; Shankoff, T.A. Preparartion of dichromated gelatin films for holography. Appl. Opt. **1969**, *8* (11), 2346–2348.

12. Ashkin, A.; Boyd, G.D.; Dziedzic, J.M.; Smith, R.G.; Ballman, A.A.; Levinstein, J.J.; Nassau, K. Optically-induced refractive index inhomogenieties in $LiNbO_3$ and LiNTaO. Appl. Phys. Lett. **1966**, *9* (1), 72–74.

13. Schmidt, R.V.; Kaminow, I.P. Appl Phys. Lett. **1974**, *25*, 458.

14. Korotky, S.K.; Alferness, R.C. $Ti:LiNbO_3$ integrated optic technology. In *Integrated Optical Circuits and Comp.*; Hutcheson, L.D., Ed.; Marcel Dekker: New York, 1987; Chap. 6, and extensive references contained.

15. Pepper, D.M.; Feinberg, J.; Kukhtarev, N. The photorefractive effect. Sci. Am. **1990,** Oct, 62–74.

16. Gunter, P.; Huignard, J.-P. *Photorefractive Materials and Their Applications*; Springer: Berlin, 1988, 1989; Vols. 1 and 2.

17. Chen, F.S. Optically induced changes in refractive index in $LiNbO_3$. J. Appl. Phys. **1969**, *40*, 3389–3396.

18. Moerner, E.; Grunnet-Jepsen, A.; Thompson, C.L. Photorefractive polymers. Annu. Rev. Mater. Sc. **1997**, *27*, 585–623.

19. Meerholz, K.; Volodin, B.L.; Sadalphon; Kippelen, B.; Peyghambarian, N. A photorefractive polymer with high gain and diffraction efficiency. Nature **1994**, *371*, 497–500.

20. Astakhova, V.V.; Nikonorov, N.V.; Panysheva, E.L.; Savvin, V.V.; Tunimanova, I.V.; Tsekhomskii, V.A. A study on the

kinetics of photoinduced crystallization in polychromatic glass. Sov. J. Glass Phys. Chem. **1992**, *18*, 152–156.

20a. Efimov, O.M. et al. High-Efficiency Bragg gratings in photo-thermorefractive glass. Appl. Opt. **1999**, *38* (4), 619–627.

20b. Efimov, O.M. et al. Measurement of the induced refractive index in a photothermorefractive glass by a liquid-cell shearing interferometer. Appl. Opt. **2001**, *41* (10), 1864–71.

21. Borrelli, N.F.; Morse, D.L. Photosensitive impregnated porous glass. Appl. Phys. Lett. **1983**, *43*, 992–993.

22. Borrelli, N.F.; Cotter, M.D.; Luong, J.C. Photochemical method to produce waveguiding in glass. IEEE J. Quantum Electronics **1986**, *QE-26*, 896–901.

23. Kuchinskii, S.A.; Sukhanov, V.I.; Khazova, M.V. The principles of hologram formation in capillary composites. Laser Phys. **1993**, *3*, 1114–1123.

24. Sukhanov, V.L. Porous glass as a storage medium. Opt. Appl. **1994**, *24* (1–2), 13–26.

25. Hill, K.O.; Fuji, Y.; Johnson, D.C.; Kawasaki, B.S. Photosensitivity in optical fiber waveguides. Appl. Phys. Lett. **1978**, *32*, 647–649.

26. Meltz, G.; Morey, W.W.; Glenn, W.H. Formation of Bragg gratings in optical fibers by a transverse holographic method. Opt. Lett., **1989**, *14*, 823–825.

27. Neustruev, V.B. Colour centers in germanosilicate glass and glass fibers. Condens. Matter **1994**, *6*, 6901–6936.

28. Tsai, T.E.; Askins, C.G.; Friebele, E.J. Pulse energy dependence of defect generation in Bragg grating optical fiber materials. Mat. Res. Soc. Symp. **1992**, *224*, 47–52.

29. Lemaire, P.J.; Atkins, R.M.; Mizrahi, V.; Reed, W.A. High pressure H_2 loading as a technique for achieving ultrahigh UV photosensitivity and thermal sensitivity in GeO_2 doped optical fibers. Electr. Lett. **1993**, *29*, 1191–1193.

30. Atkins, R.M.; Lemaire, P.J.; Erdogan, T.; Mizrahi, V. Mechanisms of enhanced UV photosensitivity via hydrogen loading in germanosilicate glasses. Electr. Lett. **1993**, *29*, 1234–1235.

31. Albert, J.; Maio, B.; Bilodeau, F.; Johnson, D.C.; Hill, K.O.; Hibino, Y.; Kawachi, M. Photosensitivity in Ge-doped silica optical waveguides and fibers with 193 nm light from an ArF laser. Opt. Lett. **1994**, *19*, 387–389.

32. Hand, D.P.; Russell, P.St.J. Photoinduced refractive-index changes in germanosilicate fibers. Opt. Lett. **1990**, *15*, 102–104.

33. Dong, L.; Archambault, J.L.; Reekie, L.; Russell, P.; Payne, D.N. Single pulse Bragg gratings written during fiber drawing. Electr. Lett. **1993**, *29*, 1577–1578.

34. Erdogan, T.; Mizrahi, V.; Memaire, P.J.; Monroe, D. Decay of UV-induced fiber Bragg gratings. J. Appl. Phys. **1994**, *76*, 73–80.

34a. Borrelli, N.F. et al. photosensitive Ge-doped borosilicate glasses. Proc. SPIE **2000**, *4102*, 242–254.

35. De Neufville, J.P.; Moss, S.C.; Ovshinsky, S.R. Photostructural transformations in amorphous As_2S_3 and As_2Se_3. J. Non Cryst. Sol. **1973**, *13*, 191–223.

36. Owen, A.E.; Firth, A.P.; Ewen, P.J.S. Photo-induced structural and physicochemical changes in amorphous chalcogenide semiconductors. Phil. Mag. B **1985**, *52*, 347–362.

37. Bertolotti, M.; Michelotti, F.; Chumash, V.; Cherbari, P.; Popescu, M.; Zamfira, S. The kinetics of the laser induced structural changes in As_2S_3 amorphous films. J. Non-Cryst. Sol. **1995**, *192–193*, 657–660.

38. Keneman, S.A. Holographic storage in arsenic trisulfide thin films. Appl. Phys. Lett. **1971**, *19*, 205–207.

39. Kostuk, R.K.; Ramsey, D.L.; Kim, T.-J. Connection cube modules for optical backplanes and fiber networks. Appl. Opt. **1997**, *36*, 4722–4728.

40. Sinzinger, S.; Jahns, J. Integrated micro-optical imaging system with a high inter-connection capacity fabricated in planar optics. Appl. Opt. **1997**, *36*, 4729–4735.

41. Song, S.H.; Joung, J.-S.; Lee, E.-H. Beam array combination with planar integrated optics for three dimensional multistage interconnection networks. Appl. Opt. **1997**, *30*, 5728–5731.

42. Kurata, Y.; Ishikawa, T. CD pickup using a holographic optical element. Int. J. Jpn. Soc. Prec. Eng. **1991**, *25* (2), 89–92.

43. Vengsarkar, A.M.; Pedrazzani, J.R.; Judkins, J.B.; Lemaire, P.J.; Bergano, N.S.; Davidson, C.R. Long-period fiber-grating-based gain equalizers. Opt. Lett. **1996**, *21*, 336–338.

44. Wysocki, P.F.; Judkins, J.B.; Espindola, R.P.; Andrejco, M.; Vengsarkar, A.M. Broad-band erbium-doped fiber amplifier flattened beyond 40 nm using long-period grating filter. IEEE Phot. Tech. Lett. **1997**, *9*, 1343–1345.

45. Laming, R.I.; Ibsen, M.; Cole, M.J.; Zervas, M.N.; Ennser, K.E.; Gusmerol, V. Dispersion compensation gratings. In *Bragg, Gratings, Photosensitivity, and Poling in Glass Fibers and Waveguides: Applications, and Fundamentals*, 197 OSA Technical Digest Series 17, 1997; 271–273.

46. Liu, K.; Tong, F.; Boyd, S.W. Planar grating wavelength demultiplexer. *Multigigabit Fiber Communication Systems*; SPIE, **1993**; Vol. 2024, 278–283.

47. Madsen, C.K.; Wageener, J.; Strasser, T.A.; Milbrod, M.A. Planar waveguide grating spectrum analyzer. In *Opt. Soc. Am. Tech. Digest*, Integrated Photonics Research Conference, Victoria, Canada, March 29, 1998; *4*, 99–10.

48. Mathews, J.; Walker, R.L. *Mathematical Methods of Physics;* Benjamin: NY, 1964, Chap. 4, 235.

8

Optical Isolators

8.1. BACKGROUND

Small-sized optical isolators are a relatively recent invention brought about by the need for protection of optical devices from stray reflected signals [1,2]. The two most common examples are laser diodes and optical amplifiers. The reflected light destabilizes the output of laser diodes, whereas with optical amplifiers, the reflected light could prematurely "dump" the inverted population. The optical isolator package must always be small to fit into optoelectronic device structures; therefore, they require a rather ingenious arrangement of micro-optical components.

The elements that make up the optical isolator depend on the specific kind of isolation required. This distinction is usually made by requirements on the state of polarization of the input beam and the source of the reflections. There are essentially three types of isolator constructions that we show in Fig. 8.1.

The simplest is shown in Fig. 8.1a; it consists of a polarizer aligned with the polarization direction of the optical

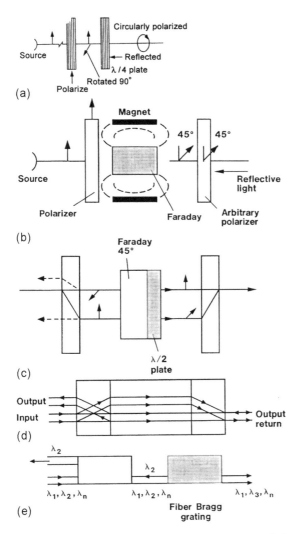

Figure 8.1   Schematic representations of three types of optical isolators. (a) Simple circular polarizer type; (b) more general Faraday effect type where all reflections, irrespective of the source are rejected; (c) polarization independent type, meaning that it rejects backward traveling light irrespective of input polarization (polarization direction through the device are indicated by the arrows); (d) circular device showing how the return beam is captured at the upper port; (e) application of the circulator where it is used to capture the reflected beam, λ_2 in the drawing, from a Bragg grating.

source, followed by a quarter-wave plate with its optic axis aligned at an angle 45° to the polarizer direction. For this arrangement to work, the backward-directed light would have to come from a direct reflection. The initial light is made circularly polarized by the action of the $\lambda/4$ (90°-phase shift). After the reflection, the phase is advanced (or retarded) by 180°, and then it passes through the waveplate again, and a −90° shift is produced. The net effect then is a 180°-phase shift that corresponds to a 90° rotation. This is crossed to the polarizer thereby blocking the reflected light. The limitations of this type of isolator are obvious: first, the source must have a fixed polarization direction; second, and much more restrictive, is that the only blocked light is from direct reflections that do not suffer any additional phase shift.

The structure shown in Fig. 8.1b is a much more general isolator in that all light returning, regardless of its origin, will be blocked. This is because it uses the Faraday effect that acts on the phase of the light in a nonreciprocal fashion. The *Faraday effect* is the rotation of the axis of polarization of light with an applied longitudinal magnetic field.

$$\Theta = VHL \tag{8.1}$$

where Θ is the rotation, H is the applied magnetic field in the direction of the light, L is the sample path length in that direction, and V is the constant of proportionality, called the "Verdet" constant. The nonreciprocity means that the rotation will continue in the same sense independent of the direction of the light. This is in contrast to say a half-waveplate that will also rotate the polarization, but in the opposite sense, depending on whether the direction of travel is forward or backward. The origin of this nonreciprocity stems from the fact that the magnetic field direction does not change, and the sense of rotation derives from its direction, not the light direction. In referring to Fig. 8.1b, one, sees that the light from the polarized source is rotated 45° by the Faraday element. It then passes another polarizer that is aligned in this direction. Any light returning will be polarized at 45° by this polarizer that rotates the polarization by *an additional 45°* in the same direction. This makes the polarization 90° relative to the initial polarizer; hence, it is blocked. The one major limitation of this type of isolator is that it requires a linear-polarized source. Its major use is for laser diodes in which this criterion is met.

The third type of isolator is the so-called polarization-independent type. This structure removes the limitation of the previous two isolators in that it blocks return light irrespective of the initial state of polarization of the source. This is an important condition because in single-mode optical fibers, the state of polarization is not constant owing to mode coupling between the TM and TE modes. Not only is the state of polarization unknown, but it is changing with time. The structure is shown in Fig. 8.1c. The polarization independence derives from the splitting of input beam into the two orthogonal polarizations. This is accomplished by using the different ray paths of the ordinary and extraordinary rays of a birefringent material. The Faraday element is used to provide the nonreciprocity. We have shown in Fig. 8.1c a representation of the polarization condition after each element. This has been done for both the forward and backward direction. It should be clear that the rejection arises by having the polarization of the backward-traveling beams arrive at the input birefringent prism, reversed by 90° from what they were at the beginning.

In the Appendix we will give the Jones matrix form [3] for each of the elements. What this will allow one to do is to quickly determine the state of polarization as a consequence of passing through any sequence of these elements in any order.

A related device is called a circulator [4]. It has the same exact structure as the polarization-insensitive isolator; however, its function is to separate and capture the backward beam as shown in Fig. 8.1d. A common application is to place it before a reflecting fiber Bragg grating, thus routing the reflected beam out. This is shown in Fig. 8.1e.

In the next subsections we will discuss the elements that go into the isolator. For example, we will cover polarizers, waveplates, polarization separators, and Faraday elements.

Following that we will discuss actual optical isolator structures.

8.2. POLARIZERS

The familiar polarizing material called Polaroid is not adequate for highly reliable micro-optic devices because it is made in a plastic base. Moreover, that limits its use for the telecommunication wavelengths window. There are various inorganic materials that can be used as polarizers in this region. It would be worthwhile to make the following distinctions for the type of polarizers. There are essentially three types of polarizers, (1) dichroic, (2) wire-grid, and (3) beam separators. We schematically represent the three effects in Fig. 8.2.

In the dichroic type, one is dealing with a material that absorbs the light strongly in one polarization orientation,

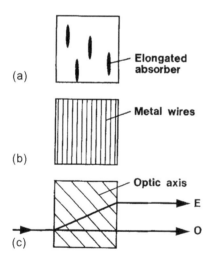

Figure 8.2 Three types of polarizers: dichroic, which works on the principle of an anisotropic absorption; wire-grid that is based on the behavior of the reflection of light from a pattern whose spacing is comparable to the wavelength of light; and birefringent crystals where the polarization components are separated by double refraction.

and not in the other. The polaroid is such a material in which the anisotropic absorber is an aligned molecule within the plastic strip. The stretching of the plastic leads to the alignment. We shall see that there is a glass analogue to this approach called Polarcor.[TM]

The wire-grid type relies on reflection to differentiate the respective polarizations. A good example of this is the polarizer used to polarize light in the near infrared (IR) spectrum. It is based on the property that light polarized parallel to the wires is strongly reflected, whereas that perpendicular is not.

A common example of the beam separator type is the birefringent crystal polarizers in which the linear polarized components refract differently as a consequence of the anisotropic refractive index. As a result of this, for light entering at an angle to the optic axis, the two polarizations will see different refractive indices and travel different paths. This is discussed in a later section.

In the following sections we will go through examples of each of these types, together with some discussion of how they are made and their level of performance.

8.2.1. Dichroic Polarizers

The dominant IR polarizer is the "stretched" metal–halide glass type. They are made from special alkali aluminoborosilicate glasses that contain Ag and or Cu, together with halides, such as Cl and Br [5]. The glasses are subsequently heat-treated to develop a phase-separated droplet metal–halide phase within the glass. The glass is then heated to near its softening temperature and then stretched, which provides elongation of the metal–halide droplets in the process. The schematic of the process is shown in Fig. 8.3a, together with an electron photomicrograph of the elongated metal–halide phase (Fig. 8.3b). At this stage, the glass is birefringent, but not dichroic. The glass is now in the form of a strip 2–4 cm wide and 3 mm thick. To produce the dichroic property, the stretched glass is treated in an atmosphere of pure hydrogen at 400°C for sufficient time to chemically reduce the metal–halide phase to the metal. The optical dichroism now comes

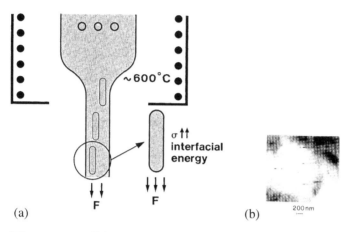

(a)   (b)

Figure 8.3 Schematic drawing of the method used to make polarizing glass. (a) Glass containing a droplet phase is stretched above the glass softening temperature that produces a force that elongates the droplet. The forces involved are the shear force on the particle as a consequence of the applied stress, and the restoring force related to the interfacial energy. (b) An electron micrograph of the resulting stretched glass.

from the anisotropic interaction of the metal needle with the incident radiation. The absorption in the direction parallel and perpendicular to the long direction of the particle can be expressed as Eq. (8.2) [6]:

$$\gamma = \frac{NVk\varepsilon_2}{[(L_1(\varepsilon_1 - 1) + 1]^2 + [L_1\varepsilon_2]^2} \tag{8.2}$$

where N is the number of particles per unit volume, V is the particle volume, k is the wavenumber $(2\pi/\lambda)$, and ε_1, ε_2, are the real and imaginary parts of the dielectric constant of the metal normalized to the optical dielectric constant of the glass. The L_1 refers to the depolarization factors for the respective symmetry directions derived from potential theory. The depolarization terms define the local field in the principal directions of the particle

$$E_{\text{loc}} = E_{\text{ext}} + 4\pi LP \tag{8.3a}$$

where E_{ext} is the external field. The anisotropy develops as a consequence of the way the external field redistributes (E_{loc}) around the particle. One notes that if the first term in the denominator goes to zero, then a resonant-like behavior is observed. This behavior is assured if the following is true:

$$\varepsilon_1 = \frac{L-1}{L} \tag{8.3b}$$

When for a given aspect ratio—measure of the major to minor axes—of the particle, the Ls are determined, the term will become zero if at some wavelength ε_1 achieves this value. For free-electron-like metals, such as Ag, Au, and Cu, the real part of the dielectric constants is negative, thus the condition in Eq. (8.3b) is always met for some wavelength. For relatively simple shapes, such as prolate spheriods, the depolarization terms are easily obtained. The transmittances in the stretched direction and perpendicular to that direction for a Polarcor material are shown in Fig. 8.4. Contrasts of more than 50 dB are common. Another desirable feature is the breadth of the high-contrast region. This is likely due to a distribution of particle aspect ratios.

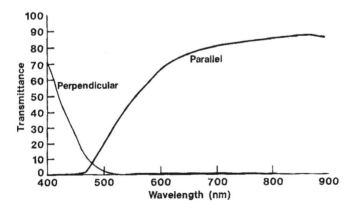

Figure 8.4 Typical transmittance spectra of the polarizing glass in the direction parallel to the stretched direction, and perpendicular to that direction.

8.2.2. Wire-Grid Polarizers

The wire-grid polarizer is based on the fact that, if the spacing of a parallel array of long conducting wires is small enough, then it will reflect radiation for which polarization is in the direction parallel to the wires, and transmit radiation with the polarization perpendicular to the wire array direction. The transmittance and the contrast a as function of the ratio of the wavelength to the wire spacing, where it is assumed that wire width is equal to half the wire spacing, is shown in Fig. 8.5 for three different substrate indices, n_0 [7]. The expression for this case in the two-dimensional limit are the following:

$$T_j = \frac{4n_0 A_j^2}{1 + (1 + n_0)^2 A_j^2}, \quad j = 1 \text{ is perpendicular}$$

$$j = 2 \text{ is parallel} \tag{8.4}$$

where

$$A_1 = \frac{1}{4B}, \quad B = (d/\lambda) \left[\frac{0.3466 + 0.25Q}{(1 + 0.25Q) + 0.0039(d/\lambda)^2} \right]$$

$$Q = \frac{1}{[1 - (d/\lambda)^2]^{1/2} - 1} \tag{8.5}$$

One can see from Fig. 8.5 that to obtain usable contrast one needs to have the ratio λ/d be greater than 2 and preferably greater than 8, to obtain a contrast of 20 dB. For the near-infrared wavelengths this means a separation of 200 nm. which is about at the present photolithographic limit. The contrast is relatively flat with wavelength because the contrast curve slopes are not steep.

One method of making such devices is the straightforward use of photolithography with which one would pattern the metal lines at the desired separation. This is pushing the limit of resolution of the optical method for the shorter wavelengths. One additional possibility is to pattern a larger period and reduce the separation by a high-temperature redraw process, similar to that described earlier for the polarcor material.

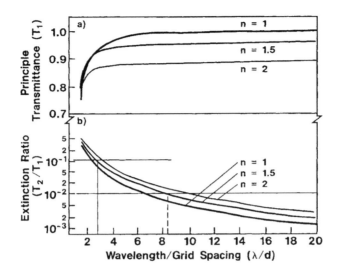

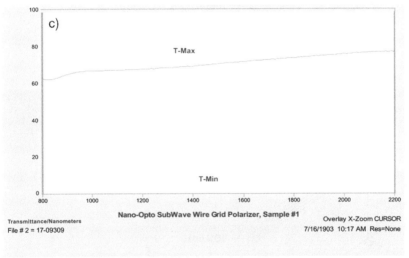

Figure 8.5 (a,b) Graphs of the computed transmittance and the contrast of a wire-grid polarizer as a function of the grid pitch normalized the wavelength. (From Ref. [7].) (c) Transmission spectrum of commercial wire-grid polarizer made by "nano-opto", Somerset NJ, 08873.

Another technique is to prepare the structure by depositing alternate layers of a metal and dielectric film of the thickness corresponding to the separation for the desired wavelength. The number of layers are built up to a thickness of the about 100 μm. The wire-grid structure is formed from a cross-sectional slice, as shown in Fig. 8.6. This is a process used to produce polarizers under the trade name Lamipol™. The reported extinction ratio is higher than 45 dB [8].

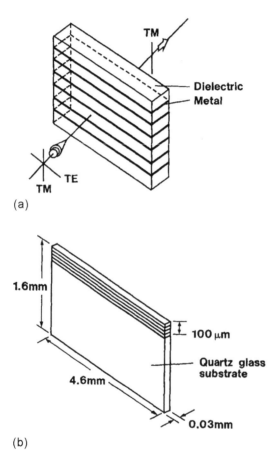

(a)

(b)

Figure 8.6 Representation of Lamipol™ wire-grid type polarizer. (a) The schematic of the alternate metal dielectric structure and (b) the actual dimensions of the finished element.

8.2.3. Gratings

Gratings also have a polarization dependence, that is, the diffraction efficiency can be different for light polarized along the grating relative to across the grating [7]. However, most of the large effects are associated with reflection gratings and are angle- and wavelength-sensitive. For the micro-optic application, a normal incidence transmission grating is preferred, if not required.

An example of a normal incidence transmission grating that produces a high polarizing is one that is based on single-crystal $LiNbO_3$. It is based on the interesting property of the induced index change that occurs on ion exchange of H^+ into $LiNbO_3$. The H^+ ion exchange is used to increase the refractive index and is a standard way in which waveguiding structures are made in $LiNbO_3$ [9]. The incorporation of the Ti^{+3} into the lattice produces an anisotropic refractive index change, changing more in the crystallographic c-direction than in the basal plane. If a pattern of equally spaced lines were produced by the ion exchange by the standard photolithographic method, then as a consequence, the grating will have a different efficiency for light polarized along the grating than for the perpendicular. This follows because the phase of the grating $\varphi = 2\pi \Delta n L / \lambda$ will be different for the two directions; $\Delta n(\text{par}) = n_c - n_0$, and $\Delta n(\text{per}) = n_a - n_0$. It actually turns out that $\Delta n(\text{per})$ is negative. The polarizer action occurs by applying a patterned film of a given index of sufficient thickness to exactly ballance the $\Delta \varphi(\text{per})$, then there will be no phase shift for light polarized perpendicular to the grating direction, and have $\Delta \varphi(\text{par}) + \Delta \varphi(\text{per})$ for the direction parallel to the grating (Fig. 8.7). The efficiency can be quite high. The limitation is in the spacing of the ion-exchange lines produced that will determine the diffracting angle.

8.2.4. Reflection from Thin Metal Layers

It is clear from the solution of Maxwell's equation at the boundary between a metal and a dielectric that the reflectivity will be different for light for which polarization is perpendicular to the plane of incidence, s-polarized, from that for

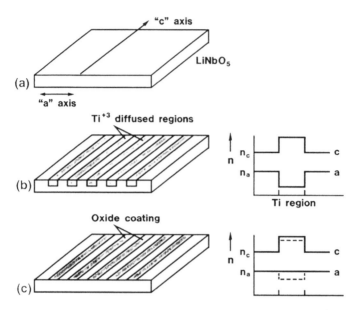

Figure 8.7 Schematic representation of the LiNbO₃ grating polarizer. Initially, H⁺ is diffused in the alternating regions. This produces the index profiles shown at the right for the *a* and *c* direction. The last step is to give the diffused region as oxide coating to eliminate the index step in the a direction as shown on the right.

which the polarization lies within the plane, p-polarized [10]. Consequently, it is possible to produce a polarization separation as a result of reflection from a metal—dielectric interface. One of the more dramatic examples of this effect is the coupling to the surface plasmon of the metal [11]. Here, the reflectivity for the s- and p-polarized component can be quite large for a specific wavelength, as shown in Fig. 8.8a. The arrangement is that of light traveling in a medium of refractive index n making contact with an A1 film an angle θ. The calculated reflectivity as a function of the incident angle for 633 nm is shown in Fig. 8.8b,c for two different metal thicknesses. For this arrangement, the metal must be very thin to allow the electromagnetic wave to set up the surface plasmon mode. The word *plasmon* is used to represent a quantized inhomogeneous electron density wave.

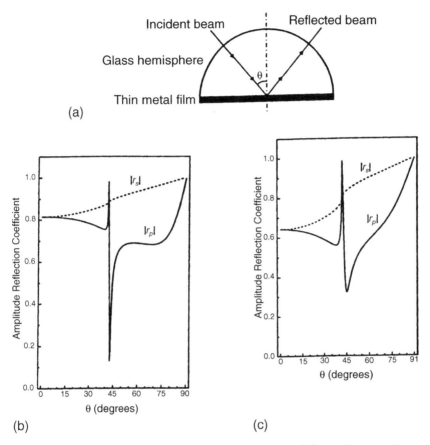

Figure 8.8 Diagrams defining the behavior of the reflection from a thin metal film. In the two graphs the amplitude reflection is shown as a function of incident angle for the *s* and *p* states of the polarizations; this is shown for two different metal layer thicknesses. (From Ref. [11].)

8.2.5. Optical Waveguide Polarizers

From the solution of Maxwell's equations for single-mode guided waves, one naturally obtains two solutions, usually referred to as the TE and the TM modes. Considering $\boldsymbol{E} = (E_x, E_y, E_z)$ and $\boldsymbol{H} = (H_x, H_y, H_z)$, then z-propagation the TE mode is defined by the nonzero components (H_x, E_y, H_z), whereas the TM mode is defined by (E_x, H_y, E_z). It is clear from this that the polarizations of these two modes are orthogonal. The propagation constant is different for these two modes; however, the difference is usually not large. One can take advantage of this difference to create a polarizer. A straightforward way to accomplish this is to use a Mach–Zehnder configuration. This is shown in Fig. 8.9. Directional couplers are guides brought sufficiently close to together to overlap their modal fields for a length long enough to allow that coupling to proceed to whatever extent is desired. There are a few ways this can be accomplished, one of which is to draw down a rod that has two cores to waveguiding dimensions. It is conceptionally easier to imagine this is a planar fabrication, although the operation is the same.

One attaches unequal arms to each leg as shown. By unequal is meant that the propagation constants, or lengths, or both are not the same. One then completes the structure with another directional coupler. The idea is the following: for an arbitrary polarization input into one arm, the respective βLs are chosen such that the TM is made to come out

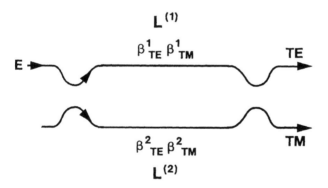

Figure 8.9 Representation of a single mode fiber Mach–Zehnder structure that would separate the input polarization into TE and TM at the output. β^1 represents the propagation constant in each arm for the respective polarization. L represents the length of the respective arm. Choosing the correct value for β and L make the phase shift for the TE mode in the upper arm an even multiple of π, and in the lower an odd multiple. The opposite should be true for the TM mode.

one arm and TE out the other. If there is no phase shift between the arms, light entering, say the upper arm, will always exit the lower arm. If one can introduce a π-phase shift, then light in the upper arm will exit the upper arm. If in the upper arm one wants the phase shift for the TE to be an even multiple of π, it will exit the lower arm. Likewise one designs the βL_{TE}, so that this will produce a phase shift of an odd multiple of π; consequently, the TM mode will exit the upper arm. One designs the lower arm to do just the opposite. It should be clear that the guides in the two arms should be reasonably anisotropic to allow for the different phase shifts for the TE and TM components.

8.2.6. Birefringent Crystals

The method of using the difference in the propagation of orthogonally polarized components of a ray in birefringent, single crystals to produce a polarizer is well known. One can understand this behavior referring to the *indicatrix*, which is a way to represent the refractive index for crystals of arbitrary crystallographic symmetry [12]. One writes it in the form,

$$\left(\frac{a}{n_1}\right)^2 + \left(\frac{b}{n_2}\right)^2 + \left(\frac{c}{n_3}\right)^2 = 1 \tag{8.6}$$

where a, b, and c are the principal axes of the crystal, and n_j is the values of the refractive index along this direction. This is the general case of a biaxial crystal. With no loss of generality, one can discuss a uniaxial crystal for which a n is equivalent to b, and $n_1 = n_2 = n_0$. Now we have a two-dimensional (2-D) representation that we can show graphically in Fig. 8.10a. A ray traveling in any direction other than along the two axes will have a different refraction, depending on the polarization of the ray. Consider the ray shown in Fig. 8.6b. One polarization component will see the refractive index n_1, no matter what the ray direction, called the ordinary ray, whereas the other component will see an index somewhere between n_0 and n_3, depending on the exact angle. It is this situation that is represented by the ellipse in Fig. 8.10a and

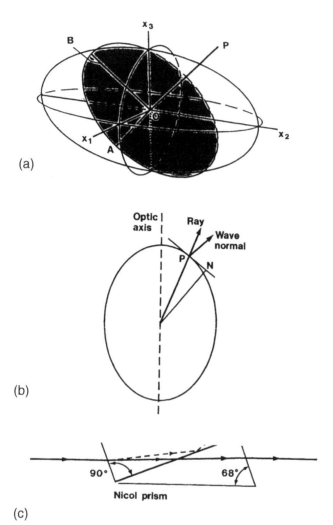

(a)

(b)

(c)

Figure 8.10 The upper figure is a representation of the index ellipsoid with principal axes x_1 where the dark region represents the plane normal to the direction of incidence as indicated by the line OP. The middle figure shows that the wave normal does not follow the same direction as the ray normal for the extraordinary which accounts for the double refraction phenomenon. The lower figure shows how one can use this effect to create a polarizer called a Nicol prism. The crystal is cut to make the extraordinary ray totally internally reflected when the ordinary ray is transmitted.

represents the condition for the extraordinary ray. The direction this ray will take is determined by the wave normal, as shown in Fig. 8.10b. The wave normal is perpendicular to the tangent to the curve at the point P. One sees that the wave normal makes an angle with the ray direction that represents the difference in refraction relative to the ordinary ray, the wave normal of which would always be in the ray direction.

There are several variations of the method, but they are all based on separation of the two rays and then, ultimately, rejecting the extraordinary ray. A simple example of this is the so-called Nicol prism. Here the cut of the crystal combined with its orientation relative to the input beam, totally internally reflects the extraordinary ray (see Fig. 8.10c).

These are obviously all derived from single crystals, but there is an artificial way to create birefringence that we discuss later in the beam separator section. The reason for this is that it is used in devices as a beam separator, rather than as a polarizer, although it would be possible to fashion it into a polarizer.

8.3. WAVEPLATES

Waveplates are optical elements that provide specific phase shifts between the orthogonal polarization components of a light beam. In general, one can write the expression for a light beam of arbitrary state of polarization relative to an x–y reference frame,

$$\boldsymbol{E} = E_x \boldsymbol{x} + E_y e^{-j\delta} \boldsymbol{y} \tag{8.7}$$

where $\boldsymbol{x}$ and $\boldsymbol{y}$ are the unit vectors in the given direction, E_x, E_y, the real amplitudes, and δ is a phase shift. If $\delta = 0$ one has linearly polarized light making an angle $\tan^{-1}(E_y/E_x)$ with the x-axis. If $E_x = E_y$ and $\delta = \pi/2$, one would have circularly polarized light. There is a convenient way to present the function of a waveplate using the Jones matrix formalism that we show in the appendix. One can refer to Fig. 8.11 to obtain the result of the action of a waveplate of phase shift equal to $\delta = 2\pi\Delta n L/\lambda$, the optic axis of which makes an angle ϕ relative

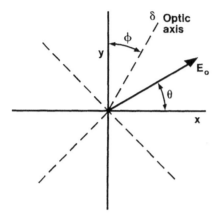

Figure 8.11 Reference system showing a δ phase shift waveplate whose optic axis makes an angle ϕ with the reference axis. The incident beam makes an angle $(90 - \theta)$ with the reference axis.

to the reference axis. An arbitrary polarization can be decomposed into E_x and E_y, which represents the input to the matrix equation

$$
\begin{pmatrix} E_x^{\text{out}} \\ E_y^{\text{out}} \end{pmatrix} = \begin{pmatrix} \cos\phi & -\sin\phi \\ \sin\phi & \cos\phi \end{pmatrix} \begin{pmatrix} e^{j\delta} & 0 \\ 0 & 1 \end{pmatrix}
$$
$$
\times \begin{pmatrix} \cos\phi & \sin\phi \\ -\sin\phi & \cos\phi \end{pmatrix} \begin{pmatrix} E_x^{\text{in}} \\ E_y^{\text{in}} \end{pmatrix}
$$

$$(8.8)$$

One obtains the state of polarization of the beam after having passed through the δ phase-shift waveplate. Some simple examples that can be easily verified from Eq. (8.8) are the following. If $\delta = \pi$ (called a half waveplate, or $\lambda/2$), then for a linearly polarized input, say $E_x = 1$, $E_y = 0$, the result will be linearly polarized beam rotated 2φ from the x-direction. Clearly, if φ were equal to $45°$, the effect of the half waveplate would rotate the polarization direction of a linearly polarized beam by $90°$. The use of half waveplates to rotate the polarization is the common. Another common use is to create circularly polarized light from a linearly polarized state. This is done by letting $\delta = \pi/4$, and $\varphi = 45$.

As we have mentioned in most applications, the role of the waveplate is to produce one state of polarization from another. More commonly, to convert linearly polarized light into another state of polarization, or vice versa. From the discussion in the previous section it is easy to see that birefringent materials provide the simplest way to produce the desired phase shifts. Referring to Fig. 8.10a, it is seen that by putting OP in either the x_1 or x_2 direction will give rise to the condition that the vibrating component in the x_3 direction will travel at a different uniform speed relative to that in either the x_1 or x_2 directions. In this orientation there is no double refraction.

8.3.1. Single Crystals

The common waveplate materials are made from slices of uniaxial single crystals cut in such a way that they contain the c-axis. Crystals such as quartz, calcite, rutile, and mica are often used because of their relatively large birefringence. The phase shift is calculated from the expression.

$$\delta = \frac{2\pi(n_e - n_0)L}{\lambda} \tag{8.9}$$

where $(n_e - n_0)$ is the difference in refractive index along the respective optical axes, and L is the thickness of the plate. The refractive indices for the aforementioned crystals are listed in Table 8.1.

Clearly, these waveplates are fabricated to a give thickness to produce the desired phase shift at a given wavelength.

Table 8.1 Birefringence of Common Crystals at 589 nm

Crystal	n_0	n_e
Calcite	1.658	1.486
Quartz	1.544	1.553
Rutile	2.616	2.903
Mica	1.56	1.59
$CaWO_3$	1.92	1.936

Because the birefringence is quite large (0.01–0.1) the phase shift is multiple order; that is, $\delta + m2\pi$, m being an integer. There are certain advantages to what are called zero-order waveplates $(m = 0)$ because of their reduced sensitivity to wavelength, thickness, and angle of incidence [13]. To make a zero-order plate of materials with large birefringence would make the thickness inpractically thin. One method to make a zero-order waveplate with a reasonable thickness is to glue two pieces together with the optic axis rotated 90°. The difference in the thickness then constitutes the phase shift, which can be zero-order.

8.3.2. Stretched Glass

The stretched metal–halide glasses discussed earlier are transparent before the halide phase is chemically reduced to the metal. Consequently, they exhibit birefringence derived from the anisotropic local fields. Similar to the foregoing expression for the absorption cross-section for a collection of aligned particles, one can write the expression for the birefringence [14]

$$\Delta n = \left(\frac{V_f}{4n}\right)(\varepsilon - 1)\{[L_1(\varepsilon - 1) + 1]^{-1} - [L_2(\varepsilon - 1) + 1]^{-1}\}$$

(8.10)

where V_f is the total volume fraction of the particles, ε is the ratio of the optical dielectric constant of the particle relative to the medium $(n_p/n_o)^2$, and L_1 and L_2 are the depolarization factors along the symmetry axes of the particle that are dependent only on the aspect ratio of the particle. The phase shift as a function of thickness for various different wavelengths is shown in Fig. 8.12 [14].

8.4. BEAM SEPARATORS

In the foregoing section we discussed the use of double refraction for achieving polarization separation. It also can be used to provide a convenient way to separate the incident beam

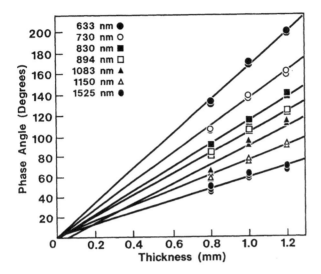

Figure 8.12   Phase shift vs. sample thickness for stretched halide glasses. Results are from a number of measurement wavelengths.

into separate paths. The advantage of this approach is that one can produce an interferometer-type structure necessary to produce an optical isolator that is independent of the input polarization state (e.g., see Fig. 8.1c). Again, there are essentially two ways to bring about this behavior; birefringent single crystals and artificial structures that utilize form birefringence.

8.4.1. Single Crystals

We have already discussed the aspect of double refraction in uniaxial single crystals relative to the polarization directions. The phenomenon of refraction is more complicated to deal with especially when the angle of incidence does not coincide with the optic axes, as shown in Fig. 8.13. For the ordinary ray, one has the ordinary Snell's law form,

$$\sin \theta_i = n_0 \sin \theta_{r0} \tag{8.11}$$

However, for the extraordinary ray, one can write a similar form, but only after recognizing that the refracting angle

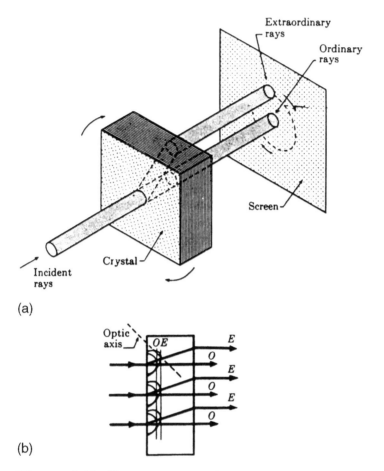

Figure 8.13 Representation showing how double refraction is used to separate the TE and TM components of an incident beam.

depends on the angle of incidence through the dependence of the velocity with angle. One could write this as,

$$\sin \theta_i = n(\theta_i) \sin \theta_e \qquad (8.12)$$

where the $n\ (\theta_1)$ reflects the dependence of the velocity with angle. The closed expression for the beam-split angle will be given later.

8.4.2. Form Birefringence Structures

Alternate thin layers of dielectric materials can develop a high degree of optical anisotropy, proportional to the disparity of the refractive indices of the alternating layers [15]. The definitions of the terms are given in Fig. 8.14. Taking the ordinary wave as that for which the electric field is parallel to the layers, and the extraordinary wave as that for which electric field is perpendicular to the layers, the refractive indices are given by the following expression [16]:

$$n_0 = [n_1^2 q + n_2^2 (1-q)]^{1/2}$$

$$n_e = \left[\left(\frac{1}{n_1}\right)^2 q + \left(\frac{1}{n_2}\right)^2 (1-q) \right]^{1/2} \tag{8.13a}$$

where q is the ratio of the thickness of the n_1 layer to the period. It should be clear that for this expression to hold the period is much less than the wavelength of light. There is a closed form expression for the beam-split angle, which is given by the following:

$$\tan \phi = \frac{(n_0^2 - n_e^2) \tan \theta}{n_0^2 - n_e^2 \tan^2 \theta} \tag{8.13b}$$

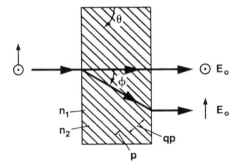

Figure 8.14 Representation of the alternate layer structure of artificial birefringent medium. θ represents the slant angle of the layers and φ is the resulting splitting angle which is dependent on the index difference of the layers and θ.

The angle θ is defined relative to the direction of the slant of the layered structure. The split angle reaches the maximum value when the slant angle is given by the expression

$$\theta = \tan^{-1}\left(\frac{n_0}{n_e}\right) \tag{8.14}$$

When this is true then the maximum split angle is just

$$\tan\phi = \frac{n_0^2 - n_e^2}{2n_0 n_e} \tag{8.15}$$

for small values of Δn, this reduces to the split angle which is equal to Δn in radians. The case of the stretched glass discussed earlier would only split beams by the order of 1.5 min of are. On the other hand, the splitting angle for the alternate dielectric layer structure can be as large as 30°. The comparison between the two common single crystal materials is shown in Fig. 8.15.

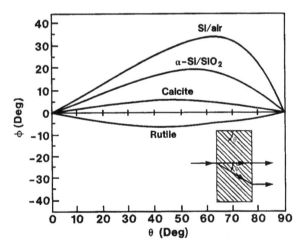

Figure 8.15 Splitting angle as a function of slant angle for two layer structures. A comparison is made to calcite and rutile. (From Ref. [16])

8.5. FARADAY EFFECT

The optical isolator is essentially based on the nonreciprocity of the magneto-optic Faraday effect. The Faraday effect in a material is the rotation of the plane of polarization with a longitudinally applied magnetic field

$$\theta = VHL \tag{8.16}$$

Here, θ is the rotation of the polarization direction, H is the magnitude of the magnetic field in the direction of the light propagation, L is the sample path length, and V is the constant of proportionality called the *Verdet constant*. All materials exhibit such an effect to some degree. The microscopic origin of the effect is beyond the scope of this book. The reader is referred elsewhere [17] for a discussion of the quantum mechanical nature of the Faraday effect. It will have to suffice to say here that the effect originates from the magnetic field splitting of the atomic energy levels of the constituent atoms (ions), and the corresponding differences that arise in the oscillator strengths of the transitions to the split levels. The transitions in question turn out to have different oscillator strengths for left and right circularly polarized light. One can schematically represent the situation as shown in Fig. 8.16. One can see how the circular birefringence arises from the splitting and the corresponding selection rule for the transitions relative to the left or right circular polarization [17]. The example depicted here is for the so-called diamagnetic effect. This simply means that the ground state is nondegenerate; and thus, only the excited states can be split by the external magnetic field. When the ground state is degenerate, (paramagnetic case), then a similar picture arises, but now, the ground state is also split. The circular dichroim is much larger, however, because the populations of the two split ground states are different as a consequence of the temperature.

The macroscopic phenomenological treatment of the Faraday effect shows that it arises from a nonlinear optical effect from a nonlinear term in the optical frequency polarization as indicated In Eq. (8.17)

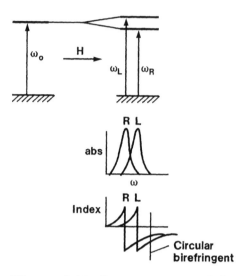

Figure 8.16 Representation of the origin of the Faraday effect. A magnetic field splits the degenerate excited state (diamagnetic effect). The absorption band splits into two states according to left and right circular polarized light. From the dispersion of these two absorption bands, the refractive index difference for left and right circularly polarized light arises.

$$P_i(\omega) = \chi_{ij}E_j(\omega) + \chi_{ijk}(\omega; \omega, 0)E_jE_k$$
$$+ \cdots + \gamma_{ijk}(\omega; \omega, 0)E_jH_k + \cdots \qquad (8.17)$$

The first term is the ordinary optical dielectric tensor refractive index, the second is the next nonlinear term in the electric field that would lead to such phenomena as the linear electro-optic effect. One can add cubic terms, another such leading to other effects. The next written term is a mixed field term involving both the electric field and the magnetic field. And is responsible for the Faraday effect. One can see this in an approximate way be realizing that the refractive index $(\varepsilon[\omega])^{1/2}$ is proportional to the derivative of the polarization relative to the electric field.

$$\frac{dP(\omega)}{dE(\omega)} = \chi = \frac{(\varepsilon - 1)}{4\pi} \qquad (8.18)$$

To evaluate the nonzero tensor elements one must use the symmetry operations consistent with the overall symmetry of the crystal. It is at this point that the difference between the way the electric field transforms compared with the magnetic field under certain symmetry operations comes into play. The most important and illustrative is the behavior of the respective fields under the inversion operator. For the electric field one uses the simple cartesian unit vectors to demonstrate the inversion, as shown in Fig. 8.17. Because the magnetic field is associated with the spin of the electron rather than its vibration, one uses clockwise and counterclockwise spin directions around the unit vector directions to ascertain the behavior of the magnetic field under any given symmetry operation. As one can see from comparing the result for inversion in Fig. 8.17a and b, the electric field is characterized by $E = -E$, whereas $H = H$ under inversion

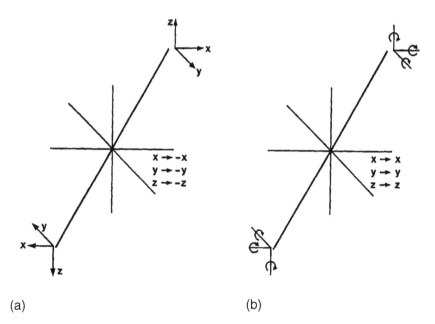

(a)   (b)

Figure 8.17   The transformation of a polar vector similar to the electric field under the symmetry operation of inversion compared to that of an axial vector similar to the magnetic field.

For the case of cubic or isotropic symmetry, one can show that the dielectric tensor for the case of including the first and third terms of Eq. (8.16), has the form [18]

$$
\varepsilon = \begin{pmatrix} \varepsilon_0 & -4\pi j\gamma H & 0 \\ 4\pi j\gamma H & \varepsilon_0 & 0 \\ 0 & 0 & \varepsilon_0 \end{pmatrix} \tag{8.19}
$$

Importantly, the Faraday effect term appears only in the off-diagonal elements. This is directly due to the way the magnetic field transforms under the symmetry operations. The practical significance of these off-diagonal terms is that they influence, not the linearly polarized light, but rather, the circularly polarized light. As we discussed earlier in microscopic explanation, the Faraday effect is really a magnetic-field-induced circular birefringence. In other words, right circularly polarized light travels with a different speed through the material in the presence of a magnetic field than does left circularly polarized light. It is simple to image linearly polarized light as made up of two in-phase circular-polarized components, that is $E = (1-j, 1+j)$. If one adds a phase shift δ to one component relative to the other, as the Faraday effect would do, then the result would be a linearly polarized beam rotated at an angle half that of the phase shift.

8.5.1. Reciprocity

Reciprocity is the word often used to describe the symmetry relation between the cause and effect in a medium [19]. If we put it in terms of the present context, it would state if one propagates light in the forward direction through an optical medium, the resultant state of the light will be the same as that if the light had been propagating in the reverse direction. As an example, consider the action of a birefringent sample, the thickness of which makes it a half-waveplate at some wavelength. Consider now that a polarized beam is incident from the left, making an angle φ with the optic axis. The effect would be that the polarization direction would rotate to an angle 2φ in say the clockwise direction *when viewed in the direction of propagation*. (The actual sense of rotation would depend on whether the optic axis was the fast or slow axis.) Invocation of reciprocity would say that light traveling through the sample from the left would produce the exact same effect. The consequence of this would be that the rotation is just reversed from the forward-traveling effect, when the sense or rotation is reckoned from the original propagating direction. The formal presentation of this problem would consider how the dielectric tensor transforms on the reversal of the propagation direction. The Faraday effect represents the classic nonreciprocal effect because Eq. (8.19) has the dielectric tensor elements that contain the magnetic field. The explanation of the lack of reciprocity can be argued in two ways. The first way is merely to invoke the fact that because the magnetic field is not reversed for the backward-traveling wave, the situations are not equivalent. The second way to understand the lack of reciprocity is outlined in the preceding section where we described the different transformation properties of H as compared with E. In either case one is left with the situation that for the backward-traveling wave the form of Eq. (8.19) would be such as to reverse the signs in the off-diagonal element.

8.5.2. Verdet Constants

One can classify the nature of the Faraday effect into two categories: diamagnetic and paramagnetic. Moreover, one can distinguish between paramagnetic and ferrimagnetic in the sense that the ferrimagnetic materials are invariably crystals and large magnetization occurs with a relatively low-applied field. In other words, the Faraday effect will be hysteric and saturate at some value of the applied field. In Table 8.2 we will list a representative sampling of the Verdet constant for the various classes of materials. We will break it down into both diamagnetic and paramagnetic glasses or crystals. The final category will be the ferrimagnetic crystals. The units of the Verdet constant will be in degrees of rotation per tesla (10^4 Oe) of field per centimeter of length. Similar to all magnetic effects, the units are as diverse as the number of people who write papers. The Verdet constant is certainly no

Table 8.2 Representative Verdet Constant for Various Materials

Materials	V (deg/T cm)	Wavelength (nm)
Glasses		
FR-123 (Tb-borate)[a]	−67	633
FR-5[b]	−68	633
Pr-phosphate[c]	−38	633
FR-7 (Tb-fluorophos)[c]	−33	633
SF-59[c]	27	633
Tl-gallate[d]	52	633
Paramagnetic crystals		
CdMnTe	2000	633
EuF$_3$	−250	633
Tb$_3$Al$_5$O$_{12}$	−170	633
Bi$_4$Ge$_3$O$_{12}$	29	633
ZnSe	112	633
Ferrimagnetic crystals[e] (degrees/cm)		
YIG	870	633
	280	1064
	210	1300
Bi$_3$Fe$_5$O$_{12}$	−55,000	633
Gd$_{1.6}$Bi$_{1.4}$Fe$_5$O$_{12}$	−28,500	633
	−1800	1300
Gd$_2$Bi$_1$	−3300	1064
	−2100	1300
Y$_{1.7}$Bi$_{1.3}$Fe$_5$O$_{12}$	−2100	1300
Yb$_{0.7}$Tb$_{1.7}$Bi$_1$Fe$_5$O$_{12}$	−1800	1300
	−1200	1550

[a] Kigre.

[b] Hoya.

[c] Schott.

[d] Aitken, B.G.; Borrelli, N.F. Thallium-containing glasses with high optical nonlinearity, Proceedings International Symposium on Glass Problems, Vol 1, 527–533, Sept. 4–6 Istanbul, Turkey, 1996.

[e] Weber, M. J. Ed. *CRC Handbook of Laser Science and Technology*, Suppl 2, *Optical Materials*; CRC Press: Boca Ratan, FL, 1995.

Sources: From SPIE 30th Annual International Technical Symposium on Optical and Optoelectronic Applied Sciences and Engineering. Conference Proceedings 681, Laser and Nonlinear Optical materials; "Faraday Rotator Materials for Laser Systems," M. J. Weber, July 1996.

exception. In the early literature, one often sees min/Oe cm, and lately one sees radians/T m. They are all relatively easily converted to one another. In the ferrimagnetic crystals, the Verdet constant will be given as the saturated value. That is, it will be at the value achieved at sufficient field to saturate the hystersis loop. The field is not often listed, but is the order of a few kOe for most of the ferrimagnetic crystals. It is desirable, on the one hand, to have them of sufficiently hard magnetic materials, a high coercive field, stable against stray magnetic fields, and on the other hand, not to require a large magnet to provide saturation. As we will shortly see, the optical isolators that employ the Faraday element should be as compact as possible. This favors the high Verdet constant ferrimagnetic crystalline materials, because less than a millimeter of path length is required to produce 45° of rotation.

8.5.3. Other Nonreciprocal Effects

There have been two reported nonreciprocal effects, in addition to the Faraday effect. The first is based on a nonreciprocal phase shift experienced by light propagating in the TM mode of a magnetic waveguide with the external magnetic field *perpendicular* to the propagation direction [20]. The TM mode is the only state that experiences the nonreciprocal phase shift because it has a component of the electric field in the propagation direction. The extent of the phase shift is proportional to the Verdet constant, but much smaller in magnitude. Because E_z changes sign at the center of guide, the waveguide structure must somehow be asymmetrical to avoid cancellation. This is shown in Fig. 8.18. There are various ways that this is suggested both in a planar format as well as a cylindrical fiber structure.

The relevant literature citations that deal with other nonreciprocal structures are Refs. [21a–b]. They are all based on planar waveguide structures utilizing some aspect of off-diagonal elements of the dielectric tensor to effect nonreciprocal behavior. Because each is unusual and unique, it is difficult to deal with them here without extensive background. If the reader is interested, consult the original articles.

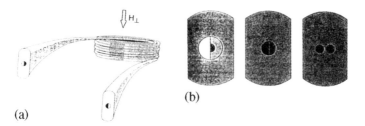

(a)

(b)

Figure 8.18 (a) Coiled optical fiber with asymmetric core. Magnetic field is applied perpendicular to the axis of the fiber. (b) Three possible core designs to avoid cancellation. In the first, half of the core is air; in the second, the core is made of two different materials with the opposite sign of the Verdet constant; and the third, two coupled cores of different Verdet constants are used. (From Ref. [20] with permission.)

8.6. ACTUAL OPTICAL ISOLATOR STRUCTURES

The need for ways to eliminate, or capture light traveling back along an optical path has become great in laser diode and fiber-based amplifier systems. Because these are commercial devices that have many other components, to a large extent the isolator function has to fit the device. As a good example of this, the total thickness of the isolator, or circulator package is a major issue, the thinner the better. This means that if the Faraday effect is being used, then the rotation of 45° must be achieved in as thin a sample thickness as possible. This is true for all the functions, beam separation, waveplates, polarizers, or others. This is why the isolators in use predominantly use ferrimagnetic crystal slices as the Faraday rotator because the rotation per unit path length is so much larger.

In this section we will deal with two important structures. The first will be the polarization-insensitive fiber pigtailed device, and the other will be the planar waveguide device.

8.6.1. Polarization-Insensitive Isolator

This is the isolator of choice for optical fiber amplifiers because the state of polarization is not fixed in optical fibers owing to TE–TM mode coupling. We showed this type in Fig. 8.1c. The structure is shown in more detail in Fig. 8.19. Here, we see the lenses that couple the light from the pigtail fibers. The lenses are gradiant index (GRIN) microlenses as described in Chapter 3. An actual 60 × SEM photomicrograph of an isolator is shown in Fig. 8.20. Each section is identified under the micrograph with the appropriate length scale. The beam separation is effected by a slice of a single crystal of rutile. The half waveplate is quartz, and its optic axis is oriented 22.5°, so that it rotates the polarization by 45°. The Faraday rotator is the ferrimagnetic garnet called BIG ($TbYb_xBi_{3-x}Fe_5O_{12}$) at a thickness of about 0.38 mm. The Verdet constant must be 118°/mm because 45° is required. The closest material shown in Table 8.2 is the YBi-garnet. The magnetic field can be either from small external permanent

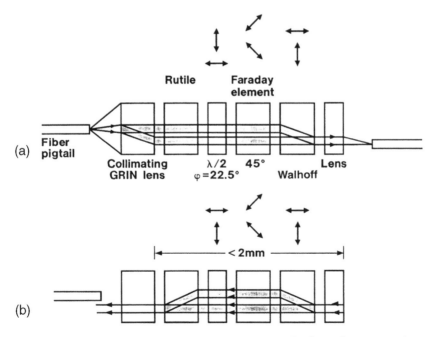

Figure 8.19 Schematic representation of the elements that constitute a polarization independent isolator. The beam paths are shown for the forward and the backward paths. The polarization direction is indicated above the elements.

| GRIN lens
doped
Borosilicate
glass,
n=1.59
on axis | TiO2 (Rutile or Titania)
Birefringent "walkoff"
polarizing beam splitter
ne=2.548, no=2.872 (visible) | Soda-Lime
(Na, Ca)Glass
Spacer | TbYbBiFeO   (TbYb BiG)
Tb, Yb doped
Bi substituted Iron Garnet
Faraday rotator (saturable) | TiO2 (Rutile or Titania)
Birefringent "walkoff"
polarizing beam splitter
ne=2.548, no=2.872 (visible) | |
| | | Birefringent   SiO2
Waveplate   λ/2
ne=1.56, no=1.55 (vis) | | | Fiber ferr |

Figure 8.20 SEM photograph of an isolator of the design shown in Fig. 8.19. The elements are indicated below the photo.

magnets mounted in the package, or with sufficiently hard magnetic materials, the fully magntized state of the crystal is used. In this case no external magnets are needed. The actual function of each element is shown in the appendix using the Jones matrix method.

The isolation of these isolators (amount of backward light that leaks through) is over 50 dB. Typical insertion losses are 1 dB, or less. Fig. 8.21 shows the actual positions of the isolators in the schematic of a forward pumped Er-doped fiber amplifier structure. In the more sophisticated WDM designs circulators are extensively used to separate out the backward-traveling diffracted wave.

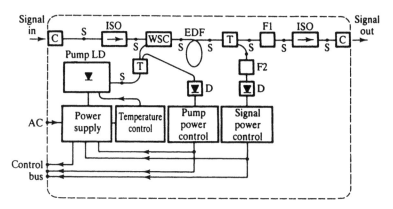

Figure 8.21 Schematic of an Er optical amplifier showing the position of the isolators, marked as ISO.

What are the disadvantages. At first glance, it would appear then the extreme hybrid construction of the assembly would make it prohibitively expensive. However, the price has decreased as the volume has increased; hence, this may not be a problem, at least for the present applications. The performance is more than adequate, so the move to some other approach is not likely unless the alternative performs at equivalent, or better levels of isolation: It would certainly be desirable to have a more integrated design, such as an all-fiber structure, but this is not likely to happen soon. The more integrated isolators will likely be of a planar nature as described next.

8.6.2. Integrated Planar Isolator

The approach to a fully integrated planar optic circuit, that is an optical equivalent to the silicon chip, will require the ability to fabricate planar optical isolators. The simplest way to approach this is to have the Faraday element act as a guided wave. An example of such a scheme is shown schematically in Fig. 8.22. The idea is to fabricate a waveguide in the magnetic film in such a way that it mates with low loss to the incoming and outgoing waveguides that are carrying the signal. For a ridge waveguide in the magnetic film, it is to be mated with

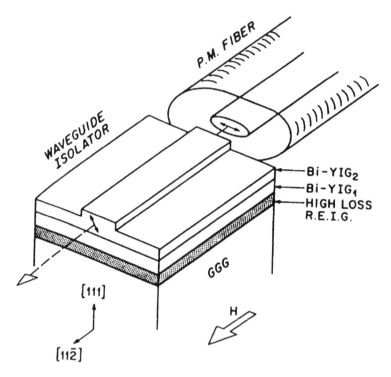

Figure 8.22 Planar waveguide isolator. The Faraday film is fashioned into a channel waveguide as shown. The film is epitaxially grown on GGG and the other layers establish a guiding structure to provide the correct modal field matching the input and output cylindrical guides. (From Ref. [22].)

a cylindrical guide, but it could also be planar. Structures such as this have been made and tested. Wolfe et al. [22] have fabricated the channel guide in a $(BiY)_3(FeGa)_5O_{12}$ material exhibiting a rotation of $133°$/cm at 1550 nm. This layer, as well as the other single-crystal layers shown, was epitaxially deposited by LPE on GGG. Isolation of greater than 40 dB was reported. It should be borne in mind that this is not the polarization-insensitive design. To do this, one would have to implement a separate path, as in the device shown in Fig. 8.1c. This would likely entail some sort of yet unspecified polarization splitter into the beam to create two paths.

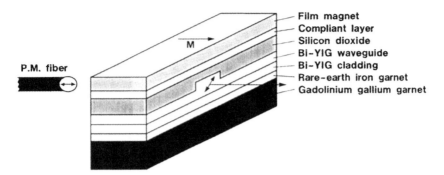

P.M. fiber

M

Film magnet
Compliant layer
Silicon dioxide
Bi–YIG waveguide
Bi–YIG cladding
Rare–earth iron garnet
Gadolinium gallium garnet

Figure 8.23 Fully implemented planar isolator where even the magnet is deposited as a film. (From Ref. [23])

A fully implemented design [23] for which even the magnet is deposited as a thin film, is shown in Fig. 8.23. Here too, the garnet layers are epitaxially prepared on a GGG (gadolinium gallium garnet) substrate. To facilitate the waveguiding of the Faraday film, in this case the bismuth-doped YIG, other layers of garnet are deposited. The purpose here is to lower the refractive index between the core and the cladding so that the single-mode thickness of the guiding layer is the order of 4–5 μm. The magnet was a TbCu-type material deposited by sputtering.

8.6.3. Other Isolator/Filter Applications

8.6.3.1. Array Isolator

The idea of integration is as important in the opto-electronic domain as it has been in the purely electronic area. Unfortunately, it is somewhat more difficult to implement because of the innate differences in the components. A good example of this is an attempt to integrate a function of an "array isolator" device. The objective is to have eight or more channels of data entering a structure which function as an isolator for each channel. The pitch, or separation of the inputs is to be either 0.25 or 0.5 mm. A schematic of such a device is shown in Fig. 8.24a. In Fig. 8.24b we show a photograph of an actual array isolator compared to a commercial single channel

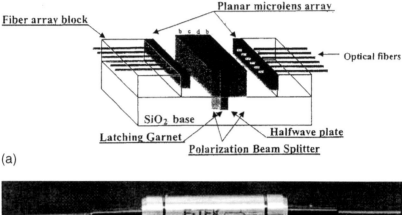

(a)

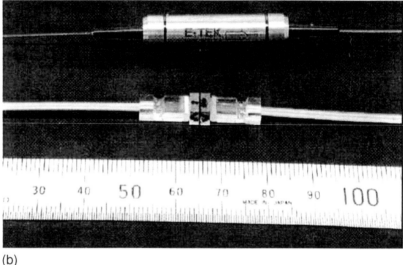

(b)

Figure 8.24 (a) Schematic representation of an array optical isolator for a 1–8 fiber array (taken from Ref [24]) and (b) a photographic comparison of the size of the arrayed isolator package to that of a commercial single channel fiber isolator.

isolator package [24]. One sees that it involves the technology of lens arrays that we discussed in Chapter 2, of fiber array holders as discussed in Chapter 9t, as well as isolator elements mentioned above. The total thickness of the device is just 4.55 mm, yet it contains two lens arrays, two polarization beam splitters, a garnet crystal and a half wave plate. The cross-section of the device is 5 mm × 5 mm.

8.6.3.2. Filters

The idea of using a series of birefringent crystal to produce a filter with a specific spectral shape (often called Solc filter) has been suggested [25–28]. There are a number of different structures but the idea is the same in that they are based on the idea that the propagation of two waves at different speeds will produce a phase difference and consequently a sinusoidal variation in the amplitude with wavelength. If the beam passes through a series of birefringent elements, phase shifts $\delta_1, \delta_2, \ldots, \delta_N$ in sequence, then there will present the permutations of all the n-fold sums and differences, up to N. Each term would be equivalent to a different spatial frequency component, so one can consider the overall spectral shape as a result of a finite-term Fourier series.

$$A(\omega) = \sum^n C_n \exp(ia_n\omega) \tag{8.20}$$

The values of a_n represent the various permutations of the phase shift. As a simple example, suppose there were three birefringent elements with all the same phase shift, then the series of Eq. (1) would be the following:

$$A(\omega) = C_0 + C_1 \exp(i\omega) + C_2 \exp(2i\omega) + C_3 \exp(3i\omega) \tag{8.21}$$

The values of the coefficients control the amount of each frequency component and are the adjustable parameters that are used to synthesize the desired spectral shape. The way this is done in the birefringent crystal case is by the choice of angle the optic axis of the crystal makes with the input since this determines the splitting ratio into the two orthogonal propagation directions.

In the birefringent crystal approach, angles that respective crystals make with the input polarization are α, β, and γ, the additional phase in each crystal over and above the integral order are signified by φ_1. The free spectral range is given by the expression,

$$f_{\mathrm{FSR}} = \frac{c}{L\Delta n} \tag{8.22}$$

L is the path length and Δn is the magnitude of the birefringence. The total phase shift in each crystal is given by the following:

$$\delta_i = \frac{2\pi}{\lambda} L\Delta n + \phi_i = \frac{2\pi f}{f_0} + \phi_i \qquad (8.23)$$

One can symbolically write the sequence of operations as Jones Matrix elements (see Appendix) in the following form. The symbol R stands for the rotation and P for the phase shift

$$\begin{pmatrix} e_1 \\ e_2 \end{pmatrix}_{out} = A(\phi)P(\delta_3)R(\gamma)P(\delta_2)R(\beta)P(\delta_1)R(\alpha)\begin{pmatrix} e_1 \\ e_2 \end{pmatrix}_{in}$$
$$(8.24)$$

The angle ϕ is the analyzer angle (projection of the output of the last crystal back onto the original axis defined by the input polarization) and is related to the crystal angles by the expression

$$\varphi = 90 - (\alpha + \beta + \gamma) \qquad (8.25)$$

An example of a 100 GHz filter fashioned from three slices of the birefringent crystal yttrium vanadate is shown in Fig. 8.25.

There is another interferometric scheme that utilizes a Mach–Zehnder fiber structure as its phase shifting element. This is shown in Fig. 8.26. The design is the same as above. The same formalism is used as in Eq. (8.24) where the rotation matrices are replaced by the coupling matrices. The coupler matrix is given by the following expression where α is the coupling constant of the Mach–Zehnder structure

$$C(\alpha_i) = \begin{pmatrix} \sqrt{1-\alpha_i} & i\sqrt{\alpha_i} \\ i\sqrt{\alpha_i} & \sqrt{1-\alpha_i} \end{pmatrix} \qquad (8.26)$$

8.6.3.3. Tunable Gain-Flattening Filter

The amplification produced by a EDFA (erbium-doped fiber amplifier) is not uniform across the telecom spectrum of

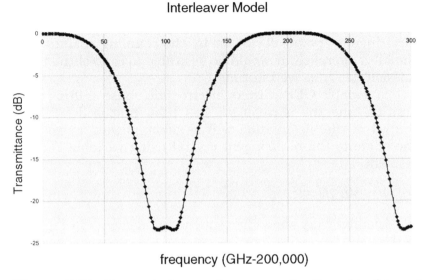

Figure 8.25 Transmission vs. frequency of a three birefringent crystal filter.

1530–1560 nm. To compensate for this condition, use is made of a static GFF (gain-flattening filter). There are a number of approaches to provide such a function and they are all inter-ferometrically based.

However, in the dynamic situations, it is desirable to be able to tune the filtering function in real time, both in amplitude and frequency. In Eq. (1), the added phase term is fixed where it would be desirable to be able to adjust it in real time. The amplitude or depth of the transmission is provided by the splitting ratio into the two paths. Again, in the static devices,

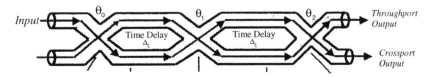

Figure 8.26 Fiber Mach–Zehnder interferometer.

this is a fixed ratio controlled either by the angle the polarization makes with the optic axis of the crystal in the case of the birefringent crystal device, or by the coupling coefficient in the M–Z approach. It would be desirable to control the transmission depth as well in time.

A tunable GFF with an electrooptic phase shifting property to tune the wavelength has been proposed [29]. The control of the attenuation will be accomplished by rotating the angle to the birefringent crystal using variable Faraday element

$$\theta = VL_M H \qquad (8.27)$$

We schematically show the sequence of elements in Fig. 8.27 [24].

The overall function can be described as follows. The polarized light input from the fiber is incident on a Faraday element. The rotation produced by this element will alter the angle that the polarization will make with the birefringent crystal.

As mentioned above, the rotation produced by the Faraday element ultimately determines the angle the polarization makes with the birefringent crystal optic axes as shown in Fig. 8.27. The beams are then incident on an electro-optic crystal whose optic axes are aligned to that of the birefringent crystal. The electro-optic effect can be either linear or quadratic in the applied electric field

$$\Delta nr \begin{pmatrix} E \\ E^2 \end{pmatrix}$$
$$\delta = \frac{2\pi}{\lambda} \Delta nL \qquad (8.28)$$

The total thickness and net static birefringence of these two crystals determines the free spectral range. The variable phase derives from the electro-optic crystal portion of the birefringent crystal. The sample then goes through another Faraday rotation equal and opposite to the first in order to restore the alignment of the polarization direction for the output analyzer.

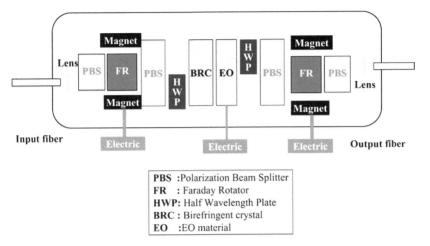

Figure 8.27 Schematic representation of a tunable gain-flattening filter using a Faraday and electro-optic material as the tunable elements.

The state of the polarization under the respective operations is most easily described by using the Jones matrix formulation where each matrix represents the function of each element as shown in the Appendix.

8.9. APPENDIX A.: JONES MATRIX

There is a convenient and useful formalism that helps one deal with the sequence of actions that different optical elements perform along a light path. These elements produce things ranging from phase shifts to axis rotations. The basic idea is to define a 2×2 matrix for each operation. The sequence of operations becomes nothing more than the product of the matrices in the order that the element is encountered along the path. Since optical isolators contain a number of elements, use of the Jones matrix formalism proves quite useful in describing their overall function.

The light beam is represented by a column vector, the components being the x and y components of the field.

The output from the element is then related to the input by the matrix expression,

$$\begin{pmatrix} E_x^{\text{out}} \\ E_y^{\text{out}} \end{pmatrix} = \begin{pmatrix} \alpha_{11} & \alpha_{12} \\ \alpha_{21} & \alpha_{22} \end{pmatrix} \begin{pmatrix} E_x^{\text{in}} \\ E_y^{\text{in}} \end{pmatrix} \tag{A8.1}$$

The matrices for the most common operations will be listed in the following:

Rotation (clockwise angle θ)

$$R = \begin{pmatrix} \cos\theta & -\sin\theta \\ \sin\theta & \cos\theta \end{pmatrix} \tag{A8.2}$$

Phase shift δ with optic axis making an angle φ

$$S = \begin{pmatrix} \cos\phi & -\sin\phi \\ \sin\phi & \cos\phi \end{pmatrix} \begin{pmatrix} \exp(j\delta) & 0 \\ 0 & 1 \end{pmatrix} \begin{pmatrix} \cos\phi & \sin\phi \\ -\sin\phi & \cos\phi \end{pmatrix} \tag{A8.3}$$

Polarizer at angle γ

$$P = \begin{pmatrix} \cos\gamma & -\sin\gamma \\ \sin\gamma & \cos\gamma \end{pmatrix} \begin{pmatrix} p_1 & 0 \\ 0 & 1-p_1 \end{pmatrix} \begin{pmatrix} \cos\gamma & \sin\gamma \\ -\sin\gamma & \cos\gamma \end{pmatrix} \tag{A8.4}$$

The overall effect of a series of optical elements is just the product of the particular matrices representing the function in the order they appear. Reciprocity can be expressed as the equality of the matrix multiplication, applied forward, or in reverse.

For the Faraday rotator, which will have the form of Eq. (A8.2), because of the nonreciprocity attributable to the magnetic field, light traveling in the opposite direction must be represented by the matrix

$$\begin{pmatrix} \cos\theta & \sin\theta \\ -\sin\theta & \cos\theta \end{pmatrix} \tag{A8.5}$$

where $\theta = VHL$.

REFERENCES

1. *Handbook of Laser Science and Technology,* suppl 0002, Optical Materials, Weber, M.J., Ed., section 9, Magnetooptic Materials; Deeter, M.N., Day, G.W., Rose, A.H., Eds.; CRC Press: Boca Raton, FL, 1995, 367–411.

2. Wilson, D.K. Optical isolators cut feedback in visible and near-IR lasers. Laser Focus **1990**, *24*, 103.

3. Clark Jones, R. A new calculus for the treatment of optical systems. JOSA **1941**, *31*, 488–493.

4. Fujii, Y. High-isolation polarization-independent optical circulators. J. Lightwave Technol. **1991**, *9*, 1238–1243.

5. Borrelli, N.F. Optical elements derived form stretched glass. Proc. SPIE **1992**, *1761*, 202–212.

6. POLARCOR Product sheet, Specialty Optical Products, Corning Inc, Corning, NY.

7. *Handbook of Optics*, II, *Devices, Measurements and Properties*, 2nd Ed.; Bass, M. Ed.; McGraw Hill: New York, 1995.

8. Lamipol product sheet, Sumitomo Osaka Cement Co, Optoelectronics and Electronics Div., Tokyo, Japan.

9. Nishino, S.; Yamamoto, H.; Kasazumi, K.; Wada, H.; Sano, K.; Saimi, T. Application of a polarizing holographic optical element to a recordable optical head. Jpn. J. Appl. Phys. **1996**, *35* (1B), 357–361.

10. Born, M.; Wolfe, E. *Principles of Optics*; Macmillan: New York, 1964.

11. Mansuripur, M.; Li, L. What in the world are surface plasmons. Opt. Photonic News, May, **1997**, 50–57.

12. Nye, J.F. *Physical Properties of Crystals*; Oxford Press: London, 1957.

13. Hale, P.D.; Day, G.W. Stability of birefringence. Appl. Opt. **1988**, *27*, 5146–5153.

14. Borrelli, N.F. Waveplates from stretched glass. Proc. SPIE **1992**, *1746*, 336–342.

15. Shiraishi, K.; Sato, T.; Kawakami, S. Experimental verification of a form–bire-fringent polarization splitter. Appl. Phys. Lett. **1990**, *58*, 211–212.

16. Shiraishi, K.; Kawakami, S. Spatial walk-off polarizer utilizing artificial anisotropic dielectrics. Opt. Lett. **1990**, *15*, 516–518.

17. Ballhausen, C.J. *Introduction to Ligand Field Theory*, Chap. 9; McGraw-Hill: New York, 1962.

18. Borrelli, N.F. Magnetooptic properties of magnetite films. J. Appl. Phys. **1971**, *42*, 1120–1123.

19. Mathieu, J.P. *Optics*, parts 1 and 2; Pergamon Press: London, 1975.

20. Wolfe, R.; Wang, W.-K.; DiGiovanni, D.J.; Vengsarkar, A.M. An all-fiber magnetooptic isolator based on non-reciprocal phase shift in asymmetric fiber. Opt. Lett. **1995**, *20* (16), 1740–1742.

21. (a) Yamamoto, S.; Makimoto, T. Circuit theory for a class of anisotropic and gyrotropic thin-film optical waveguides and design of nonreciprocal devices for integrated optics. J. Appl. Phys. **1974**, *46* (2), 882–889. (b) Auracher, F.; Witt, H.H. A new design for an integrated optical isolator. Optics Comm. **1975**, *13* (4), 435–438.

22. Wolfe, R.; Lieberman, R.A.; Fratello, V.J.; Scott, R.E.; Kopylov, N. Etched-tuned ridged waveguide magneto-optic isolator. Appl. Phys. Lett. **1990**, *56*, 426–430.

23. Levy, M.; Osgood, R.M.; Hegde, H.; Cadieu, F.J.; Wolfe, R.; Fratello, V.J. Integrated optical isolator with Sputter-deposited thin film magnets. IEEE Photonics Technol. Lett. **1996**, *8*, 903–905.

24. Takeuchi, Y. Corning Inc. private data.

25. Evans, J.W. The birefringent crystal. JOSA **1949**, *39* (3), 229–242.

26. Ammann, E.O.; Yarborough, J.M. Optical network synthesis using BR crystals. V. JOSA **1966**, *56* (12), 1746–1754.

27. Harris, S.E.; Ammann, E.O.; Chang, I.C. Optical network synthesis using BR crystals. I. JOSA **1964**, *54* (10), 1267–1278.

28. Evans, J.W. Solc birefringent filter. JOSA **1958**, *48* (5), 142–149.

29. Frisken, S.; Abakoumov, D.; Bartus, A. Low-loss polarization independent dynamic gain equaliser filter. OFC paper WM14–1/251, **1999**.

9

Photonic Crystals

9.1. BACKGROUND AND INTRODUCTION

Over the past few years, more attention has been paid to an approach to light propagation in dense media based on a some what different way of looking at the solutions to Maxwell's equations [1,2]. The different look is based on the realization that one can cast Maxwell's equations as a mathematical linear operator and, thereby view the solutions in different terms. This is the familiar quantum mechanical formulation for which one defines the hamiltonian operator [3]. The hamiltonian operator acts on a fictitious set of functions called the wavefunctions, in the form

$$\boldsymbol{H}\Psi = E\Psi \tag{9.1}$$

$$\boldsymbol{H} = \frac{-(h/2\pi)^2}{2m}\nabla^2 + V(x) \tag{9.2}$$

Here the periodic potential that the electron sees is contained in the $V(x)$. The particular linear operator $\boldsymbol{H}$ represented by

Eq. (9.2) has a number of useful and important properties [2,3]. It is sufficient here to mention that, in general, the set of eigenfunctions H is orthogonal and complete, allowing one to express any state of the electron as an infinite linear combination of these unique functions. The concept is similar to being able to represent any vector in three spaces as a linear combination of the unit vectors in the x, y, and z directions. When Eqs. (9.1) and (9.2) are applied to electrons, say in a crystal where the electron experiences a periodic potential owing to the charged lattice, one then obtains a map of allowable bands of energy as a function of the wavevector k. For a simple illustrative example in one dimension, one writes the wavefunction in the form

$$\Psi(x) = u(x) \ \exp(jkx) \tag{9.3}$$

Here the periodicity is contained in the property of $u(x)$ such that $u(x+a) = u(x)$ where a is the period of the potential that the electron is in. The result of this formulation results in an energy (frequency) vs. k diagram as shown in Fig. 9.1. What one sees is the parabolic shape of the energy bands based on what one would expect for a free electron. [This corresponds to $V(x) = 0$ in Eq. (9.2), and $u(x) = 1$ in Eq. (9.3).] The symmetry of the structure allows one to draw the extended picture of the bands as shown. For example, the energies must be the same as $k = 0$ and all multiples $2\pi/a$, where a is the period of the lattice. This will be true for any value of k within the range $-\pi/a < k < \pi/a$ which is termed the reduced Brillouin zone [3]. What is important to see, and forms the particular emphasis in what follows, is that in the presence of a non-zero potential, the bands split as shown. What this means is that there will be forbidden ranges of energies for the electron. These so-called bandgaps play the crucial role in all semiconductor devices. It is the judicious placing of impurities or defects in these materials that provide localized energy levels within the gap. It is the electrons and holes from these localized levels that are manipulated to produce the transistor.

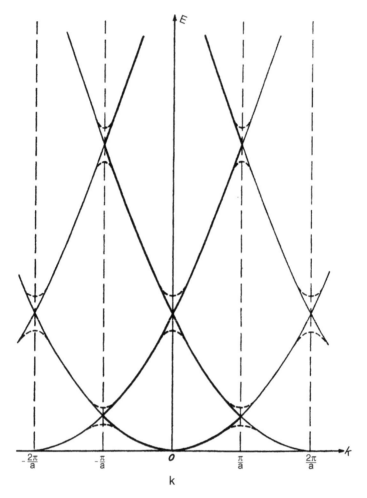

Figure 9.1 Simple one-dimensional reduced Brillouin representation of energy bands for free electrons and the effect of the modifications produced as a consequence of the periodic potential of the nuclei as a solid lattice. The effect is the opening of "forbidden gaps", regions of energy within which the electron cannot reside.

One can write Maxwell's equation in terms of the magnetic field **H** in this form as well as shown in Ref. [2].

$$\nabla x \left[\frac{1}{\varepsilon(r)} \nabla x H \right] = \left(\frac{\omega}{c} \right)^2 H(r) \tag{9.4}$$

where ε is the dielectric constant. Following the notation of Ref. [2], one can define the operator Θ as

$$\Theta = \nabla \mathbf{x} \left[\frac{1}{\varepsilon(r)} \nabla \mathbf{x} \right] \tag{9.5}$$

so that the operator form is

$$\Theta \boldsymbol{H} = \left(\frac{\omega}{c} \right)^2 \boldsymbol{H} \tag{9.6}$$

Aside from the fact that one can now apply the considerable and powerful computation methods developed over the years for large-scale quantum mechanical calculations, based on Eq. (9.1), to an entirely new area, the significance of this formulation has a more conceptual benefit. The solution of Eq. (9.4) for any given spatial variation in the dielectric constant $\varepsilon(x, y, z)$ (here we mean the optical dielectric constant that is equal to the square of the refractive index) gives rise to an energy vs. propagation constant diagram similar to that in Fig. 9.1. We will outline an algorithm for the calculation in the next section, but here it is sufficient to realize that the eigenvectors of Eq. (9.4) will be of the form

$$\boldsymbol{H}_n = \boldsymbol{A}_n \exp[j(\omega t + k_n \cdot \boldsymbol{r})] \tag{9.7}$$

where k_n characterizes the nth propagation mode. We will show a number of these for different geometries in the following sections.

In one sense, it should come as no surprise that a forbidden energy or frequency gap would occur as a result of a periodic pattern in the refractive index. The dielectric multilayer thin film stack, as represented in Fig. 9.2a, is an example of this phenomenon in a more familiar context. Normally, one would see the transmission, or reflection, plotted against the wavelength for a stack of alternating layers of refractive indicies [4] as shown in Fig. 9.3a–c. We have shown the reflection vs. wavelength as a function of the number layers to indicate how the reflection maximum begins to rise and flatten as the number of layers increases. It is only in the limit of an infinite number of layers that a flat, totally reflecting spectral region,

(a)

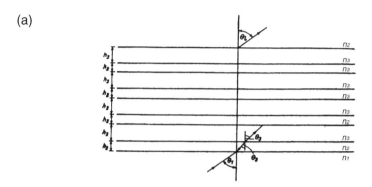

(b)

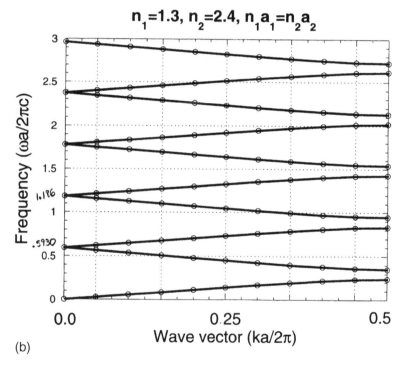

Figure 9.2 (a) Multilayer dielectric stack of thin films. Refractive index alternates from n_2 to n_3 in each layer. (b) The calculated infinite layer one-dimensional photonic bandgap; here, the condition of equal optical thickness was used.

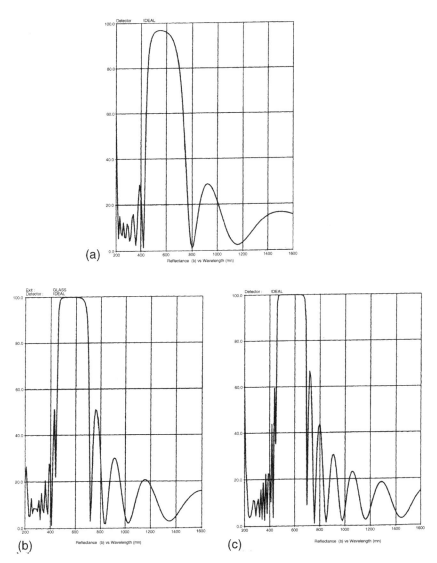

Figure 9.3 Computed reflection spectra for the case given in Fig. 9.2 for 9, 15, and 21 layers, respectively.

or gap, would result. It is this condition that corresponds to the photonic bandgap depiction, described in the foregoing, at which frequencies within a certain region are excluded. Plotting the behavior of the multilayer stack in this way leads to the diagram shown in Fig. 9.2b. Here one sees that there is a gap in the possible propagating frequencies indicated by the separation on the *y*-axis. One can compare the calculated gap to that predicted by the conventional approach (for example, see Ref. [4], p. 69) where a transfer matrix approach is used. It can be shown that the reflection will increase exponentially with the number of layers when the argument of the polynomial has an exponential behavior, rather than a sinusoidal one. The condition is thus expressed as the following equation:

$$\cos^2 \beta - (1/2)\left[\frac{1}{n_1} + \frac{1}{n_2}\right] \sin^2 \beta < -1 \tag{9.8}$$

Here, $\beta = (2\pi/\lambda)a[(1/n_1) + (1/n_2)]$, a is the period and the *ns* are the indices of the layers. The gap is then the values of a/λ for which Eq. (9.8) is met. We will see from a practical point of view, the gap is to be viewed as existing for some arbitrarily defined reflectivity, as close to 100% as desired.

It is the extension of this forbidden gap concept into two and three dimensions, in particular, the structures that will be required, that will provide the interesting microoptic applications that we will try to cover in the succeeding sections.

9.2. FREQUENCY VS. *K* DIAGRAMS

Following the method described in the previous section, the code written by Joannopoulos et al. [7] was used to calculate the photonic bandgap for a few structures that have two-dimensional periodicity. This is just to familiarize the reader with the types of behavior that can arise from different geometries. The first is that of a square array of dielectric cylinders, as shown in Fig. 9.4. Here, there is no overall bad-gap in the sense that the no mode is everywhere forbidden. This is referred to as a full gap. Such a full gap is exhibited for a structure shown in Fig. 9.4b where cylindrical holes are hexangonally arranged. Note that there is a normalized frequency region, roughly between 0.43 and 0.5, at which no

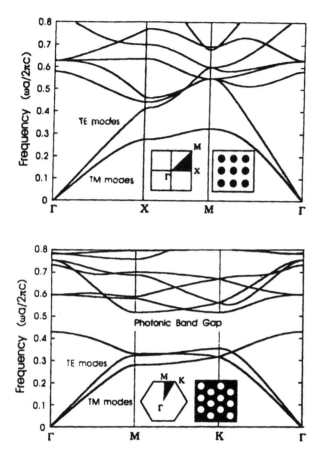

Figure 9.4   Calculated energy diagrams for two simple 2D lattices as shown. (From Ref. [20].)

modes exist for any direction. The width of this gap is a function of the ratio of the refractive index of the material that constitutes the cylinders and that of the surrounding material. In the example shown, the ratio is 3.6. The dependence of the width of the gap as a function of this ratio is shown in Fig. 9.5. This presents an interesting departure from the single-dimension case where there was no threshold for the appearance of a gap. One can understand this difference, at least qualitatively, in that in two dimensions one has both TE and TM modes possible. One can see on inspection of

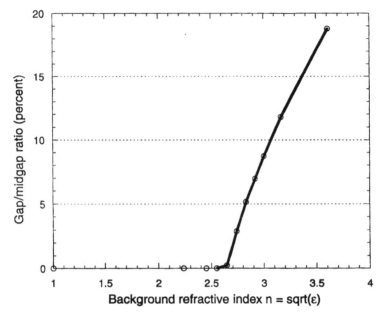

Figure 9.5 Computed value of the refractive index contrast required to produce a full gap for a triangular pattern of holes.

Fig. 9.4 (top) that there is a gap for the TM mode, but not for TE. Similarly, if one made the inverse structure holes, rather than posts, one would have seen the opposite situation occur. Suffice to say here that with sufficient disparity between the dielectric constant, one can force an opening of a full gap.

There is another very important aspect of the photonic crystal approach. That is the role the defect in the structure plays in the allowable energy states. Identical with the role defects play in the electronic band structure, the result is the formation of localized states that may appear within the forbidden gap. The way that one calculates the consequence of defects is to proceed with the calculation of the infinite periodic array except to remove, one or more of the periodic elements as in the example shown in Fig. 9.6a, or it could be a whole line of holes, as shown in Fig. 9.6b. The extent that the translational symmetry is broken by the presence of the defect determines the way one can represent the results. If,

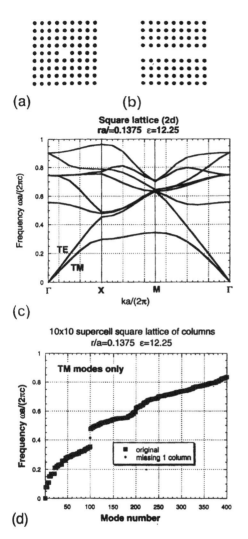

(a)   (b)

(c)

(d)

Figure 9.6 Representation of defect states (a) and (b) in a 2D array. In the former, one element is missing, and in the latter, a whole row is missing. (c) The standard energy vs. k diagram for the perfect array. (d) A representation that shows the consequence of defect structures such as (a) and (b). The energy of the first four TM modes for a 10×10 array is plotted in ascending order for both the perfect array and one with a single defect. In this representation, not only is the gap evident, but also the localized state caused by the defect.

for example, the translational symmetry is maintained in the propagation, as would be true if the propagation were along the row of missing elements in Fig. 9.6b, then one can still represent the modes relative to the k corresponding to this direction. However, if the propagation were perpendicular to the line, the translational symmetry is broken, and there is no way to define k. There are several ways to display the allowable energy of defect structures. One way to do this is to plot the modes in ascending order of energy. We have done this for the example discussed earlier for which one defect is located in the middle of a 10×10 array for the first four TM modes, as shown in Fig. 9.6a. We have done this because a gap exists for the TM modes, as seen from the conventional energy diagram for the infinite (Fig. 9.6c). Because there is no way to represent the situation of a defect on this diagram, for there is no k, one can circumvent this difficulty; we replot results in the way shown in Fig. 9.6d. This is done for the defect-free case where the gap is still a feature of this representation. For the defect case, one sees not only the gap, but also the mode in the gap, which is the desired result. In general, then, the description of cases with defects will have to be represented in this way, or a similar way, or directly in the time domain.

Again, it should be borne in mind that these calculations are based on a 2D model and thus infinitely long cylinders or holes in the z-direction are assumed, as well as infinite extended in the x–y plane. The diagrams we have shown are for the case in which the light propagation is confined to the x–y plane, that is, perpendicular to the cylinder direction. Joannopoulos et al. [2] have shown that for propagation out of the plane, k_z not equal to zero, the gaps tend to close. We will see in a later section that gaps can open up in the z-direction for certain structures.

9.3. STRUCTURES

We have shown some simple photonic crystal structures in the previous section. Although they provide some practical properties themselves and can be used to demonstrate certain unique properties, it will be the more complicated 3D structures that will likely be important. We will briefly discuss these prototype structures so that the reader will have a better appreciation of the fabrication challenges.

9.3.1. Two-Dimensional

What are meant here are structures that have a two-dimensional periodicity, but exist in three dimensions. The structures shown in Fig. 9.4 are two of such structures. One can imagine other arrangements of holes or posts, but the energy diagrams are not expected to be significantly different. The truncated version of these idealized structures is what one can really produce. In other words, the posts or holes will not extend to infinity, but will be limited by the fabrication process. As we will see, the fabrication of some of these structures is well suited to conventional high-resolution e-beam lithography. Others will require new and unique fabrication techniques.

9.3.2. Three-Dimensional

The extension of periodic dielectric structures into three dimensions is not hard to imagine, and actually the number of possibilities grows. As Joannopoulus et al. [2] have pointed out, just take any three-dimensional lattice of points and place a dielectric sphere at the points. A hexagonal close-packed arrangement of spheres would be a simple example of this. One can also use the lattice to construct a photonic crystal by imagining dielectric cylinders connecting the points, as shown in Fig. 9.7. They do not necessarily have energy gaps of consequence. The Brillouin zone is now in three dimensions and the full gap criterion takes on a more demanding role. Nonetheless, such structures have been proposed. One must also consider the possibility of fabrication. There are some structures that can be made, at least in principle, and the reader is referred elsewhere ([8], and the references contained therein) for a more complete discussion.

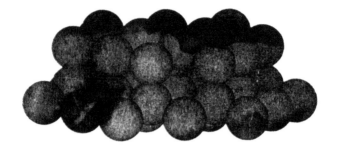

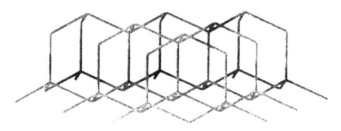

Figure 9.7 Representation of 3D structures arising from point symmetry arrangements: in the upper figure one puts a sphere where the point would be and in lower one runs tubes from the points. (From Ref. [2].)

We will primarily be dealing with the 2D structures and fabrication methods in this book. This is simply because the fabrication of the three-dimensional structure is much more complicated and is not yet to a point at which one can write any synopsis or comparison of them.

9.3.3. Magnetic Photonic Crystals

We saw before that one can write the Maxwell operator in the following way, analogous to the Hamiltonian in quantum mechanics:

$$\Theta = \nabla x \left(\frac{1}{\varepsilon(r)} \nabla \right) x \qquad (9.9)$$

The tensor way of writing this would be the following:

$$\Theta_q = \delta_{qml} \nabla_m [\varepsilon_{lk} \delta_{kij} \nabla_i H_j] \tag{9.10}$$

where $\delta_{kij} = 1$ for 123 and any single permutation, $= -1$ for any two permutations, and $= 0$, otherwise.

One can easily show that if the dielectric tensor "ε" is of the form εI, where I is the identity tensor then the following is true:

$$\omega(k) = \omega(-k) \tag{9.11}$$

Clearly, this is not the most general case. Consider the situation of an otherwise isotropic medium in a static magnetic field. One can now define the dielectric tensor from the following [25]:

$$D = \varepsilon E + i\gamma BxE \tag{9.12}$$

The γ for an isotropic material is a scalar, but in the more general case, it is a tensor whose form depends on the symmetry of the material.

If one expands Eq. (9.12) for the case of a static magnetic field in the z-direction, one has the following result:

$$D = \begin{pmatrix} \varepsilon & -i\gamma B_z & 0 \\ i\gamma B_z & \varepsilon & 0 \\ 0 & 0 & \varepsilon \end{pmatrix} \begin{pmatrix} E_x \\ E_y \\ E_z \end{pmatrix} \tag{9.13}$$

In the case of no external magnetic field, one has the symmetric dielectric tensor, but the general condition that is required is that it be Hermitian

$$\varepsilon_{ij} = [\varepsilon_{ji}^*] \quad \text{but } \varepsilon_{ij} \neq \varepsilon_{ij}^* \tag{9.14}$$

The expression for the dielectric tensor in Eq. (9.14) is now the one we will use in Eq. (9.9) or (9.10). One finds now that although Θ is still Hermitian (conjugate transpose is equal) nonetheless $\Theta \neq \Theta^*$. The latter is the condition that determines time reversal property. This means that Eq. (9.11) is no longer necessarily true. It was true in the special case where the dielectric tensor was real and symmetric. Consequently in the situation described above where the external magnetic produces a dielectric tensor as in Eq. (9.12) leaves open the possibility that in certain cases (symmetries) time reversal is indeed not holding.

To consider time reversal in the more general case, and in particular the consequence it may have on spectral asymmetry, one must consider its properties in more detail. It is clear that for the anti-symmetric $\omega(k) = -\omega(-k)$ condition to hold then both time reversal and spatial inversion must not be elements of the symmetry group of the crystal. The situation has been considered for the case of the degeneracy of energy bands in magnetically ordered crystals. Time reversal, in addition to $t \rightarrow -t$, has the effect of reversing the direction of the spin. For crystals exhibiting magnetic ordering (either internal or through an external magnetic field), the symmetry classification has to be expanded to take this condition into account. The fact that the time invariance operator is non-unitary does not allow for a simple representation, as exists for the spatial symmetry operations.

For the higher dimensional structures, one must satisfy the requirement that there is no element of the group (group "$G_k(R)$" is made up of the symmetry operations, R_i of the particular point in the Brillouin zone) which takes k to $-k$. This condition must be true for any arbitrary point of the BZ. The next condition that should be met to insure a strong asymmetry follows from Ref. [26] which states that there is some direction of the wave vector where there is a symmetry operation R such that the following is true:

$$Rk \rightarrow k \quad \text{for some } R \in G \tag{9.15}$$

Figotin and Vitebsky [27] show some computed examples for a simple 1D structure. This is reproduced here in Fig. 9.8a. Here they use alternate layers of magnetic and non-magnetic materials to create a dielectric reflector. To have spectral asymmetry in a 1D structure requires anisotropy of the material that makes up the non-magnetic layer. It is important to realize that this is not required for higher dimensional structures. In other words, in 2D photonic crystal structures, the materials that make up the structure, both magnetic and non-magnetic can be isotropic.

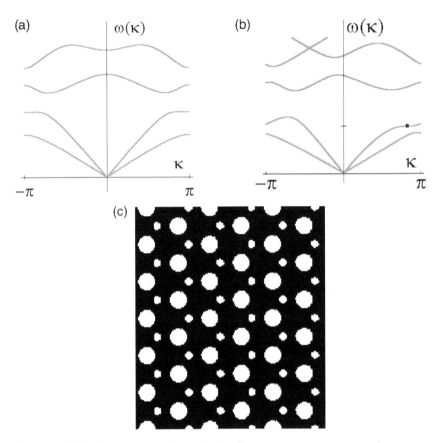

Figure 9.8 Representation of the frequency vs. propagation constant diagrams for (a) a conventional three layer dielectric stack displaying the normal spectral symmetry $\omega(k) = \omega(-k)$ and (b) where one layer is a magnetic, and the other two are anisotropic dielectrics leading to a spectral asymmetric behavior as shown. (Taken from Ref. [27].) (c) Cross-section of photonic crystal structure with point group that would permit spectral asymmetry.

The important practical thing to note here is the possibility of having a structure where the propagation characteristics of a given mode can be radically different in the forward and backward directions by a change in the direction of a magnetic field. In a real way, this is related to the discussion in Sec. 8.5 concerning the Faraday effect. In that case, the characteristic that changed with propagation direction was the sense of the polarization rotation. In the situation here, we are extending the propagation characteristic to the group velocity. In the example shown, it would be possible at some frequency to essentially have a zero group velocity in the $(-k)$ direction, while having a non-zero in the $(+k)$ direction. Even more intriguing is the device idea where one could switch from one to the other group velocity by reversing the direction of an external magnetic field.

The desired property derives from only certain symmetries. The inclusion of time invariance as a symmetry element increases the number of point group symmetry groups from 32 to 122. They break down into three classes:

a. the original 32 applicable to non-magnetic crystals;
b. 32 where T is a symmetry element of the group; and
c. 58 where T is not a member of the group, but is in combination with one or more of the other elements of the group, RT.

In order to achieve the spectral asymmetry condition one needs to choose the 2D structures from the last class since the time inversion operator must not be an explicit member of the point group for spectral asymmetry to exist.

An example is shown by the Brillouin zone for the modified honeycomb structure in Fig. 9.8c. In this case, G represents the group C_3. We have satisfied the requirement that there in no element of group C_3 which takes k to $-k$. This is true for any arbitrary point of the BZ. The next condition that should be met to insure a strong asymmetry states that there is some direction of the wave vector for Eq. (9.15) to be true. We see that the direction K to M satisfies this condition under the threefold rotations.

9.4. PHOTONIC CRYSTAL FIBERS

The unique aspects of the photonic crystal structure that has been demonstrated thus far have been in photonic crystal waveguide fibers. This is in large part due to the existing and available capability to fabricate such periodic structures in fiber form. Similarly it can be stated that the slower progress in the general area of both 2D and 3D photonic crystals has been due to the lack of such readily available and accessible fabrication techniques. For the fibers, there was a technology in place for the fabrication of fiber-optic faceplates that utilized the bundling of capillary tubes in a hexagonal close packed pattern and then re-drawing them to the appropriate dimension. We will show this technique, as well as others, in a subsequent section.

Russell et al. [9–11] were the first to report the results on the propagation in a micro-structured fiber with a periodic dielectric strucure shown in Fig. 9.9. It might appear surprising to see an effect of a periodic dielectric structure in the plane perpendicular to the axis of the fiber on the propagation in the fiber axis direction. In the prior section, we have been concerned with propagation in the plane of the periodic structure as shown in Fig. 9.4. However, as described above, the modes that exist are described by a vector, "k". One can write the expression for the electric field for a given mode propagating in the axial, or z-direction in the following form:

$$\mathbf{E}(r,z) = \mathbf{A}(r,k_\perp) \exp(ik_z z) \tag{9.16}$$

The component of this vector in the axial, or z-direction, k_z, is used to characterize the any given fiber mode. The component of k in the radial direction, $k_\perp$, is the one that samples the periodic structure, and as we will see gives rise to energy gaps, just as we have seen in the in-plane propagation. The momentum conserving condition provides us with the following relationship between these quantities:

$$(nk_0)^2 = k_z^2 + k_\perp^2 \tag{9.17}$$

The refractive index of the medium of the propagation is denoted by n, and k_0 is the free-space propagation constant, $2\pi/\lambda_0$.

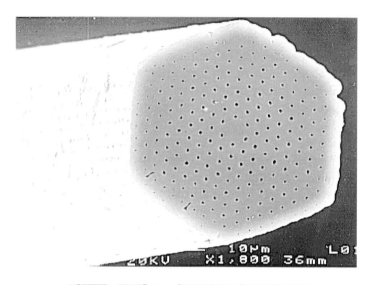

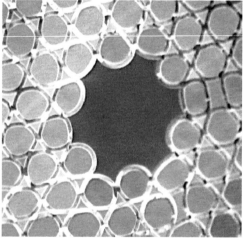

Figure 9.9 Photographs of photonic crystal fibers produced by Russell and his group, upper is a solid core and the bottom is a hollow core.

9.4.1. Generalized Description of Propagation of Light in Fiber

Before we discuss the various kinds of photonic crystal fibers and their properties and applications in subsequent sections, it is useful to first consider a unified way to look at light propagation in a fiber consistent with what we might already be familiar within the broad optical waveguide literature. In other words, in the normal description of waveguide propagation, one hardly ever sees this expressed on a frequency vs. propagation constant diagram as we see used above for the description of light propagation in photonic crystal structures. In will be instructive to have a description where we can see the behavior of the more familiar conventional optical waveguide as well as that of photonic crystal waveguides.

The easiest way to start is to draw a frequency vs. z-propagation constant diagram for an infinite block of glass with a refractive index of n_0 as shown in Fig. 9.10a. (These are projection diagrams in that they include all allowed values of $k_\perp$.) All modes are allowed above the line representing the reciprocal of the refractive index of the glass. Below this boundary line modes are forbidden (shaded area) since on that line $k_z = n k_0$. We can consider the space below this line as a "forbidden gap". Now, let us introduce a rod of glass into this infinite glass whose refractive index is $n_1 > n_0$. We can now draw a dashed line with the slope of the reciprocal of its rod index, to indicate its position in the forbidden gap as shown in Fig. 9.10b. The presence of the rod of higher refractive index glass creates a defect state in the gap. The defect state permits localized modes to exist which would be otherwise forbidden. By being localized we mean they can propagate in the z-direction modes with decaying field in the radial direction. This is nothing more than the description of a conventional step-index optical waveguide. We sketch in Fig. 9.10b what

Figure 9.10 Schematic representations of normalized frequency vs. propagation constant for (a) solid piece of glass of refractive index n_g. Forbidden gap (shaded region) is defined by the line with slope $1/n_g$; (b) rod of glass with a larger refractive, n_c, index is inserted into infinite glass. Dashed line corresponds to this with slope $1/n_{eff}$. All of the conventional step-index guided modes fall between this line and the boundary line. The fundamental mode, is sketched in as the solid line.

(a)

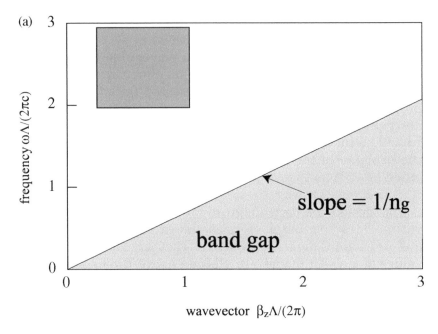

(b)

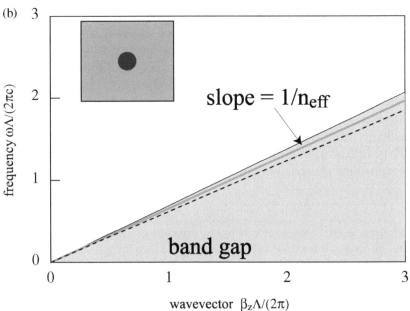

the dispersion of these modes would look like on this diagram. As expected, for the lowest order mode, at the large value of k_z, the mode is essentially all in the core and its dispersion hugs the line representing the core refractive index. As one goes to smaller k_z, the mode must eventually spread into the matrix glass (cladding) and thus become asymptotic to the line corresponding to the index of the matrix glass. The slope of this curve is the group velocity $d\omega/dk_z$. The derivative of the reciprocal of the group velocity is the waveguide dispersion term. One can see that the zero-dispersion point $(d^2k_z/d\omega^2 = 0)$ will occur in the region of the transition from the light essentially traveling at the velocity of the core index to that of traveling at the velocity of the cladding.

We now look at the situation where into the solid piece of glass, index n_0, we produce a periodic array of holes, refractive index $= 1$, as shown in the inset to Fig. 9.11a. (In actuality, we use "holes" here to mean any material with an index less than the matrix glass.) Two new things appear in the resulting ω vs. k_z diagram. First, we see that energy gaps (shaded areas) now appear, and some above the light line of the glass. (The light line for air divides the diagram into the region above where propagation in air is allowed and the region below where it decays in air.) The second difference is that the limiting index line (often referred to as the fundamental space-filling mode or FSM) is no longer a straight line with a slope of the reciprocal glass index. The first effect occurs as a result of interference phenomena analogous to, but clearly much more complicated than the 1D dielectric stack case. Where the gaps appear and how wide they are is a function of the exact geometric pattern of the holes, and

---

Figure 9.11 Schematic representation of normalized frequency vs. propagation constant for (a) a photonic crystal structure as shown in the inset. Energy gaps now appear above the FSM and vacuum light line. The FSM is now a curved line. (b) Solid-core defect introduced as shown. Defect modes appear below the FSM. (c) Hollow-core defect as shown in inset. Defect line can now intersect a gap that lies above the vacuum light line as shown.

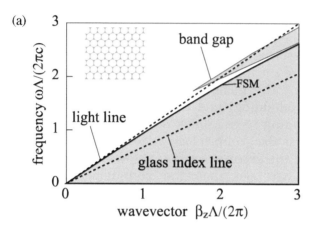

(a)

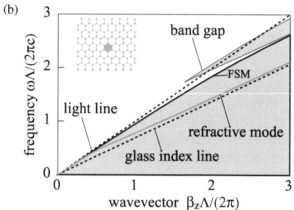

(b)

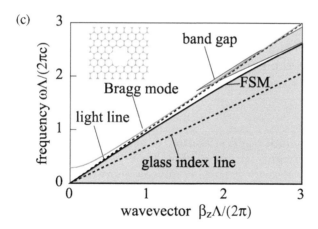

(c)

their geometry, as well as the refractive index difference between the matrix material and that of the holes. For example, the air-fill fraction, that is the volume fraction of the air holes to the total volume, plays a critical and rather sensitive role in producing wider gaps. The reason for this is not really understood.

The non-straight gap boundary mode line, or FSM, is a result of the fact that light can redistribute between the holes and the glass in a continuous manner as a function of wavelength. An easy way to imagine the reason for the curvature of this line which divides the allowed modes from the forbidden ones is to consider the behavior at very small k_z and very large k_z. At large values of k_z, the line has the expected slope corresponding to the $1/n_0$. However, as we go to very small values of k_z, the wavelength is now very large compared to the hole spacing and it essentially sees an average refractive index. One would see a steeper slope depending on the air fraction. The steeper slope gradually merges with the lesser slope giving rise to the curved boundary line. What is important about this feature is that one can control to some extent the curvature of this line by the specific geometry and air fraction.

The next structures to consider are the ones shown in Fig. 9.11b and c. In the first, (b) we remove a hole and create a solid glass region which will act in a similar way to a core in a step-index fiber. In other words, the localized propagating modes will still appear in the forbidden gap below the FSM; however, some interesting situations occur that are unique to the photonic crystal version. This type of photonic crystal fiber is often referred to "solid-core" fiber, or sometimes more simply as a PCF. In Sec. 9.4.2, we will discuss in more detail the most interesting consequence of the solid-core photonic crystal fiber.

In Fig. 9.11c, we have created a larger central hole among the periodic array of holes. This enlarged hole constitutes a defect, that is it represents a break in the perfect periodic structure. If the structure has forbidden gaps above, the FSM, or more importantly, above the vacuum line, and the dispersion line of the defect passes through one of these gaps then a localized mode can propagate. As we will see below, the size of the unique hole will determine whether its dispersion will intercept one or more of the gaps.

It is of interest to note that there is not any meaningful distinction to be made between the guiding mechanism for localized modes propagating above or below the vacuum line, or in any gap. Often people refer to the conventional wave guiding mechanism as due to total internal reflection. This ray-based terminology is not appropriate for structures where the dimensions are comparable to the wavelength of light. What is really being referred to is that the refractive index of the core is larger than the index of the surrounding cladding. But, this condition is always met for any guided mode. This can be seen from the momentum matching conditions expressed in Eq. (9.17) for the core and the cladding. For a localized mode to propagate, one must have $k_{\perp}(\text{cl})$ to be imaginary, thus by subtracting the momentum matching equation for the cladding from that of the core, the quantity $n_c{}^2 - n_{cl}{}^2$ must always be > 0. The solution of Maxwell's equation for a given structure produces the allowed modes, some of which will be localized. The mechanism is always the same in all cases; simply put it involves the destructive interference of the modes in the outgoing radial direction leading ultimately to the localization in the core. We may see reference to "diffractive" modes, but in the same way, this is a needless specification.

9.4.2. Solid-Core Photonic Crystal Fibers

9.4.2.1. Endlessly Single Mode Behavior

The unique aspect of the solid core PCF is the ability to control and manipulate "waveguide" dispersion. The waveguide dispersion of an ordinary step-index optical fiber can be understood from the consideration of the effective index of the mode. We can express this by the following limiting condition for the allowable values of the effective index:

$$n\,(\text{cladding}) < n_{\text{eff}} < n\,(\text{core}) \tag{9.18}$$

The wavelength dependence of this mode index varies with wavelength through the way the light distributes itself in the two media as a function of the wavelength. In other words, one can think of the effective mode refractive index as some weighted value between the core and cladding index, and the weighting factor is dependent on the actual light distribution between the two.

In a micro-structured fiber or PCF, there is another parameter that enters in and that is how much of the light in the cladding itself is distributed in the glass webbing relative in the air holes. As mentioned above, this distribution between air holes and webbing makes the boundary line between allowed modes and forbidden modes, curved not straight. Furthermore, the degree of curvature is a function of the ratio of the hole size to the pitch, in addition to the overall geometric pattern of holes. In general, then one has control of the waveguide dispersion through geometric parameters. The simplest way to think of this is that now the effective cladding refractive index depends on the way the light is distributed between the air holes and the glass webbing at any given wavelength.

This behavior leads to a very interesting feature of the "micro-structured fibers" with a solid core. This feature is called "endlessly single mode" behavior [9,10,27]. To understand this behavior, the reader is referred to Fig. 9.12. In (a), we show a schematic diagram of the defect mode structure of a conventional step-index fiber. It is single mode only below the the dashed vertical line. From the momentum conservation equation (9.17), one can write the condition for single mode as the following:

$$\frac{2\pi}{\lambda}r\sqrt{(n_c^2 - n_{cl}^2)} < 2.4 \qquad (9.19)$$

Here, we have assumed that the effective index is essentially equal to the cladding index. In the solid-core micro-structured fiber, the cladding index in the LHS of the above equation is now a function of the wavelength

$$\frac{2\pi}{\lambda}r\sqrt{[n_c^2 - n_{eff}(\lambda)]} \qquad (9.20)$$

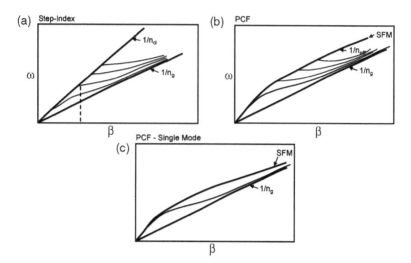

Figure 9.12 Schematic representation of how endlessly single mode behavior occurs. (a) The situation in a conventional step-index waveguide. The dashed line indicates the point in which the structure supports just one mode; (b) photonic crystal in the multimoded condition; (c) optimum design of periodic structure to trap the single mode excluding all the other higher modes out of the gap.

This is not to be confused with material dispersion. Rather it is what is usually termed as "waveguide dispersion" since it derives from the relative distribution of the light in the media in which the light is traveling. It appears in a step-index fiber from the distribution of the light between the core and cladding as a function of wavelength. In the micro-structured fiber, an additional distribution factor comes into play in the way the light distributes itself within the air holes and glass that make up the "cladding" region. How this is manifest on the ω vs. k diagram is in the bending of the lowest line dividing the allowed modes from the forbidden modes. This is shown in Fig. 9.12b. In the step-index fiber, this line is straight as shown in Fig. 9.12a. (This ignores material dispersion.) The bending in the micro-structured fiber allows one by appropriate choice of the geometry of the holes to essentially "trap" the fundamental mode between the defect line defined by the index of the core, and the boundary line so that the mode can propagate as a single mode at all wavelengths.

We show this schematically in Fig. 9.12C. The computed modal behavior as a function of the geometric parameters of the hexagonal structure depicted diagram is shown in Fig. 9.13a. The computed modal profile of a micro-structured fiber displaying the endlessly single mode property, comparing the profile at long and short wavelength, is shown in Fig. 9.13b. One can see clearly how the light in the cladding is redistributing as a function of wavelength. One distinct advantage of this behavior is that one can obtain single mode performance with a large effective area fiber.

9.4.2.2. Dispersion

A related consequence of the ability to produce large waveguide dispersion in solid-core micro-structured fibers is the ability to position the zero-dispersion point to wavelengths as short as 800 nm through structural changes. An example of this is the triangular lattice geometry of air holes in silica, where the change in the hole to pitch ratio is used to position the wavelength of the zero-dispersion point. This is shown in Fig. 9.14. In the upper graph, the hole diameter to pitch ratio is 0.3, and the dispersion parameter, "D", is plotted against wavelength for the values of the pitch that are listed. In the lower curve, the same is done for a hole to pitch ratio of 0.4. In this case, the index of the glass is 1.5 and the holes are air. One can see that one can move the zero-dispersion point from roughly 700 nm out to 1700 nm for the case of silica with air holes.

One can imagine how this arises by considering the effect that the particular geometry has on the nature of the curvature of the line defining the gap as was discussed above. The zero-point of the dispersion comes from the inflection point of the defect dispersion (the lines drawn between the gap line and the line representing the core). It is where the mode makes its transition between being tightly bound to the core to that of extending into the cladding. We shall discuss the applications of this unique feature later on in the section on Applications.

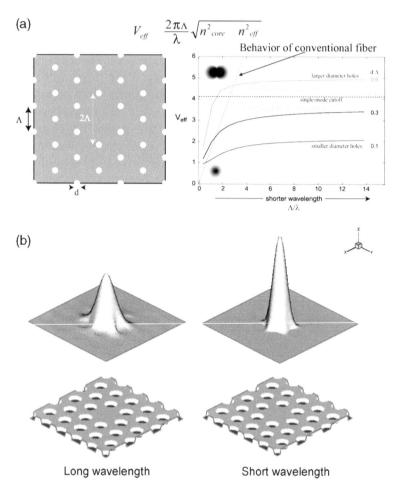

Figure 9.13 (a) Calculated modal behavior as a function of normalized wavelength with the diameter of the holes to the pitch in a triangular structure as shown. Comparison to conventional step-index fiber is shown by the dotted line. (b) Calculated modal profile in the endlessly single mode condition comparing the long wavelength intensity distribution to the short wavelength.

9.4.3. Hollow-Core Photonic Crystal Fibers

In the above discussion, we have described photonic crystal waveguides that guide light in the lowest gap. In this regime, the gap in which the defect mode is propagating is below the

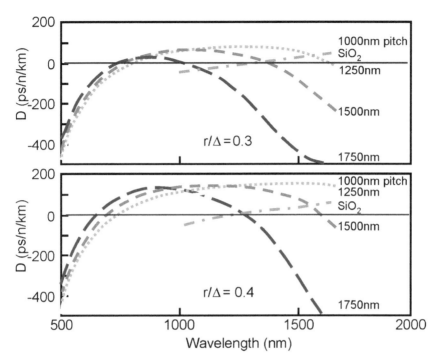

Figure 9.14 Calculated dispersion parameter D vs. wavelength for triangular lattice as a function of pitch. This is done for two different hole sizes. The structure is air/holes in silica. The typical step-index silica waveguide dispersion is also shown for a comparison.

space-filling mode as shown in Fig. 9.11B. Now we are considering modes that are in gaps above this line and in particular above the vacuum line as shown in Fig. 9.11C.

The simplest example of this guiding structure is the Bragg configuration shown in Fig. 9.15a. This structure was first examined by Yeh and Yariv [28]. The simplest way to look at this as if it were a dielectric stack wrapped around to a cylinder. There is another structure that is somewhat less intuitive that also is capable of producing propagating modes above the vacuum line and that is the one shown in Fig. 9.15b [14]. It is an array of holes on a triangular lattice with a larger hole cut out to serve as the core. To make this all

(a)

(b)

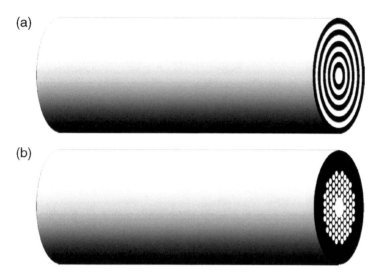

Figure 9.15 (a) Schematic drawing of a hollow-core Bragg waveguide. (b) Schematic drawing of a hollow-core waveguide in a triangular lattice.

understandable in a more quantitative way, one must appeal to the computation codes that are available. (For example, see Appendix A.) What is done is to solve Maxwell's equations in the form of Eqs (9.4) with the particular periodicity of the dielectric structure represented by the function $\varepsilon(\mathbf{r})$. One projects all the allowed solutions onto a "ω vs. k_z diagram" (These are the diagrams we schematically represented above in Fig. 9.10 and 9.11.) What is important is the appearance of energy *gaps above the vacuum light line*. As mentioned above, the so-called "light lines" are lines with slope $1/n$. They divide the diagram into an upper portion where modes with effective index of less than refractive index "n" can propagate, and the lower where they decay exponentially, and therefore do not propagate. When we consider the light line with slope of unity, we are speaking about the modes that can propagate in vacuum, or air. An actual experimental result for light propagating in a triangular structure of holes in silica with a hollow core is shown in Fig. 9.16.

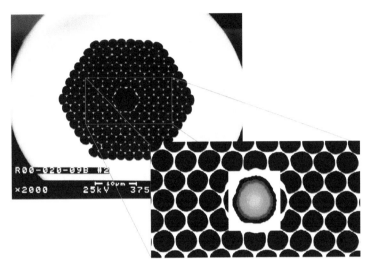

Figure 9.16 Actual SEM picture of a hollow-core waveguide with an inset of a photograph of light propagating in the core.

The way one can produce a localized mode above the vacuum light line is to produce some sort of defect, or designed disruption in the periodic structure whose dispersion intersects the forbidden gap. The easiest realization of this is to enlarge one hole of the array. One can approximate the dispersion of the defect by using the dispersion equation of the perfectly reflecting cylinder [29]. By perfectly reflecting cylinder is meant the solution for the propagation in a cylinder of radius "r_d" where the field is made to vanish at the boundary

$$\left(\frac{\omega}{c}\right)^2 = \beta^2 + \left(\frac{\phi_{nm}}{r_d}\right)^2 \tag{9.21}$$

Here r_d is the radius of the circular defect, and ϕ_{nm} are the eigenvalues of the mth zero of the Bessel function of order n for the TM modes. (For the TE modes, it is the derivative of the Bessel function.) The first TM mode would correspond to $\phi_{01} = 2.4$ in Eq. (9.21). Knowing the frequency position of the lowest order gap, one calculates the approximate radius of the defect in order to have it intercept the defect dispersion

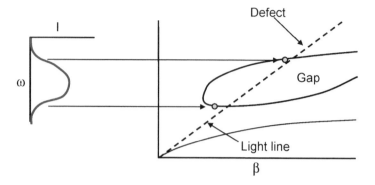

$$\text{Defect dispersion} \approx \left(\frac{\omega}{c}\right)^2 = \beta^2 + \left(\frac{\phi_n}{r_d}\right)^2$$

Figure 9.17 Schematic representation of pass band of a hollow-core waveguide. The defect line is governed by the expression given.

line. This is what is schematically drawn in Fig. 9.17 to show how it determines the effective bandwidth of the localized mode. Calculations of the modal pattern of the first two modes of the structure shown are depicted in Fig. 9.18.

The confinement of the mode is determined by the number of layers surrounding the hollow core as well as the contrast in refractive index [30]. This is similar to the concept that one uses to determine the reflectivity of a dielectric stack. Often the word tunneling is used to describe the extent of radial flow of energy permitted by a given structure. This

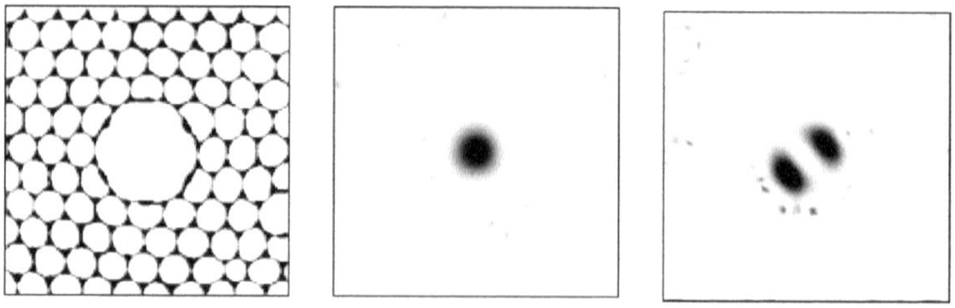

Figure 9.18 Computed modes for real structure to SEM bitmap was used as input to the code for the computation.

analogy to a dielectric stack is less obvious in the case of structures made up of a triangular pattern of holes. This just points out that the actual destructive interference pattern that is set up in the propagation of the modes in this structure is a truly complicated 3D problem capable of being dealt with only through the use of Maxwell's equations numerical mode solvers.

9.5. FABRICATION TECHNIQUES

We will briefly review some of the more important fabrication methods that have been used thus far in the construction of photonic crystal structures. We will cover the simpler 2D methods as well as the photonic crystal fiber. For the 3D methods, the reader is referred to Ref. [31].

9.5.1. E-Beam Lithography

The major additional problem in the use of the standard state-of-the-art e-beam lithographic method to create a simple two-dimensional photonic crystal pattern is the depth. The required separation distances are of the order of, $a/\lambda = 0.4 - 0.5$ as indicated from the energy vs. k diagrams of the simple structure. If one is considering telecommunication devices, this translates into separations of 600 nm, which present no real resolution problem. However, the stipulation of how deep is more difficult to answer. As we pointed out in the earlier sections, the 2D structure, which is periodic structure in the x–y plane, is assumed to be infinite in the z direction. The measure of how deep in the z-direction, as well as the related issue of how extended in the x–y direction, must come from some estimate of how fast the field decays. As we will see later on in the application, the devices will all be based on creating a defect state thereby localizing the light to a given region. It will be the field decay in this arrangement that will be the measure of the required dimensions.

In any event, one might anticipate a dimension of something close to 10 periods, which would translate into the order of 5 μm if the wavelength regimen were 1500 nm. This is quite deep for conventional resist. In other words, what is important is the etching contrast between the resist and the substrate. Thus, it will also depend on what type of material one is trying to use for the pattern.

An example of a technique that can be used to enhance the contrast of patterning in silicon is to use the resist to pattern a thermal oxide grown on top of the wafer [14]. The oxide material is then used as the resist for the subsequent deep reactive ion-etching step.

There is a novel process involving the use of an electron beam deposition in a scanning electron microscope. Molecules are injected in a gaseous phase and aggregated through the action of the electronic beam. As a result small posts of the organic composite can be precisely positioned [17].

It is too early to tell how the less than "perfect" structures will correspond in performance to the idealized mathematical structures. Issues such as roughness of the holes, radius variation, perpendicularity, and other natural consequences of the fabrication process are likely to influence the actual propagation, but to what extent has yet to be determined.

9.5.2. Nanochannel Arrays

An interesting approach to making a two-dimensional array of channels is to use a precision-patterned texturing treatment of the surface of polished Al [30]. This is done just before an anodization step. The molded depressions serve as initiation points and guide the growth of the channels in the oxide film. The initial texturing was accomplished by a special molding process. The process is schematically shown in Fig. 9.19a–e. Note that the last step removes the Al, and one obtains holes that go all the way through. Scanning electron microscopic (SEM) pictures of some of the patterns that were produced by the method are shown in Fig. 9.19. The authors claim that the method can be applied to other materials, such as semiconductors, by using a two-step process.

9.5.3. Other Methods

There have been a number of novel methods that have been applied to the fabrication of periodic 2D arrays. One involves

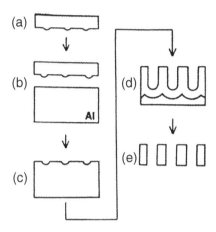

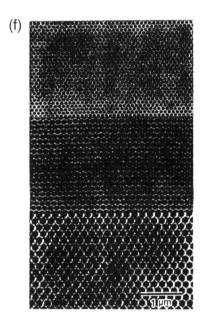

Figure 9.19 Texturizing treatment of Al to produce nanochannel patterns. (a) Mold used to pattern Al; (c) on anodization the textured regions etch differentially; (b,d) the channels; (e) when the Al is removed the holes go all the way through; (f) The lower panels are SEM micrographs of the hole patterns that could be produced. (From Ref. [18].)

the use of unique exposure schemes applied to photopolymeric materials. An example is shown in Fig. 9.20. Here the interference of multiple beams is used to produce a hexagonal 3D pattern as shown in a photopolymerizable resin [32]. There are related techniques using two-photon photochemistry [33,34]. The essential advantage here is to be able to

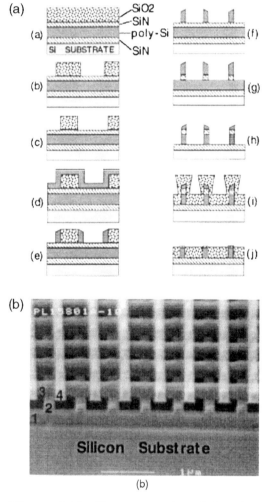

Figure 9.20 Representation of photolithographic steps to produce a "pile of logs" photonic crystal structure. (Taken from Ref. [32].)

expose a region in a 3D structure. Another method is to utilize self-assembly of colloidal particles. A schematic is shown in Fig. 9.21 as well as an SEM photo of an array of polystyrene spheres [35].

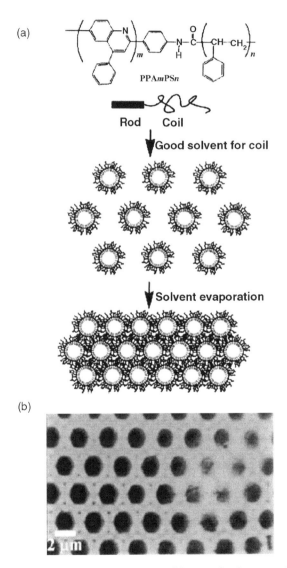

Figure 9.21   Self-assembly method to produce a photonic crystal structure for colloidal particles. (Taken from Ref. [35]).

9.5.4. Photonic Crystal Fiber Fabrication

There are available approaches to the fabrication of fibers that have a periodic structure of refractive index in the direction normal to the axial direction. The advantage of this structure is that it is compatible with existing optical fiber technology. We will cover the major methods in the making of photonic crystal fibers, both solid and hollow core.

9.5.4.1. Capillary "Stack-and-Draw"

The process starts with a tube with a desired ratio of wall thickness to inside diameter [36]. The hole may either be drilled to establish the correct radius to wall thickness, or may be formed by the process by which the tube was made. Schematically the process is shown in Fig. 9.22. Here a hole is drilled into a rod, which is then finished to a hexagon to allow the stacking. The rods are usually redrawn to reduce their size and then stacked as shown to the close-packed

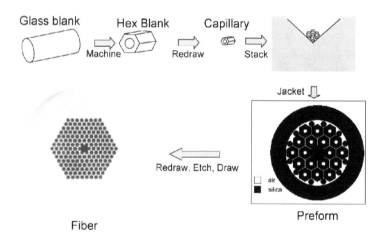

Figure 9.22 Schematic drawing of the "stack-and-draw" technique to make a photonic crystal fiber. Glass is formed into a hexagonal-shaped tube. Tube is redrawn to intermediate size and sectioned into smaller pieces and then stacked as shown. A jacket is placed around the assembly and then this perform is redrawn to fiber.

structure. The bundled structure is then redrawn until the desired pitch is achieved. This may be done in more than one step. The intermediate pitch is more than likely the order of 50 μm. The final redraw takes the preform to the fiber form with the desired range of 1–5 μm. For the solid-core version, the defect is introduced by replacing the center element of a bundle with a fiber that did not contain a hole. For the hollow-core fiber, one or more rows of fibers are removed from the center to produce the hollow core.

There are a number of critical process features. One must maintain the geometric integrity of the structure, that is, all the dimensions and shapes. This is especially true around the center core region. All these dimensional variations have an effect on the position of the gaps and their widths.

9.5.4.2. Bragg-type Waveguides

This fabrication of this fiber structure has roots in the proposal of a structure that guides light by diffraction put forth by Yeh and Yariv [28]. The present form is shown in Fig. 9.15a, and is sometimes referred to as a Bragg fiber. In the original structure, the contrast was quite low consistent with the ability at the time to produce rings with high contrast. The MIT group [37] has pioneered a method of construction which essentially rolls a high contrast layered structure into the desired cylindrical form. The refractive index of the layers is of the order of 1.5 and 3. The thickness of the layers is also designed to meet a kind of broadband quarter wave condition. They design the radius of the center hole to be consistent with the propagation of the low loss TE_{01} mode. It appears that in this type of Bragg structure the tunneling loss (confinement due to destructive interference) has a strong dependence on the particular mode. The reader is directed to the detailed analysis in the literature [28a]. The single-mode behavior of such guides is a function of mode stripping due to loss. In other words, the structure is multimoded but only one mode has sufficiently low loss to propagate over any long distance. This behavior is in contrast to the diffractive waveguide fibers that are based on a periodic array of holes where the lowest order mode seems to propagate with similar loss.

9.5.4.3. Multiclad

There is no reason that the index of the included phase of the array has to be 1; in other words, there need not necessarily be holes. In terms of strength and structural integrity, it would be much easier if the included phase were another glass. Clearly, this presents a lower contrast case, but as far as the extended single-mode behavior indicated by Fig. 9.11B, it still persists. The effect of the lower refractive index contrast is to move the flattened portion of the curve to larger values of Λ/λ and d/λ. There are methods to make optical fiber couplers by a technique in which multiple cores within a common cladding tube are redrawn to separation distances that are sufficiently close to that of the photonic crystal structures. This process can be extended to make arrays of cores within the same cladding host.

For the reader not familiar with the multicore process, one is referred to Fig. 9.23a,b in which the process is shown for a four-core structure [20,21]. There are four preforms shown. These would be made by any of the waveguide processes such as MCVD. For the photonic crystal, they are rods with a core of a different refractive index. The desired spacing would be provided by the starting core to rod diameters. The assembly shown in Fig. 9.23a is collapsed under vacuum so that all the gaps disappear, ultimately yielding a structure shown in Fig. 9.23b. Bundling these together using a square rod would produce the repeating extended structure. One should be able to see how this process can be extended to other geometries.

Lucent has used a variety of micro-structured designs like the one shown in Fig. 9.23c that reduce environmental sensitivity, increase acceptance angle for fiber laser applications, and capitalize on unique dispersive properties.

9.5.4.4. Extrusion

This is a method that involves the extrusion of a powder in a binder through a metal die. The die geometry contains the

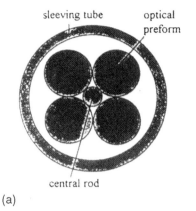

(a)

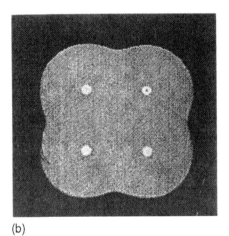

(b)

Figure 9.23 Representation of the multicore process: (a) The upper curve shows the initial structure for a four-core device. (b) The lower shows the result after drawing down to fiber size. (c) Cross-section of a solid-core photonic crystal fiber with microstructured cladding.

desired spacing and shape characteristics. One can then heat the extruded body to drive off the binder and further heat it at a higher temperature to consolidate the body to theoretical density. This process was originally developed to produce ceramic honeycomb structures [22], but it can be applied to such materials as glass or silica. Depending on the specific material, the preform can be redrawn to reduce its size.

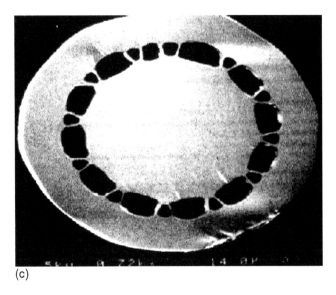

(c)

Figure 23 Continued

Bundling and further redrawing would be required to reduce it to the 1- to 10-µm range. Example of extrusion process is shown in Fig. 9.24a. An example of a photonic crystal perform made by extrusion is shown in Fig. 9.24b.

9.6. APPLICATIONS

It is a bit premature to talk about actual applications of photonic crystal structures because few, if any, devices have actually been fabricated for any kind of real evaluation. However, there have been a number of ideas put forth based on the mathematically modeled behavior, and these have been interesting enough to warrant the discussion that follows.

9.6.1. Filters

One can consider the use of the photonic crystal structure as being manifest as narrow-band multilayer optical filters. It employs the defect structure discussed in the foregoing, one period missing, or skipped in the dielectric stack, to create

(a)

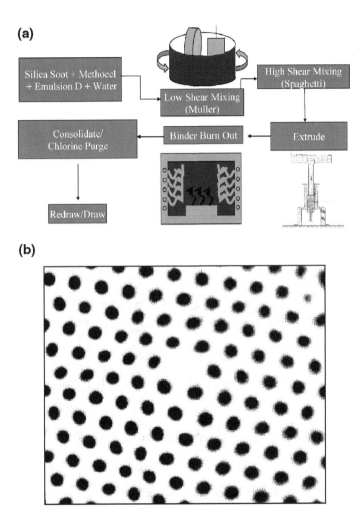

(b)

Figure 9.24 (a) Schematic of the extrusion process, (b) Photograph of a photonic crystal perform made by extrusion.

the allowable mode. This is commonly referred to as a Fabry–Perot filter. One can easily understand the name because it is the conventional Faby–Perot etalon structure with the mirrors stemming from the multilayer stacks. The behavior is shown in Fig. 9.25 where we have reproduced the reflection spectrum of the 15- and 21-layer stack of Fig. 9.3 with the defect (skip one-layer sequence). The spectra are shown in

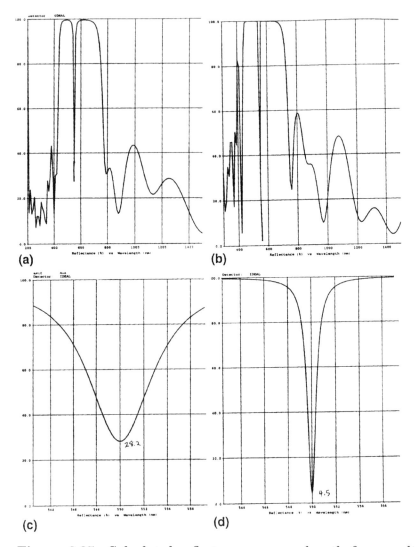

Figure 9.25 Calculated reflectance vs. wavelength for a multi-layer dielectric stack where one layer is skipped to produce the defect, a Fabry–Perot filter: (a) 15 layers and (b) 21 layers. The lower two traces are a blowup of the cases showing how the width and depth of the transmission dip depend on the number of layers.

Fig. 9.25c and d. The transmission spike is a consequence of the allowed propagation through the localized state in the gap. The position of the transmission maximum, the width of the transmission peak $\Delta\lambda$, and the peak transmission are functions of the mirror reflectivity, hence, of the number of layers, and index contrast and thickness of the alternating layers that make up the stack.

9.6.2. Microcavities

In two dimensions, one can have a similar behavior. Here, the defect is referred to as a *microcavity* [23] because the defect is surrounded by the undisturbed periodic structure. Villeneuve et al. [23] show the computed spectral behavior for a simple dielectric structure, as shown in Fig. 9.26a. They compute the output transmittance spectrum in Fig. 9.26c at a single point, labeled detector in the drawing, for a polarized input pulse, the spectral width of which is shown in Fig. 9.26b. An important performance property, namely the quality factor $\omega/\Lambda\omega$ was calculated as a function of the size of the crystal, as is shown in Fig. 9.27. It is also important from another point of view: the physical size required for the device. The typical pitch is roughly less than a micrometer, so the total dimension of the device in the x–y plane is less than 10 µm. The rods here are assumed infinite for the calculation. If one assumes that the performance as stated will not suffer by truncation to the same level as in the x–y plane, then the entire filter is a 10-µm cube. This issue notwithstanding, one could imagine a spatial array of such elements, each tuned to a different wavelength by the particular choice between the rod diameter and the spacing.

Foersi et al. [38] have demonstrated a microcavity in a waveguide structure. The structure consisted of spaced holes

Figure 9.26 Two-dimensional photonic crystal filter arrangement: (a) schematic of the measurement; (b) input pulse; (c) measured transmission of structures. SEM photo of an actual waveguide structure made in silicon. (From Ref. [38]).

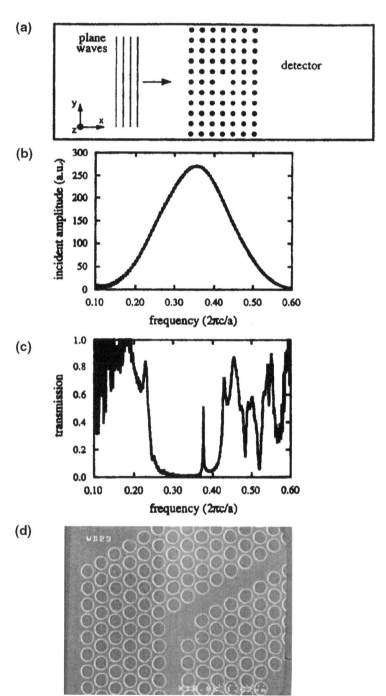

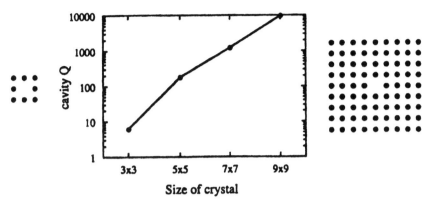

Figure 9.27 Quality factor $\omega/\Delta\omega$ vs. number of elements in the array for the structure shown. (From Ref. [38].)

in a ridge silicon waveguide. They reported measurements of a microcavity resonance in a PBG structure integrated directly into a sub-micron-scale waveguide of 1.56 μm with a quality factor of 265. This is an analog of a distributed feedback device.

9.6.3. Waveguides

If one extends the defect along a whole line, rather than just a point, one can then have a structure that guides light by diffraction. For example, consider when light is forbidden to propagate in the energy region, as indicated in Fig. 9.28a for the TM modes of a square array of dielectric rods. By removing a row of the rods, there now appears allowable states within the forbidden gap, as indicated by the solid line [2]. For this particular structure, the simple removal of a single row is sufficient to produce a single-mode condition. The propagation around a bend is simulated in Fig. 9.28b [8]. This is essentially an illustrative example, rather than a representation of an actual structure. It is after all a two-dimensional structure. Nonetheless, it does indicate the possibility of building a true three-dimensional waveguiding structure.

One might question the advantage of guiding light by diffraction, rather than refraction. There certainly is no lack

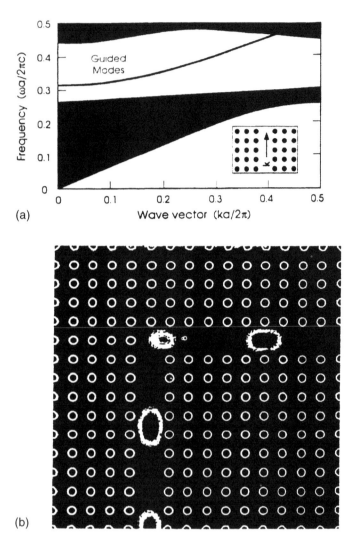

(a)

(b)

Figure 9.28 Representation of the dispersion of the defect within the forbidden gap of the 2D structure shown. The lower figure shows the time domain simulation of the propagation around a right-angle bend. (From Ref. [8].)

of technology available to construct extensive planar-waveguiding structures, and this has been amply demonstrated [24]. One could argue that the photonic crystal offers possibly smaller structures. However, small is only as good as it allows efficient light coupling in and out. Once one arrives at submicron dimensions, this becomes the major problem. However, if the photonic crystal structure were to be totally integrated, source pathways, detectors, all fabricated in the same volume, then the overall size would be an advantage, and coupling once in and once out, even if it were inefficient would not be a severe drawback. If GaAs were the material, this would be within the realm of possibility.

Kuchinsky et al. [39a] have produced 3D localization in a channel waveguide of a photonic crystal with 2D periodicity. The structure is schematically shown in the inset to Fig. 9.29. The localization in the plane of the structure is produced through diffraction, while confinement in the *z*-direction (thickness direction) is through the conventional

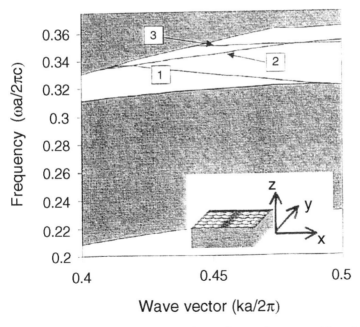

Figure 9.29 Band diagram for a channel waveguide with a structure shown in the inset. Guidance is considered refractive in the plane and refractive in the *z*-direction. The labels refer to the primary guiding nature, 1 refractive-like, and 2,3 diffractive-like. (Taken from Ref. [39a,b]).

silicon–air interface. Three-dimensional band structure calculations showed that with an appropriate thicknes, one could support a small number of modes. These modes were shown to be of the two kinds indicated, diffractive and refractive. The dispersion curves of these modes are labeled by the numbers 1, 2 and 3 in Fig. 9.29. From the shape of the curves, it was concluded that curve 1 corresponds to a refractive mode, whereas curves 2 and 3 are diffractive modes because their group velocity approached 0 at the BZ boundary.

Temelkuran et al. [39b] have reported a low loss structure 1D stack based on the idea of omnidirectional reflectivity for broadband propagation in the near IR. The dielectric properties of the layer materials (like polyethylene for the low index layer), and the high index layer which in this case was Te are much better suited to the longer wavelength IR.

The realization of true 3D structures is still a fabrication issue, namely the ability to produce periodic structures in all three dimensions. One example of this is shown in Fig. 9.30. What makes the problem even more difficult is the lack of ability to produce specific designed defects within the structure.

Figure 9.30 Example of a 3D photonic crystal structure. (Taken from [31].)

9.7. OPTICAL FIBERS

One area of photonic crystals that has progressed quite rapidly into possible applications is that of photonic crystal fibers. As we have discussed above, these are essentially 2D periodic structures drawn into optical fiber dimensions in the z-direction. The two categories, solid core and hollow, have distinctly different application areas. In the former, it is essentially their unique dispersion characteristics that drive the applications, whereas for the hollow core, it is the fact one can propagate in essentially vacuum that drives its applications. In the following, we will briefly review some of the important applications of both types of photonic crystal fibers.

9.7.1. Solid-Core Fiber Applications

We had shown in Fig. 9.14, the dispersion parameter "D" as a function of wavelength of a solid-core fiber for two ratios of the hole radius to the hole separation distance (pitch) in a silica/air structure. For our purposes, one can approximately write the GVD (group velocity dispersion) as the sum of the material dispersion contribution and the waveguide contribution

$$D \, (\text{ps/km} - \text{nm}) = \frac{-\lambda}{c} \frac{d^2 (n_{\text{m}} + n_{\text{eff}})}{d\lambda^2} \qquad (9.22)$$

Here, n_{eff} is defined through the relationship as $\beta = n_{\text{eff}} k_0 = (\omega/c_0) n_{\text{eff}}$. This ability to control and tailor the dispersion of a guiding structure has two major application areas. The first is that of optical fiber dispersion compensation devices. The normal spread of the signal pulses in a communication fiber through dispersion eventually has to be corrected otherwise the pulses overlap and bit information is lost. The dispersion compensation fiber module consists of an optical fiber with the opposite dispersion. For device compactness, one would do this with a fiber with a very large opposite dispersion so that the compensating fiber can be relatively short. The solid-core photonic crystal fiber affords this range of dispersion values by the appropriate design.

The second area is in the ability to fabricate a single mode fiber with a high optical nonlinear coefficient, the most important of which is being able to initiate the nonlinear response with low pump power. In most applications, the efficient utilization of the nonlinearity requires that one operates in or near the zero-dispersion wavelength regime. For most highly nonlinear glasses, this wavelength is in the wavelength region $>2\,\mu m$. The following are some examples of how by moving the zero-dispersion wavelength to the more practical wavelength region of the near IR through the use of the photonic crystal approach.

9.7.1.1. Soliton Fiber

Propagation of soliton pulses in silica-based optical waveguide fibers is well established. The phenomenon results from the balancing of the broadening of the pulse through dispersion with the narrowing of the pulse produced by self-phase modulation. The propagation is governed by the so-called nonlinear Schrodinger equation [40]

$$i\frac{\partial A}{\partial t} = \frac{\beta_2}{2}\frac{\partial^2 A}{\partial T^2} - \gamma|A|^2 A \qquad (9.23)$$

Here, $A(z,T)$ is the complex amplitude of the z-propagating wave, in the slowly varying envelope approximation, and T is reckoned from the moving pulse $T=(t-z)/v_g$, v_g being the group velocity. In this equation, β_2 represents the group velocity dispersion defined as the following:

$$\beta_2 = \frac{\partial^2 \beta}{\partial \omega^2} = \frac{-\lambda^2 D}{2\pi c} \qquad (9.24)$$

The dispersion parameter D in units of ps/km nm is more commonly used to characterize the GVD. For soliton propagation, the magnitude of β must be negative, correspondingly the value of D must be positive.

One can show that the peak power required to support the fundamental soliton is given by the following equation:

$$P(\text{Watts}) = \frac{3.11|\beta_2|}{\gamma T^2} = \frac{0.495\lambda^2|D|}{c\gamma T^2} \qquad (9.25)$$

T is the pulse width. However, if one also demands that the wavelength of the soliton be consistent with the telecommunication band, then the wavelength is restricted to the region 1500–1600 nm. It is invariably the case that materials that are highly nonlinear have large material dispersion in this wavelength range. However, it is seen from the data of Fig. 9.14 that one can move the zero-dispersion point to a more practical wavelength and in doing so take advantage of the lower power required to support a soliton in a highly nonlinear material.

9.7.1.2. Parametric Amplification

The nonlinear phenomenon can be used to amplify or otherwise modify a signal beam at one frequency through the interaction with a pump at another frequency [40]. For parametric amplification, there are two beams, a pump beam is used to transfer power to the signal beam. This can be written as the following:

$$P_s = \chi_{1111}(\omega_s = 2\omega_p - \omega_i)E_pE_pE_I^* \tag{9.26}$$

The propagation mismatch is given by

$$\Delta\beta = 2\beta_p - \beta_I - \beta_s, \tag{9.27}$$

with the frequency condition being

$$\Delta\omega = \omega_p - \omega_s = \omega_I - \omega_p \tag{9.28}$$

If one expands β in a power series in frequency $(\omega - \omega_n)$ where $n = p$, s, and substitutes the expansion into Eq. 9.28 one obtains

$$\Delta\beta = \frac{2\pi c\Delta\lambda^2}{\lambda^2}\left(D + \Delta\lambda\frac{dD}{d\lambda}\right) \tag{9.29}$$

Here, D is the measure of the GVD. The overall efficiency of the process will be diminished by the phase mismatch according to the expression

$$\frac{\sin^2 \Delta\beta L}{\Delta\beta^2} \tag{9.30}$$

It should be clear from Eq. (9.30) that to optimize the process, the wavelength should be near the zero-dispersion point.

This phenomenon is related to the four wave interaction termed optical phase conjugation. In this case, the phase conjugate of a signal beam is created from the interaction of the signal with a suitable oriented pump beam. Yariv and colleagues [41] proposed the use of phase conjugation in optical fibers as a dispersion compensation technique. He suggested that if one formed the phase conjugate of a pulse after having broadened from traveling a length of fiber L_1 and then sent the conjugate through a section of fiber L_2 such that

$$D_1 L_1 = D_2 L_2 \tag{9.31}$$

then the pulse would have resumed its initial width. In effect, the conjugate had a reversed GVD. On the other hand, the group velocity itself is not reversed in sign, so the pulses would still be maintained in the same temporal order.

9.7.1.3. White Light and Continuum Generation

This application uses the high nonlinearity produced by a small core and fs-laser pulses (see next chapter) to produce nonlinear mixing of the beams. This is essentially four wave mixing gone wild. To make this process efficient, it must be done at or near the zero-dispersion point of the pump laser to reduce the phase mismatch between the interacting beams. In a normal step-index fiber, it is very difficult to move the zero-dispersion point to the 800 nm region where the fs-laser source is readily available. With a properly designed solid-core photonic crystal fiber, this is easily accomplished. With a step-index fiber, it would require essentially a silica rod in air to achieve the same shift in zero-dispersion.

Light in the range of 390–1600 nm has been generated in a photonic crystal fiber using 100 fs pulses at 800 nm [42]. See Fig. 9.31.

9.7.2. Hollow-Core Fiber Applications

The initial interest in the hollow-core fiber was that it could yield the lowest possible loss fiber. Only about 5% of the light

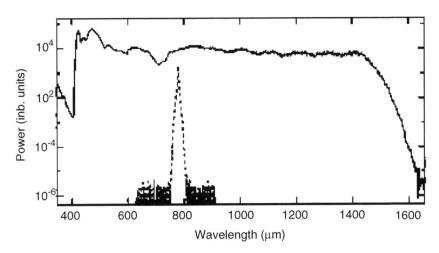

Figure 9.31 Result of continuum generation in a photonic crystal fiber. One hundred fs pulses with a peak power of 1.6 kW at 800 nm into a PCF produced a continuum from 390 to 1600 nm.

is carried in a lossy material and that material itself is of loss of the order of 0.15 db/km. We will cover this application in some detail below. Meanwhile other applications have been suggested again capitalizing on the hollow-core structure. Two of these new applications include high power light pipes for visible and uv laser light, and tunable deep uv generation through fs-laser pulse propagation through noble gas filled cores.

9.7.2.1. Ultra-low Loss Telecommunication Fiber

Assuming a non-absorbing gas fills the core, the loss of a hollow-core fiber could be 20× less than that of the best silica core fiber. This is based on the fact that about 5% of the light is carried in the glass that constitutes the webbing. The rest of the light is carried in the voids. As a consequence, the length of fiber between amplifiers could be 20× longer. The issues that will impact this reality are not insignificant by any manner of means. Notwithstanding, significant progress toward this goal has been made over the last few years. Smith et al. [43] reported a loss of 13 db/km in the

1300–1700 nm wavelength region. Their result is shown in Fig. 9.32 for the structure shown. One notes the width of the passband is from about 1300 to 1800 nm interrupted by a region of higher loss. This full passband corresponds to the wavelength difference corresponding to the intersection of the dispersion of the defect mode with the boundaries of the forbidden gap as we had indicated in Fig. 9.17. Two other groups (Blaze Photonics and Crystal Fibre) have reported of similar losses in unspecified lengths hollow-core fibers.

From a more detailed computational examination of the forbidden gap region (the geometric coordinates of the real structure are used in the computation which was obtained from an SEM photograph Fig. 9.33), one sees a number of interesting and unexpected features in Fig. 9.34. One sees the expected dispersion lines produced by the larger central hole defect, corresponding to the first two modes. The choice of the radius of the defect value was guided by Eq. (9.21).

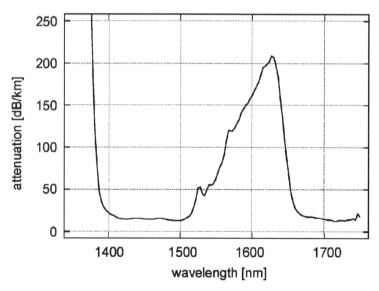

Figure 9.32 Measured loss vs. wavelength in 100 m of a hollow-core photonic crystal fiber with a structure shown in Fig. 9.33. (Taken from Ref. [43].)

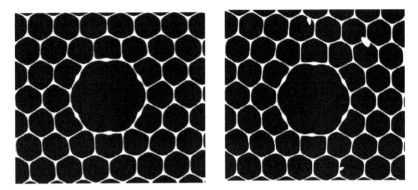

Figure 9.33 The structure has the following reported para-meters, pitch was 4.7 µm and the volume fraction of air was 94%. The average core radius was 12.7 µm. (Taken from Ref. [43]). Left from initial; Right from final.

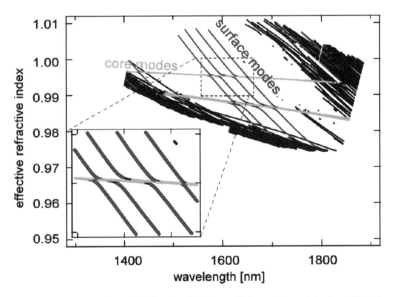

Figure 9.34 Detailed calculation of the allowable localized modes of the structure shown in Fig. 9.33. The two nearly horizontal lines are the core modes, the fundamental and next higher order mode. The four steeper lines represent the "surface" modes. The inset shows the interaction or mixing of the core modes with the funda-mental modes. (Ref. [43].)

The spectral position of these lines is in rough agreement with one would estimate from the 100% reflecting cylinder approximation.

The other modes that are calculated to exist in the gap can best be described as "surface modes." They arise from the dielectric structure that constitutes the transition from the perfect, or near perfect structure to any terminating surface like that provided by the central hole. In his book, Joannopoulos et al. [2] discuss in some detail the nature of surface modes. Depending on the specific nature of the geometric termination of the structure, these surface modes can fall within the forbidden gap and propagate as a well-defined localized mode. The unique characteristic of these localized surface modes is that they decay both in air (core) and into the photonic crystal (cladding). In the language of mode propagation, this means that both transverse wave-vector components are imaginary. This condition makes these modes tightly bound to the interface since they are decaying in both directions. These are the kinds of modes that were calculated and shown in Fig. 9.34 in addition to the core modes. The calculated mode field patterns confirm that these modes are indeed surface modes with the light confined to the boundary of the hole. (See Fig. 9.35.)

What is significant about this calculation is the prediction of an interaction of these surface modes with the core modes as indicated by the avoided crossings. These avoided crossings correspond to the mixing of the two modes, that is the core mode takes on surface mode character and vice versa.

Allan et al. [44] propose that it is through this interaction of the core mode with the surface modes that the higher loss feature from 1500 to 1600 nm arises. It is important to note that this measured higher loss region corresponds to the spectral region where the avoided crossings are computed. They proposed a coupled mode approach to simulate the situation using the calculated avoided crossing results to gain an estimate of the magnitude of the interaction. They wrote the following coupled differential equations for the core mode

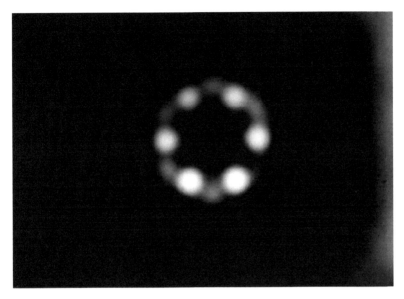

Figure 9.35 Photograph of the intensity profile of a surface mode. Mode decays into both the air core and the photonic crystal cladding.

and the surface mode:

$$\frac{\mathrm{d}A_c}{\mathrm{d}z} = -i\kappa A_K \exp(i\Delta\beta z) \tag{9.32}$$

$$\frac{\mathrm{d}A_K}{\mathrm{d}z} = -i\kappa A_c \exp\left(-\frac{i\Delta\beta}{2}z\right) - \gamma A_K \tag{9.33}$$

Here, $\Delta\beta = \beta_K - \beta_c$ and κ is the coupling coefficient which is given by

$$\kappa = \frac{\omega}{4}\varepsilon \iint E_K(x,y)E_c(x,y)\,\mathrm{d}x\,\mathrm{d}y \tag{9.34}$$

The parameter γ represents the loss. In other words, it was assumed that the surface modes were lossy. Strictly speaking, the actual loss is assumed to be from the surface mode coupling to the extended modes. They assumed this coupling was strong.

Solving the above two equations, and using the calculated avoided crossing results to estimate the value of $\beta(\lambda)$, they were able to fit the actual loss region quite well with an appropriate value of γ. (See Fig. 9.36.) The value of β controls the spectral width of overlap while γ determines the peak loss. They summed the loss estimate over wavelength for the interaction of the four surface modes with the first two core modes and obtained a reasonable fit for the high loss region. The conclusion of this analysis is that to lower the loss still further, one must find a way to fabricate the core of the fiber in such a way as to move these surface modes out of the gap.

In view of the coupling explanation of the loss, it is also important to be able to maintain the same structure over long lengths since the coupling of the surface modes to the

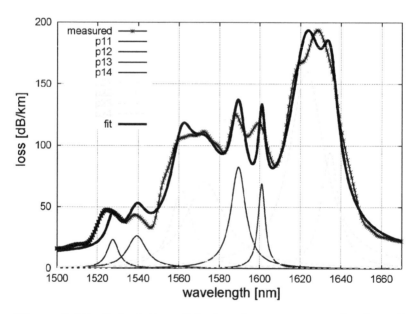

Figure 9.36 Calculation of the loss feature as a result of coupling of the core mode to the surface modes according to Eqs. (9.32)–(9.34) in the text. The estimate of the strength of the interaction was taken from the measured width of the avoided crossing shown in Fig. 9.34. The four separate peaks correspond to the four surface mode interactions. (Taken from Ref. [44]).

extended modes requires an axial variation in the refractive index. There is experimental support for this idea in the comparison of the normalized transmission of the above-mentioned low loss fiber. This is shown in Fig. 9.37 where the normalized transmission of the first and last meter of a 100 m length of fiber is compared to the transmission through the entire 100 m. The coincidence of these three curves indicates the uniformity of the structure.

The group velocity dispersion of the fundamental mode in hollow-core fiber described above, Fig. 9.33, has been measured by Ouzounov et al. [45] using a tunable femtosecond pulse source. The result is shown in Fig. 9.38. The zero dispersion point is determined to be at 1430 nm for this fiber structure.

The theoretical analysis of waveguiding in the Omniguide® hollow-core structure was extensively analyzed by Johnson et al. [37]. The bandgap diagram with the dispersion of the various modes is shown in Fig. 9.39a. They showed

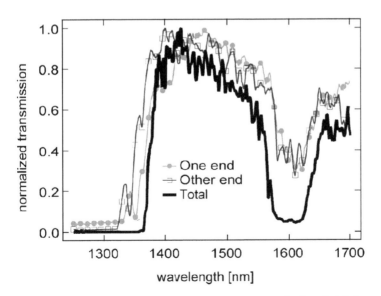

Figure 9.37 Normalized transmission of the initial and final potions of the 100 m fiber of Fig. 9.32, compared to that over the entire 100 m. The close agreement speaks to the axial uniformity of the fiber over this length. (Taken from Ref. [43]).

(a)

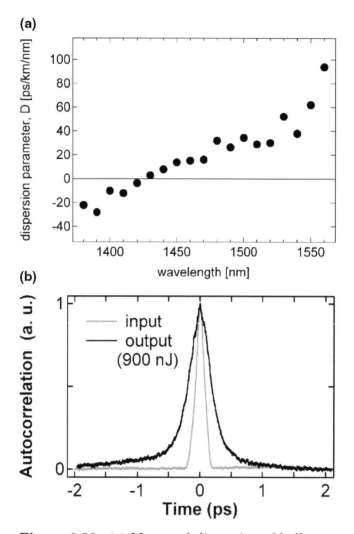

(b)

Figure 9.38 (a) Measured dispersion of hollow-core fiber of Fig. 9.33, and (b) autocorrelation measure of pulse shape at input and output of the hollow-core fiber. (Taken from Ref. [45]).

that the lowest loss mode was the TE_{01}. The computed radiation leakage (tunneling loss) is shown in Fig. 9.39b and the modal pattern of the lowest loss TE mode is shown in Fig. 9.39c along with the HE_{11} mode for comparison. The hollow-core size was chosen to be relatively large which provided

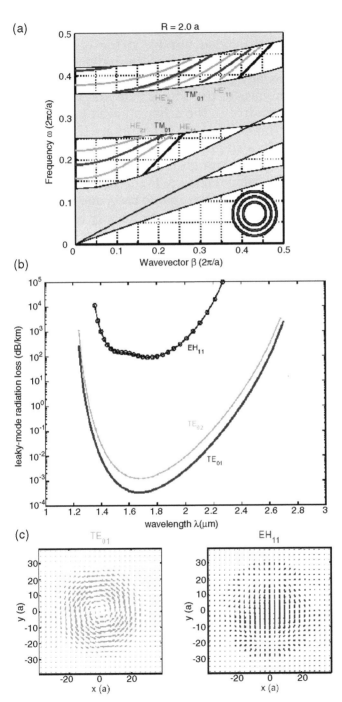

the low loss condition for the TE_{01} mode. In other words, the actual structure is multimoded but by virtue of the much lower loss of the TE_{01} mode relative to the other supported modes, the result is essentially a single mode fiber. The larger the core the lower the loss of the TE_{01} mode and the greater the disparity in loss relative to the other allowed modes. This is in contrast to the triangular lattice fiber where the core size is more in keeping with a single mode propagation. They also do an analysis on the loss through perturbation of structure such as bending a dimensional change to indicate the stability of the loss. Nonetheless in the Omniguide case, one must keep in mind that these are theoretical losses due only to radiation leakage whereas in the triangular lattice structure the loss in measured loss and as shown above is not yet to the level where one needs to worry about tunneling.

9.7.2.2. Light Pipes for High Power

One obvious advantage to having a fiber with a hollow core is that the power that can be transmitted is very high because of the significant reduction in the magnitude of nonlinear mechanisms that contribute to optical damage. Ouzounov et al. [45] have shown that in an inert gas filled core, one can propagate 100 fs pulses with mJ peak energy at 1400 nm with no evidence of any damage. Utilizing the dispersion results given above, in the case of an air-filled core, sufficient nonlinerarity is present to allow very high energy soliton propagation.

The loss required for this application is of the order of 0.1 db/m, so fibers are already meeting this criterion. There are other issues having to do with core size and numerical aperture that may not be so easy to meet, however.

The propagation of shorter wavelengths is also possible by either using a smaller pitch structure (wavelength of gaps

Figure 9.39 (a) Calculated band diagram for Omniguide structure, (b) calculated radiation leakage for different modes as labeled, (c) depiction of the low-loss TE_{01} mode compared to the HE_{11} mode. (Taken from Ref. [37]).

are scaled to pitch), or propagation in the higher order gaps. A triangular hole structure with a pitch of $2\,\mu m$ produced a gap in the 900 nm wavelength regime. The practical limit in the pitch imposed by the closing of the air holes in the redraw process is around $1\,\mu m$, so one might anticipate gaps in the 400 nm region with this approach. It is not clear at this point whether propagation in these higher order modes is more problematic than in the fundamental mode. One might argue that whatever constitutes the disorder parameter that leads to the surface modes would be more significant at the shorter wavelengths.

9.7.2.3. Deep UV Harmonic Generation

Another application that has come to the fore is one where one utilizes fs-pulse excitation into a gas filled hollow core to create extremely high peak intensity (Terawatt/cn^2). The high intensity combined with the tuning of the refractive index dispersion of the noble gas by the ambient pressure to a near zero value permits efficient harmonic generation to second and third, and higher orders. Results obtained in gas-filled capillary tubes [46–49] show efficient tunable uv generation, reproduced here in Fig. 9.40 [47].

The phase matching condition can be written by adding a term to the momentum matching condition written above for a 100% reflecting cylinder in Eq. (9.21). Here, however, we add a term corresponding to the presence of the gas in the tube with refractive index $n = 1 + P\delta(\lambda)$, with P being the pressure and δ being the conversion factor

$$k(\lambda)^2 = k_0^2[1 + P\delta(\lambda)]^2 - (\phi_n/a)^2$$

$$k(\lambda) \approx k_0[1 + P\delta] - \frac{\phi^2}{2a^2}\frac{1}{k_0} \qquad (9.35)$$

As we have seen above in Eq. (9.27) (section on Parametric amplification) for the nonlinear process to be enhanced the phase mismatch between the pump and the generated waves must be small. By using Eq. (9.35), an adjustment of the gas pressure, gas species, waveguide radius and the

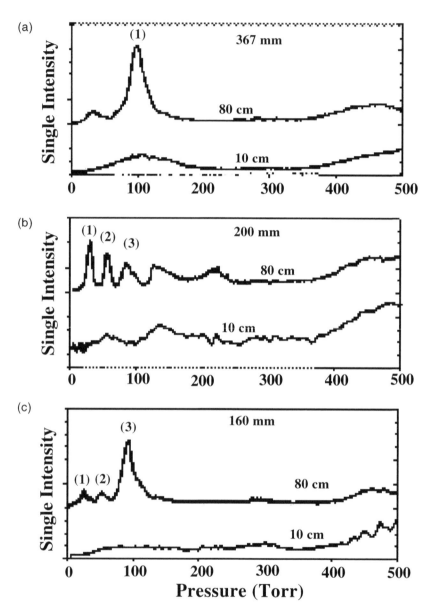

Figure 9.40 Deep uv generation of light in gas-filled capillary tube. (Taken from Ref. [46]).

spatial mode in which the light propagates, the phase mismatch can be tuned to produce the desired harmonic.

The inert gas-filled hollow-core photonic crystal fiber version offers the advantage of a lower pulse energy because of the smaller core size relative to a capillary to maintain the same intensity, and an enhancement of the nonlinearity because the pressure required to achieve zero dispersion is higher.

APPENDIX A. MATHEMATICAL FRAMEWORK

The intention is not to present any more mathematics than absolutely necessary, especially in light of several excellent articles that are available [2,5]. Nonetheless, it is valuable to see the formulation, and how the solutions are derived in the photonic crystal framework. This is especially true here because the solution of Maxwell's equations cannot be different, but the approach to obtaining the solution is quite different.

One is looking for a solution to Eq. (9.4) in terms of what are called *plane waves*. What this means is that one wants to write the solution as an infinite series of plane waves of the form $\exp i(\mathbf{k} \cdot \mathbf{r})$. Further one uses the periodic repeating nature of the dielectric structure to simplify the writing of the resulting sum. The periodicity of the structure can be characterized in real space and reciprocal space. A simple example of the relation between a structure in real space to that in reciprocal space is shown in Fig. 9.A1 for a simple two-dimensional (2D) square structure. The figure on the right is referred to as the Brillouin zone and the shaded portion, the primitive portion thereof. The capital Greek letters refer to different points on, or in, the primitive cell and represent points corresponding to the symmetry they possess [3]. The point Γ, at the center of the zone, contains the complete symmetry of the structure, whereas the other points contain less, depending on the way they transform under the complete set of symmetry operations possessed by the structure. The plane wave expansion of $\mathbf{H}$ for any given

Real lattice

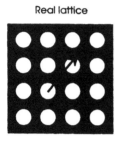

Brillouin zone of
reciprocal lattice

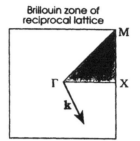

Figure 9A.1   Example of the rteal lattice on the left and the corresponding reciprocal lattice on the right. The symbols refer to positions of the Brillouin zone, center, edge corner, and so on.

value of k will be in reciprocal space in the following form [6]:

$$H_k(r) = \sum_G h_G \exp i(k + G) \cdot r \qquad (9A.1)$$

The G represents an arbitrary set of vectors in reciprocal space. On next expands the $1/\varepsilon$ term in Eq. (9.4) in a similar fashion

$$1/\varepsilon = \sum_G \varepsilon(\mathbf{G})^{-1} \exp i(\mathbf{G} \cdot r) \qquad (9A.2)$$

Substituting Eq. (9.8) and (9A.1) into (9.4) leads eventually to the expression

$$\sum_{G'} \{\varepsilon^{-1}(\mathbf{G} - \mathbf{G}')\}(k + \mathbf{G})(k + \mathbf{G}')h_{G'} = \left(\frac{\omega}{c}\right)^2 h_G \qquad (9A.3)$$

This constitutes a matrix equation with the matrix of the form

$$\Theta_{G,G'} = (k + \mathbf{G})(k + \mathbf{G}')\varepsilon^{-1}(\mathbf{G} - \mathbf{G}') \qquad (9A.4)$$

The eigenvectors are the field amplitudes and the eigenvalues are the frequencies. One can summarize the steps in the calculation of the frequency of the allowable modes as follows. Based on the symmetry of the dielectric arrangement, construct the Brillouin zone. Choose a basis, that is the number of Gs that are needed for the respective representations. Compute $\varepsilon^{-1}(\mathbf{G})$ as the Fourier transform of the spatial dielectric structure $\varepsilon^{-1}(r)$. For each value of k, selected symmetric directions in the Brillouin zone, calculate $\Theta^k_{G,G'}$. Then solve Eq. (9A.4)

ACKNOWLEDGMENT

The author is grateful to Douglas C. Allan for help in preparing the material contained in the Appendix.

REFERENCES

1. Yablonovitch, E. Photonic band-gap structures. J. Opt. Soc. Am. B **1993**, *10*, 283–295.

2. Joannopoulos, J.D.; Meade, R.D.; Winn, J.N. In *Photonic Crystals*; Princeton University Press: Princeton, NJ, 1995.

3. Tinkham, M. In *Group Theory and Quantum Mechanics*; McGraw-Hill: New York, 1964.

4. Born, M.; Wolfe, E. In *Principles of Optics*; Macmillan: New York, 1964.

5. Villeneuve, P.R.; Piche, M. Photonic band gaps in two-dimensional square and hexagonal lattices. *Phys. Rev. B.* 46, 4969–4972; **1992**; Photonic band gaps in two-dimensional square lattices, square and circular rods. Phys. Rev. B **1992**, *46*, 4672–4675.

6. Allan, D.C. Corning Inc. Private communication.

7. The codes used were provided to us by J. Joannopoulos and his group at MIT.

8. Joannopoulos, J.D.; Villeneuve, P.R.; Fan, S. Photonic crystals: putting a new list on light. *Nature* **1997**, *386*, 143–149; Fan, S.; Villeneuve, P.R.; Joannopoulos, J.D. *Phys. Rev. B* **1996**, *54*, 247–251.

9. Birks, T.A.; Roberts, P.J.; Russell, P.St.J.; Atkin, D.M.; Shepherd, T.J. Full 2-D photonic bandgaps in silica/air structures. Elect. Lett. **1995**, *31*, 1941.

10. Russell, P.St.J.; Knight, J.C.; Birks, T.A.; Cregan, R.F.; Managan, B.J. Silica/air photonic crystal fibers. In *International Workshop on Structure and Functional Optics*;

Properties of Silica and Silica-Related Glasses; Shizuoka, Japan, July 10–11, 1997.

11. Birks, T.A.; Knight, J.C.; Russell, P.St.J. Endlessly single mode photonic crystal fiber. Opt. Lett. **1997**, *22*, 961.

12. Rosenberg, A.; Tonucci, R.J.; Bolden, E.A. Photonic band-structure effects in the visible and near ultraviolet observed in solid-state dielectric arrays. Apply. Phys. Lett. **1996**, *69*, 2638.

13. Rosenberg, A.; Tonnucci, R.J.; Lin, H.-B.; Campillo, A.J. Near-infrared two-dimension photonic band-gap materials. Opt. Lett. **1996**, *21*, 830.

14. Knight, J.C.; Birks, T.A.; Russell, P.St.J.; Atkin, D.M. All-silica single-mode optical fiber with photonic crystal cladding. Opt. Lett. **1996**, *23*, 1547.

15. Marcuse, D. *Light Transmission Optics*; Van Nostrand Reinhold: New York, 1972.

16. Suzuki, Y. Cornell University, private communication.

17. Koops, H.W.P. Photonic crystals built in 3-D additive lithography. *SPIE* **1996**, *2849*, 248.

18. Masuda, H.; Yamada, H.; Satoh, M.; Asoh, H. Highly ordered nanochannel-array architecture in anodic alumina. Appl. Phys. Lett. **1997**, *71*, 2770.

19. Silvesrte, E.; Russell, P.St.J.; Birks, T.A.; Knight, J.C. Endlessly single-mode heat sink waveguide. CLEO '98 Paper CTh059, Los Angeles, CA.

20. Inno, S. High density multi-core fiber cable. IWCS'79 Proceedings, 1979; 370–384.

21. Bourhis, J.-E.; Meilleur, R.; Nouchi, P.; Tardy, A.; Orcel, G. Manufacturing and characterization of multicore fibers. Proceedings of the 46th International Wire and Cable Symposium. (IWCS), Philadelphia, PA, Nov., 1997.

22. Reed, J.S. *Principles of Ceramic Processing*; Wiley-Interscience: New York, 1995.

23. Villeneuve, P.R.; Fan, S.; Joannopoulos, J.D. Microcavities in photonic crystals: Mode symmetry, tunability, and coupling efficiency. Phys. Rev. **1996**, *54*, 7837–7842.

24. Hutcheson, L.D. Ed., *Integrated Optical Circuits and Components;* Marcel Dekker: New York, 1987.

25. Yariv, A.; Yeh, P. *Optical Waves in Crystals*; John Wiley and Sons: New York, 1984: 103, Chapter 4.

26. Vitebsky, I., et al. Electronic energy spectra in antiferromagnetic media with broken symmetry. Phys. Rev. **1997**, *55* (18), 12566–12571.

27. Figotin, A.; Vitebsky, I. Nonreciprocal magbetic photonic crystals. Phys. Rev. E **2001**, *63*, 0666091–06660917.

28. Yeh, P.; Yariv, A. Theory of Bragg fiber. J. Opt. Soc. Am. **1978**, *68* (9), 1196–1201; Electromagnetic propagation in periodic stratified medis, I General theory. J. Opt. Soc. Am. **1977**, *67* (4), 423–438.

29. West, J.A. *Practical Aspects of Photonic Crystal Fibers*; OSA, 2000.

30. de Sterke, C.M.; Bassett, I.M.; Street, A.G. Differential losses in Bragg fibers. J. Appl. Phys. **1994**, *76* (2), 680–688.

31. *Micro- and Nano-photonic Materials and Devices*, Proceedings of SPIE, Jan. 27–28, 2000, Perry, J.W.; Scherer, A. Eds.; Vol. 3937.

32. Fleming, J.G.; Lin, S.-Y. Three-dimensional photonic crystal with a stop band from 1.35–1.95 μm. Opt. Lett. **1999**, *24* (1), 49.

33. Tanaka, T.A.; Kawata, S. Three-dimensional fabrication and observation of microstructures using two-photon absorption and fluorescence. Proc. SPIE **2000**, *3937*, 92.

34. Kuebler, S.M., et al. Three-dimensional microfabrication using two-photon activated chemistry. Proc. SPIE, **2000**, *3937*, 97.

35. Jenekhe, S.A.; Chen, X.L. Self-assembly of ordered microporous materials from rod-coil block copolymers. Science **1999**, *283*, 372.

36. West, J.A., et al. Demonstration of an IR-optimized air-core photonic band-gap fiber. In Proceedings of ECOC 2000, Munich, 2000; Vol. 4, 41–42.

37. Johnson, S., et al. Low-loss asymptotically single-mode propagation in large-core Omniguide fibers. Opt. Express **2001**, *9* (13), 748–779.

38. Foresi, J.S., et al. Photonic-bandgap microcavities in optical waveguides. Nature **1997**, *391*, 143.

39a Kuchinsky, S.; Allan, D.C.; Borrelli, N.F.; Cotteverte, J.-C. 3D localization in a channel waveguide in a photonic crystal with 2D periodicity. Opt. Comm. **2000**, *175*, 147–152.

39b Temelkuran, B., et al. Low-loss infrared dielectric material system for broadband dual-range omnidirectional reflectivity. Opt. Lett. **2001**, *26* (17) 1370–1372.

40. Butcher, P.N.; Cotter, D. *The Elements of Nonlinear Optics*; Cambridge Press, 1990.

41. Ouyang, G.; Xu, Y.; Yariv, A. Theoretical study on dispersion compensation in air-core Bragg fibers. Opt. Express **2002**, *10* (17), 899–908.

42. Ranka, J.K.; Windeler, R.S.; Stentz, A.J. Visible continuum generation in air-silica microstructure optical fibers with anomalous dispersion at 800 nm. Opt. Lett. **2000**, *25* (1), 25–27.

43. Smith, C., et al. Low-loss hollow-core silica/air photonic band-gap fiber. Nature **2003**, *424* (6949), 657–659.

44. Allan, D.C., et al. Surface modes and loss in air-core photonic band-gap fibers. In Proceeding of SPIE Photonic Crystal Materials and Devices, 2003; Vol. 5000, 161.

45. Ouzounov, D.G., et al. Generation of megawatt optical solitons in hollow-core photonic band-gap fibers. Science **2003**, *301*, 1702–1704.

46. Durfee, C.G., III, Backus, S.; Murname, M.M.; Kapteyn, H.C. Ulta-broadband phase-matched optical parametric generation in ultraviolet by use of guided waves. Opt. Lett. **1997**, *22* (20), 1565–1567.

47. Misoguti, L., et al. Generation of broadband VUV light using third-order-cascaded processes. Phys. Rev. Lett. **2001**, *87* (1), 0136001–0136004.

48. Durfee, C.G., III, et al. Phase matching of high-order harmonics in hollow waveguides. Phys. Rev. Lett. **1999**, *83* (11), 2187–2190.

49. Rundquist, A., et al. Phase-matched generation of coherent soft x-rays. Science **1998**, *280*, 1412.

10

Femtosecond–Laser Interaction in Glasses

10.1. INTRODUCTION

In the last few years as a consequence of the rapid development of mid-infrared ultra-short laser pulses (in particular, Ti–sapphire laser operating at 800-nm), an entirely new way of fabricating three-dimensional photonic structures has been accomplished [1–3]. These ultra-short pulses (<100 fs) offer a way to obtain very high peak intensities of the order of 10^{15} W/cm^2. For example, a laser pulse with a pulse width of 100 fs and a pulse energy of 10 μJ focused to a spot size of 5 μm produces a peak intensity of that order. (In Appendix, a brief description of the method used to produce ultra-short pulses will be given.) A unique aspect of using <100 fs pulses is the distinct way in which the interaction with transparent and absorbing materials occurs as a consequence of the pulse duration. One example of this is the observable damage threshold increases as one uses shorter pulses as shown in Fig. 10.1. The reason for this

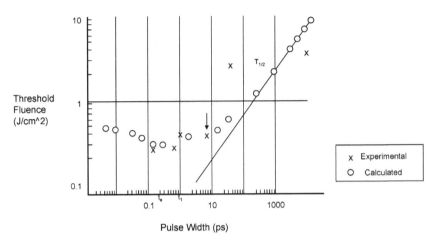

Figure 10.1 Laser damage threshold of Au as a function of pulse duration (taken from Ref. [8]).

originates from a distinctive mechanism of the interaction of short pulses with solid materials. We will discuss this point briefly in the next section.

The practical use of the ultra-short laser pulse technique in material fabrication technology falls into two broad categories. The first is the high pulse energy regime where the fs-laser is used to physically remove material through ablation as the mode of making precise patterns. This is often referred to as laser micro-machining [4–7]. Although laser ablation is not new, it will be shown how the use of short pulses avoids many of the problems heretofore associated with laser ablation. We discuss the ablation of absorbing materials and transparent dielectrics.

The second category is the low exposure intensity regime where the refractive index (or any other physical modification) can be permanently modified without causing physical damage in transparent media such as silica and other glasses [1–3]. One distinct advantage of this technique is that because of the highly nonlinear nature of the laser/material interaction, one can produce patterns inside a body on a three-dimensional way.

We will then deal with the high fluence ablation approach, and its applications. Then we will cover the lower fluence interaction with transparent media such as silica where the refractive index can be altered up to values of 0.01. We will present the results of applications and devices based on this phenomenon.

10.2. GENERAL THEORY OF OPTICAL DAMAGE IN SOLIDS

The interaction of solids with ultra-short pulses is quite different than that of longer pulses basically in that pulse duration is short compared to the transfer of the energy to the nuclei. Thus, the process is highly nonequilibrium, with electrons at a much higher temperature than the nuclei. This makes one able to produce a plasma-like electron concentration because the energy transfer to the lattice does not occur on this time scale. In other words with a longer pulse one must consider the de-excitation of the electron density which continually drains electrons. At the same time the lattice temperature is going up. In short, the ultra-short pulse regime allows one to achieve electrons concentration that can be produced no other way.

It would be helpful for the following discussion to briefly review the mechanisms by which damage is produced and how it depends on the intensity and pulse duration [8–10]. The primary damage mechanism for long pulse lasers >1 ps is called avalanche breakdown. The process begins with a stray electron in the conduction band acquiring energy through an inverse "Bremsstrahlung" mechanism. Light interacts with the electron and accelerates it when it collides with a nuclei, producing more electrons (see Fig. 10.2). The process grows in this manner until breakdown occurs. During the duration of the pulse, energy is being transferred to the lattice, thus raising the physical temperature. In general, in this regime the damage threshold goes with the square root of the pulse duration. However, when the pulse gets short enough the range of the electron is insufficient for this mechanism to be effective. The increase in the damage threshold as the pulse becomes shorter as indicated in Fig. 10.1 is the indication of the change from an avalanche mechanism to something different.

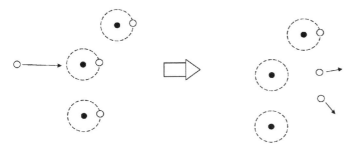

Figure 10.2 Schematic representation of the initiation of the avalanche breakdown mechanism (taken from Ref. [6]).

If the pulse is short enough, shorter that the electron–lattice relaxation time, the electron concentration can build to a density of $10^{20}/cm^3$. These electron densities could not be sustained with longer pulses because the electron is being simultaneously being de-excited. (It is like trying to fill a bucket with a big hole in the bottom.) The mechanism by which the electrons are produced is either by tunneling, or multiphoton excitation (see Fig. 10.3). Both of these

(a)

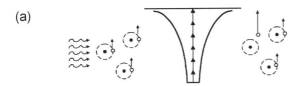

(b)

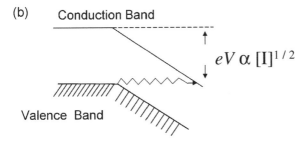

$$eV \propto [I]^{1/2}$$

Figure 10.3 Comparison of multiphoton excitation in (a) to tunneling in (b); in both cases, electron in the conduction band is produced from photons with less than band gap energy.

mechanisms require very high intensity, but the shorter pulses make this condition easier to attain. This is in contrast to the longer pulse situation where the initial electrons are essentially stray electrons. This is why to some extent the damage threshold is so variable in the long pulse regime. In the high intensity short pulse regime, the initial electrons are themselves formed by the light, so the damage event is more predictable.

How damage explicitly occurs and how to characterize it is still a matter under study. It will be discussed in the following sections as we cover laser ablation and induced refractive index change in transparent solids.

10.3. ULTRA-SHORT LASER PULSE ABLATION

We will be primarily concerned in this chapter with dielectric materials, or putting it another way, with materials whose band gap is much larger than the laser wavelength. However, the fs-laser technique can be quite useful for absorbing materials as well, so a brief discussion of this area will be given. Clearly, one will be limited to patterns on the surface of the material, however, in the case of absorbing materials an advantage of using a fs-laser over a longer pulse duration is in the reproducibility of the ablation, and the ability to obtain a smaller spot size. The latter is due to the fact that in the Gaussian beam, only the high intensity portion of beam is used. We will deal briefly with the fs-laser ablation of absorbing materials first.

10.3.1. Absorbing Materials

Metals are very problematic materials for microstructuring with laser techniques because of their high thermal conductivity, and relatively low melting temperature. With pulse duration of >1 ns, the ablation of metals is always accompanied by the formation of extensive heat-affected regions and physical material transport. This limits the achievable precision and quality of the desired structure. The improvements afforded by fs-laser pulses are (a) very rapid energy deposition resulting in plasma formation, (b) absence of molten metal, and (c) negligible heat-affected zones.

Nolte et al. [7] give a very complete description and discussion of a typical micromachining process in metals. They describe an imaging system that focuses a flat-top profile onto a copper foil which is mounted on a computer controlled three-axis translation stage inside a vacuum chamber. The beam fluence used was 140–460 mJ/cm^2; the pulse duration used was 150 fs. Use was made of a vacuum chamber (10^{-4} mbar) to avoid nonlinear effects in air and minimize re-deposition of the ablated material. Examples of patterns and the feature size that were written by this technique are shown in Figs. 10.4 and 10.5 taken from the same work. They also show experimental data for the feature depth as a function of the incident fluence, and compare it to a simple model which relates the depth to the threshold fluence and the penetration depth of the beam. Mourou et al. in their US patent [8] show a plot of the damage threshold vs. pulse width obtained on a gold film. They use a 200 fs-pulse from a 800-nm Ti–sapphire laser. The arrow on the graph indicates the significant change in behavior.

There are commercial system available that provide the laser and the three-axis micropositioning stages for

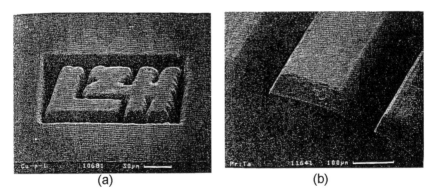

(a)   (b)

Figure 10.4 Electron micrographs pf laser ablation writing on copper. The 800 nm 150 fs-light, 700 μJ per pulse, was focused with a 150 mm lens (taken from Ref. [7]).

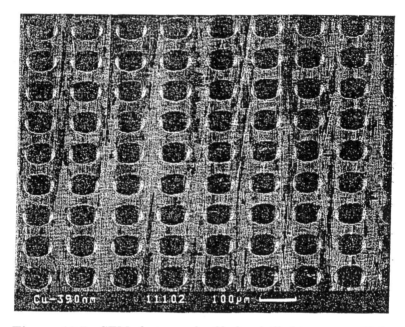

Figure 10.5 SEM photograph of holes drilled in copper. Holes are 80 μm in size, on a 150 μm pitch. In this case, the laser frequency was double to 390 nm with a pulse energy of 40 μJ. (picture taken from Ref. [7]).

patterning intricate and precise surface structures. An example of one is shown below in Table 10.1.

10.3.2. Dielectric Materials

The process of laser ablation of dielectric materials shares a lot in common with the absorbing materials discussed above except that the initial excitation occurs by a higher order process. In other words, the thermal conditions leading to ablation likely are similar, but the way the initial excitation is effected is different because the laser wavelength is below the band gap. This is borne out by similarity of the damage threshold as a function of pulse duration for silica as shown in Fig. 10.6 [8] to that shown above in Fig. 10.1 for gold.

In general, one starts the explanation of the damage process by positing the existence of a stray electron, which then

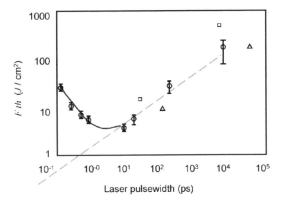

Figure 10.6 Damage threshold vs. pulse duration for silica (taken from Ref. [8]).

gets accelerated by the light. Then by collisions it produces two electrons, and the avalanche process begins as described above. This stochastic-like nature of the seed electrons is the reason why ablation, or optical damage in general can be so hard to reproduce. In the case of ultra-short pulses, the seed electrons can originate from multiphoton excitation because of the very high intensity, thus the initiation of the avalanche process is more consistent. This point is shown clearly in Fig. 10.7 where a comparison is made between the damage threshold of fused silica with 7 ns pulses, upper curve, and

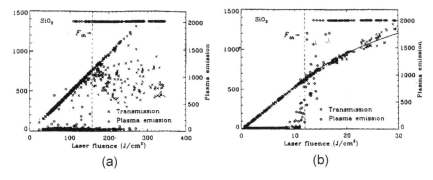

Figure 10.7 Laser-induced breakdown vs. laser fluence in silica for (a) 7 ns pulses, and (b) 170 fs pulses (taken from Ref. [12]).

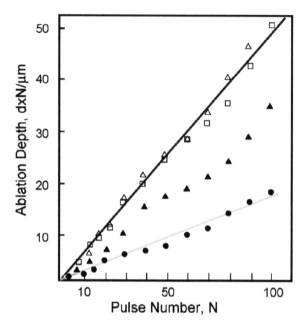

Figure 10.8 Ablation depth vs. number of pulses in barium borosilicate glass at 20, 50, 120, and 300 fs (taken from Ref. [5]).

170 fs pulses in the lower curve [12]. Both transmission and plasma emission data are plotted. The transmittance is a complicated phenomenon involving the light interacting with the plasma. The important point is that the plasma emission yields a rather distinct threshold. One can see the almost random behavior of the 7 ns data, while that of the 170 fs is quite consistent. This is directly attributable to well-defined multiphoton initiation of the plasma formation.

Kautek and Kruger [5] have measured the ablation depth as a function of the number of pulses for 20–300 fs pulse duration for Corning code 7059 glass as shown in Fig. 10.8. Similar results as a function of fluence are shown in Fig. 10.9.

10.3.3. Other Applications

Other than the ability to pattern surfaces with the use of focused fs-laser beams, there are a few new suggested

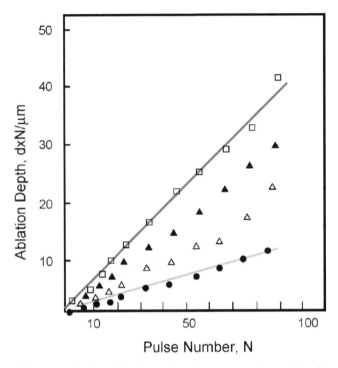

Figure 10.9 Ablation depth vs. number of pulses for different fluences/pulse duration combinations for the barium borosilicate glass. From upper line to bottom in fs/μJ, 50/15, 50/30, 120,20, 120/40 (taken from Ref. [5]).

applications in dielectric materials. One scheme is to make a 3-D memory storage device by altering the energy state of a dopant in the glass [13]. The schematic of the operation is shown in Fig. 10.10. Because of the short pulse duration, regions of the glass that are exposed to the focused beam experience intensities $>10^{12}\,\mathrm{W/cm^2}$. At this intensity, it is proposed that one can convert Sm^{+3} to Sm^{+2}, thus the absorption, as well as the fluoescence is altered dramatically in the exposed spots, allowing them to read.

Another unique idea is to manipulate the focused fs-laser generated microvoids in silica glass. Watanabe et al. [14] focused fs-laser inside the silica glass to create local structural changes or more specifically, optical damage

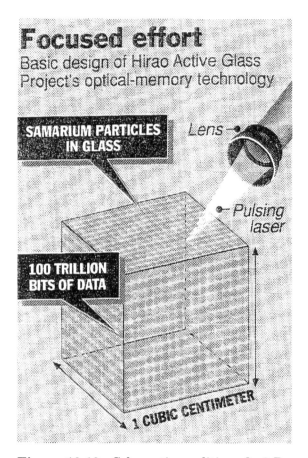

Figure 10.10 Schematic rendition of a 3-D memory based on writing small damage spots into a material (see Ref. [13]).

in the form of microvoids. They used a Ti–sapphire laser operating at 800-nm, 1 kHz, 130 fs pulse, with an energy per pulse 0.5–2.3 µJ depending on the NA of the focusing objective. What was unique was their ability to physically move the void by moving the exposure spot. This is shown in Fig. 10.11. First, they produced three microvoids as shown in (a). They then translated the focus by a distance of 0.5 µm in the negative z-direction. Continuing this process, they could step the void to a distance of 5 µm as shown in (e).

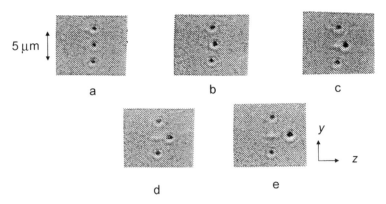

5 μm

a

b

c

d

e

y

z

Figure 10.11 SEM picture of fs-laser spots and the ability to move them, see text for explanation (taken from Ref. [14]).

10.4. INDUCED REFRACTIVE INDEX

The unique aspect of the interaction of fs-laser pulses with glass is the ability to produce permanent refractive index changes without physical damage. In other words, up till now we have been describing phenomena which cause significant physical changes, like ablation or the formation of microvoids. An example of an over exposed track where physical damage is clearly evident is shown in Fig. 10.12 [3]. For the most part, these phenomena are ultimately driven by a thermal mechanism. What we are to discuss below is not likely thermally induced, although this point has been argued. Nonetheless, there appears to be in addition some sort of structural change that is discernible only through a change in the refractive index. This ability has led to a number of very interesting suggestions and demonstrations for optical devices. The idea of a 3-D structure of coupled light guides is one of the more fascinating suggestions.

10.4.1. Typical Refractive Index Altering Exposure Set-up

An amplifier (e.g., Spectra Physics Spitfire model) seeded by the oscillator is typically the source of 800-nm, 40–100 fs pulses having a repetition rate of 1–20 kHz, depending on

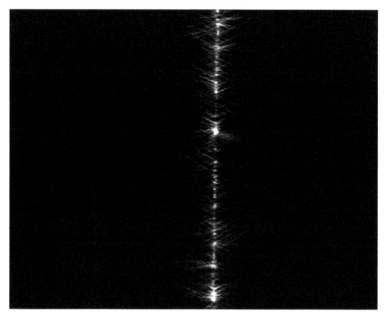

Figure 10.12 Light scattering from a damaged track in an attempt to write a waveguide.

the specific model, and energies up to 30-µJ. Mitutoyo objectives, typically 0.26 NA are used and they give a good compromise between a reasonably tight focusing and about 3-cm long working distance enabling to fabricate waveguides in thick samples. There are essentially two writing methods, the first is the "longitudinal" writing technique when the sample is traversed parallel to the optical axis of the beam, while the other is the "top writing" method. These are depicted in Fig. 10.13a and 10.13b. In either case, the positioning of the beam is accomplished by a three-dimensional translation stage usually with speeds ranging from 1 µm/s to 1 mm/s. The longitudinal writing method has the advantage of producing a cylindrical profile compared to an elliptical one in case of top-writing scheme. Multiple-scan writing can also be used.

Waveguide-like patterns written in a number of different glasses by the method shown in Fig. 10.13b are shown in Fig. 10.14 [15]. The index profile of a waveguide written in

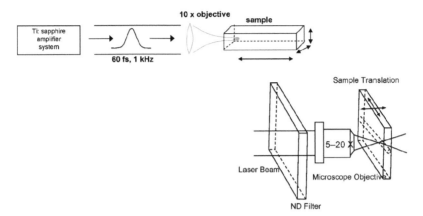

Figure 10.13 Schematic representation of the fs-laser waveguide writing technique; (a) axial or longitudinal writing, (b) top-writing.

silica by the longitudinal method is shown in Fig. 10.15 [3]. A photograph of the near field of light propagating in the waveguide is shown in Fig. 10.16. Representative dependence of the induced Δn for SiO_2 on the 800 nm laser-pulse energy has been measured and the typical curve is presented in Fig. 10.17a [3]. The functional dependence on translational velocity is shown in Fig. 10.17b [3]. There is a weak dependence of Δn vs. velocity for values higher, than 40–50 μm/s and it sharply grows with velocity decreasing below the values of 30–35 μm/s (Figs. 10.2a and 10.3a). For the fixed

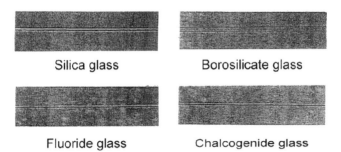

Figure 10.14 Microscope pictures of fs-laser written waveguide-like tracks in four different glasses (taken from Ref. [1]).

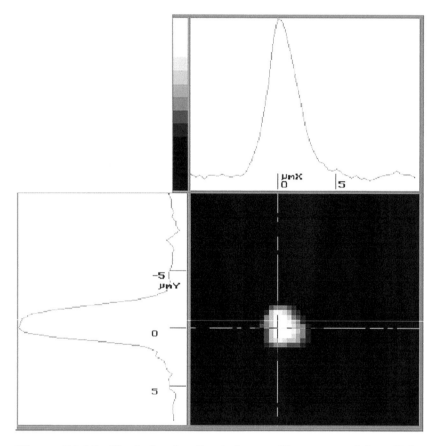

Figure 10.15 Typical refractive index profile measured for a light guiding structure written with the focused 800 nm 40 fs laser under optimum conditions [3,25].

velocity of 10 μm/s, the refractive-index increase in fused silica is saturated at pulse energies around 0.75 μJ and then decreases slightly (Fig. 10.2b). Visual inspection of light-coupling into the waveguide written either with pulse energies in excess of 1 μJ or at translation velocities below 5–10 μm/s showed the presence of scattering centers along the written track that lead to losses. This appears to be the result of physical damage similar to that shown in Fig. 10.12.

Figure 10.16 End-on microscope picture of white light propagating through fs-laser written waveguide structure.

10.5. PHYSICAL MECHANISM FOR INDEX CHANGE

The physical origin of the refractive index change that can be produced in glass by fs-laser exposure is not really well understood. There have been a number of papers relating to the "damage" of glass and crystals [2,8,9,11,16]. The word "damage" is taken to mean catastrophic mechanical damage such as breakdown, or ablation. However, at exposures well below these estimated thresholds, some sort of permanent alteration of the refractive index is occurring without any evidence of physical damage.

Below we are examining the effects that may cause the increase of the refractive index in glasses as a result of irradiation by femtosecond laser pulses. First, we discuss the role of the localized heating that accompanies the interaction of intense ultra-short pulses with the glass. Several mechanisms of nonlinear absorption are considered in the literature [17], the main being multiphoton absorption, tunneling, and

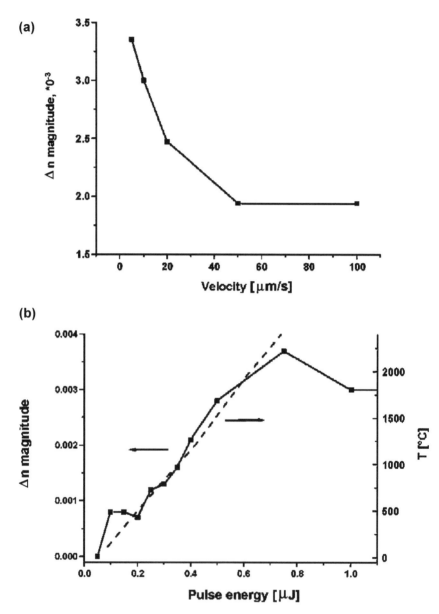

Figure 10.17 Dependence of induced refractive index in silica; (a) as a function of translational velocity at fixed pulse energy, and (b) fixed velocity of 10 µm/s as a function of pulse energy [25].

avalanche ionization. The electron excitation kinetics equation can be written in the form:

$$\frac{dn_e}{dt} = W(I, \hbar\varpi, \Delta) + \eta I n_e, \tag{1}$$

where n is the free electron density, I is the laser pulse intensity, ω is the laser frequency, Δ is the glass band gap, W is the ionization rate, and η is the avalanche coefficient. Usually, W is calculated using Keldysh formula [17]. In case of multiphoton ionization, the ionization rate can be presented as

$$W = \sigma(k)I^k, \tag{2}$$

where the value of k refers to the order of the multiphoton process.

To estimate the contribution of the multiphoton absorption and avalanche ionization one can compile Table 10.2, based on the data for the comparable experimental conditions. The data in the table indicate that for the short pulse duration of 20–40 fs the avalanche ionization has a negligible contribution.

Depending upon the electric field strength E the nonlinear light absorption may occur due to the multiphoton ionization (MPI) or due to tunneling. Tunneling mechanism is involved for higher electric field when this field distorts the boundaries of the bandgap. According to Refs. [9,18] the critical parameter γ is used to distinguish these two processes is given by the formula

$$\gamma = \frac{\Omega\sqrt{m_{\mathrm{red}}\Delta}}{eE}, \tag{3}$$

where Ω is the laser frequency, m_{red} is the reduced effective mass, and e is the electron charge. The case of $\gamma \gg 1$ (high laser frequency and/or low electric field) corresponds to multiphoton ionization while in case of $\gamma \ll 1$ tunneling is dominating. For the two regimes described above electron excitation occurs through multiphoton absorption in the low pulse energy case and through tunneling in the high energy case. However, as concerns the waveguide formation the unanswered question remains how much of the absorbed light energy is transferred to the lattice.

10.5.1. Thermal Model

Although the explicit thermal mechanism is not known, it is thought to involve the local rapid heating of a small volume of the glass at the focal spot. Heating the glass and freezing in the high temperature state could produce a lowering of an index, instead of an increase. However, one might imagine a shock wave generated by the rapid expansion leading to compression and an increase of refractive index [3].

Streltsov and Borrelli [3] measured the absorbed energy from fs-laser pulses that were focused inside the bulk of the sample, which was constantly translated so not to maintain the focus in the same spot of the sample. The energy after exiting the sample was monitored relative to the input. The results at 800-nm for both fused silica are plotted as absorbance vs. pulse energy in Fig. 10.18. They then modeled the thermal situation corresponding to the experimental exposure conditions using a finite-element heat diffusion equation with a heat source term added appropriate to the pulse energy, pulse duration, and repetition rate. In the case studied with a 20 kHz repetition rate, the time between pulses was sufficiently long relative to the thermal diffusion time (around 1 µs) so that only one pulse does the heating. Since the translation speed is slow in comparison to the repetition rate, the spot is relatively fixed.

The temperature rise was calculated using the absorbed energy, which was obtained from the measured value of the absorption coefficient. On the basis of the absorbed energy data and the thermal model, they were able to make a guess at the temperature rise associated with our experimental exposure protocols. This is shown in Fig. 10.17b for silica.

Although the predicted temperature rise is significant and seems to initially track the index change, it is nonetheless difficult to conclude on the basis of these data that the temperature is the dominant parameter.

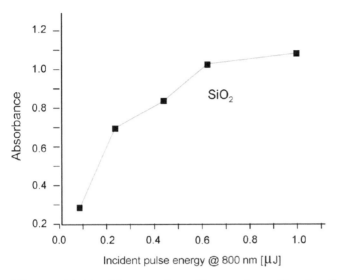

Figure 10.18 Measured nonlinear absorption in silica as a function of incident 800 nm pulse energy [25].

10.5.2. Color Center Model

It has been suggested that the possible source of the induced refractive index change is the effect of the radiation to produce color centers [19] in sufficient number and strength as to alter the index through a Kramers–Kronig mechanism. This has been a proposed mechanism for the index change produced by deep-UV excitation of the Ge-doped silica fibers resulting to fiber Bragg gratings [20]. In deep-UV irradiated Ge-doped silica fiber, the Ge E' center concentration is of the order of $5 \times 10^{17}\,\mathrm{cm}^{-3}$. The exposure was with $1\,\mathrm{MW/cm}^2$ per pulse for 10^6 pulses from a 248-nm excimer laser that resulted in induced refractive index change of the order of 10^{-4}. For pure silica, the numbers of Si E' and NBOHC that can be produced with much longer exposure at 193-nm are the order of $10^{15}\,\mathrm{cm}^{-3}$ [13].

Streltsov and Borrelli [3] measured the dominant color center produced by the fs-laser exposure (i.e., in the case of silica it was the Si E' center). The measured results indicate that color centers of the order $3 \times 10^{19}\,\mathrm{cm}^{-3}$ are being

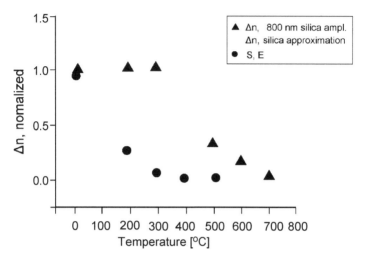

Figure 10.19 Temperature stability of the fs-laser induced refractive index change (triangles) as compared to that of the Si E′ color center (circles) [25].

produced. In comparison to the Si–Ge case mentioned above, this could correspond to refractive index changes of 10^{-3}.

They used the comparison of annealing behavior to learn if the color centers and the induced refractive index might have a common origin. In Fig. 10.19 is plotted the normalized induced refractive index change as a function of temperature for silica and compared it to the normalized E' concentration. The color center signal is seen to diminish at a faster rate with annealing temperature compared to the induced refractive index change. On the other hand, different thermal stability notwithstanding, the positive argument for the color center origin for the induced refractive index is that because of the high electron density produced by the fs-laser exposure there are sufficient trapped species (color centers) created to produce the index change via Kramers–Kronig.

10.5.3. Structural Change Model

There is the possibility that the induced refractive index change derives from some sort of structural change which in turn produces an index change. An example of this would be deep-UV induced density changes that have been observed in silica and other binary systems [21]. The signature of laser-induced densification is the stress that is produced in the surrounding unexposed medium in response to the volume changes produced in the exposed region. In the 800-nm laser case we would appeal to a multiphoton excitation of the same phenomenon as observed with deep-UV exposure. A representative set of stress birefringence patterns is shown in Fig. 10.9 a for writing speeds of 10–100 μm/s at pulse energy of 1 μJ. The scan of the measured retardance for a 0.5-mm sample thickness is shown in Fig. 10.20. This resulting refractive index change calculated from the measured birefringence as a function of the writing speed for two exposure energy is an order of magnitude smaller indicating that densification alone cannot account for the entire effect (Fig. 10.19).

There is another study alluding to fs-laser-induced structural changes evident from change in the Raman spectrum [22] (see discussion and references therein). Raman spectrum peaks at $490 \, cm^{-1}$ and $605 \, cm^{-1}$ have been correlated to structural change. For the case at hand, one must further propose the link between the observed increases in the refractive index and the growth of these peaks. It has been reported such changes occur in the Raman spectrum and indicate that

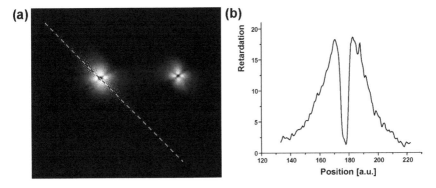

Figure 10.20 (a) Axial stress birefringence profiles of exposed tracks viewed under crossed polarizers, (b) the map of the magnitude of the induced birefringence.

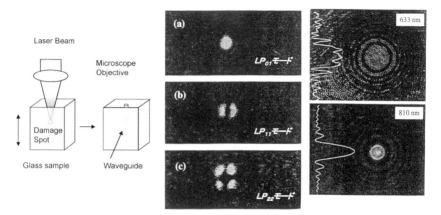

Figure 10.21 Schematic diagram of optical waveguide writing by high repetition laser pulses. Near-field patterns at 800 nm on fluoride glass waveguides where the core diameters are (a) 8 μm, and (b) 17 μm and (c) 25 μm respectively. Far-field patterns of a fluoride waveguide for different wavelengths. The sample was the same as that observed in left.

structural changes have occurred [22]. On the other hand, Stretsov and Borrelli [3] using a micro-Raman instrument directly examined four tracks each corresponding to different exposure energy ranging from 1 to 4 μJ. They found, only at the highest energy of 4 μJ was there any evidence of a change. On the other hand, measurable refractive index changes of $>10^{-3}$ were obtained for all the tracks although the track written at 4 μJ had visible mechanical damage.

10.6. DEVICE APPLICATIONS

The induced index change that can be produced by controlled fs-laser exposure has been used to produce a number of novel structures that could have photonic device implications. The general view is that it extends one's ability to create index patterns in real 3-D dimensions, like a 1 cm cube. The limitation in dimension of the path produced is imposed by the working distance of the writing lens. This is easily seen from the diagram shown in Fig. 10.13a for the axial writing case.

The thickness of the path through the sample cannot exceed the working distance of the lens. Unfortunately, one also need a reasonably high NA to produce the high intensity at the focus. Since these are opposing properties one must adopt a trade-off. The best lens used has an NA of the order of 0.25 and a working distance of around 20 mm. In the following sections, we will review the structures that have been made and their performance.

10.6.1. Waveguide Structures

The first optical structure that was made utilizing the fs-laser exposure-induced refractive index change was an optical waveguide embedded in a 3-D piece of silica, or glass. Hirao and co-workers [1] were the first to report writing such structures. They used an amplified 810-nm Ti–sapphire laser with 120 fs operating at 200 kHz with an average power of 1 W. A 5-mm beam was focused using a 5–20× microscope objective to a moving XYZ stage (speed could be controlled from 0.1 to 10 mm/s). They were able to write measurable tracks in a wide variety of glasses such as silica, fluoride glasses, lead-silicate and even a sulfide glass using this technique. The microscope pictures of the written pattern were shown in Fig. 10.14. The fact that one can see the written track as shown does not insure that a true waveguide structure has been written. However, for the fluoride glass they show a series of near-field patterns for three track diameters clearly showing the expected higher order modal behavior change. The authors are not specific to how they altered the diameter of the guide. There is some indication that they used multiple scans and in that process the altered region could be widened. They also show a far-field pattern for one of the guides at two different wavelengths. The bulls-eye pattern appears to originate from the interference of the guided mode with light from the launch that is not guided. In other words, it comes from the mismatch of the NA of the lens and the acceptance NA of the guide. Since the guides are so short, 1–2 cm, one cannot strip out the unguided light. The estimate of the induced index change is quite high, 0.01.

Homoelle et al. [23] reported on a more extensive study of the induced refractive index of silica and boron-doped silica as a function of writing fluence and speed. There results are shown in Table 10.1. It should be kept in mind that it is the intensity that is important, not the fluence per se. One must be careful to factor in the pulse duration when comparing results from different authors since they most often report the exposure in terms of the pulse energy.

The method used to estimate the refractive index of the waveguide was to measure the NA from the far-field pattern. What was interesting was the apparent correlation of the induced refractive index change with induced densification. Although it may not be the sole source of the index change, it seemed to account for the enhanced sensitivity of the B-doped silica.

Hirao and co-workers [1] reported losses as low as 0.1 db/cm, but without any description of the method by which it was obtained. Florea and Winick [24] report on the results of a standard cut-back method and obtained a value of 1.3 db/cm at 633 nm.

10.6.2. Couplers

If one were to implement a true 3-D structure using the fs-laser technique one would need to make directional couplers and gratings to direct the light from one part of the 3-D structure to another. Hoemelle et al. [23] were the first to demonstrate a Y-coupler as shown in Fig. 10.22. The structure was made using a 60 fs pulse at 1 μJ with a writing speed of 30 μ/s. An Ar-laser was used to show the relative coupling efficiency. More recently Streltsov and Borrelli [25] wrote a directional waveguide structure and measured the splitting ratio to be near 3 db. Their results are indicated in Fig. 10.23. In Fig. 10.23a is shown the actual dimensions of the beam writing paths that were used. In (b) figure, the near-field output is shown as indicated, and in (c) is shown the far-field output.

Minoshima et al. [26] also reported on the fabrication of a directional couples as shown in Fig. 10.24. This design

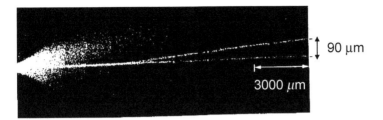

Figure 10.22 Y-coupler written by fs-laser method. Light entering from left is split in the *y*-section (taken from Ref. [23]).

provides a controllable coupling interaction distance. One problem is the rather sharp corners which could cause loss. Nonetheless, they found excellent correlation of the coupling with length as shown in the (c) through (d) figures.

A more ambitious 3-D structure is shown in Fig. 10.25 [27]. Here a central straight waveguide is coupled to two adjacent guides as shown. When light is input into the central guide one can see equal amount of light in the two adjacent guides as shown in (a). When light enters through one of the adjacent guides, one can see light coupled to the central guide and only a very weak amount to the other adjacent guide as expected. This result is shown in (b).

There are some nonobvious difficulties in creating these structures. The first is that since the induced index is a function of the writing beam speed, one must be sure that is traversing the path changes that it is done in a continuous manner at the same speed. Otherwise one develops regions of higher index and perhaps different diameter. Although this difficulty can be overcome with appropriate hardware and software, the second poses a more serious problem. This is that in the writing of the second waveguide because of the proximity to the first written guide, there is a change in the index of the first guide. This interaction is tolerable in the case of a two-guide coupler but it becomes much more serious as the number of guides increases. Each newly written guide has an effect on the characteristics of all of the other guide previously written. This makes maintaining a knowable and constant coupling ration to all the guided very difficult to obtain.

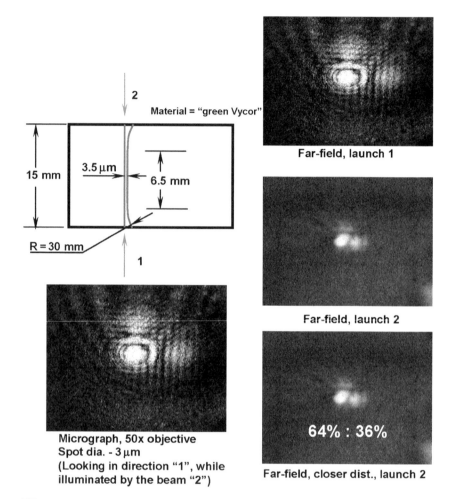

Far-field, launch 1

Far-field, launch 2

64% : 36%

Far-field, closer dist., launch 2

Micrograph, 50x objective
Spot dia. - 3 μm
(Looking in direction "1", while
illuminated by the beam "2")

Figure 10.23 Directional coupler written by 400-nm 30-fs 2.5-nJ laser pulses, speed—10-μm/s.

10.6.3. Gratings

We have discussed in some detail in Chapter 7 the methods and properties of gratings. The most important of these are the so-called fiber Bragg gratings. They are used as filters in their short period version and mode couplers in their long period state. Mention was also made of methods of making holographic gratings in bulk glasses as well. These gratings

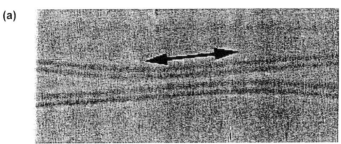

(a)

(b)

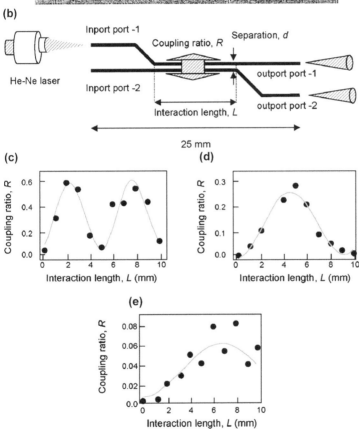

could extend through a millimeter or more of glass. In the above-mentioned scheme of an integrated 3-D structure, gratings localized to given positions within the structure would be required either to redirect the light or couple light from one waveguide to another adjacent guide. The fs-laser writing technique would allow such a positioning is a cube of glass.

Florea et al. [28] have written gratings with a 10 μm pitch into silica with a fs-laser with pulse energies of 0.5–10 μJ. The writing beam was focused inside the sample and each grating line was drawn by scanning the focused laser into parallel lines perpendicular to the eventual beam propagation direction. They used a 50× objective to form the ultimate writing beam. The scan rate was reported to be 25 μm/s. Since the top-writing method was being used, the depth of the grating was determined by the Rayleigh range of the focused beam and the threshold intensity of the writing process. What this means is that the actual affected area is smaller than the actual focal spot size. In Fig. 10.26 is shown the measurement of the diffraction efficiency vs. pulse energy. This direct writing of gratings has also been done for making long-period gratings in optical fibers by the ERATO group in Japan.

Kawamura et al. [29] used a more sophisticated writing method based on the "colliding" of two short pulses inside a

---

Figure 10.24 (a) Phase contrast microscopic image of one of the directional couplers. The aspect ratio of the image is compressed by a factor of 5 in the horizontal direction in order to better visualize the directional coupler. The arrow in the image indicates the interaction region. (b) Schematic of the coupler. Separation, d, and interaction length, L, are varied with the fixed total length, 25 mm. He–Ne laser is guided into one of the input ports and the output power at two ports is measured. The coupling ratio between the two waveguide ports, R, is obtained. (c) Interaction length dependence of the coupling ratio for waveguide separations $d = 8$-μm (d) $d = 10$-μm and (e) $d = 2$-μm. Experimental results (dots) and their best-fit results to sinusoidal curves (lines) are shown. The period of the oscillation increases from (c) 5, to (d) 9, to (e) 14 mm which is consistent with the coupled mode theory.

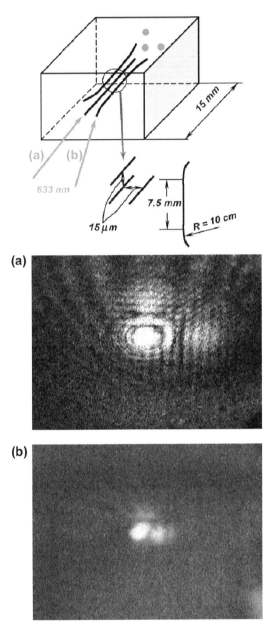

Figure 10.25   Schematic of 3-D coupler with dimensions. (a) Photo of output when light is initially in the center guide, (b) when light is initially in one of the outer guides.

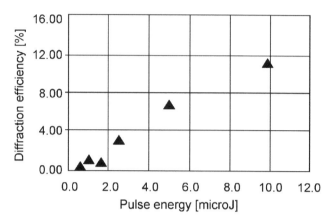

Figure 10.26 Diffraction efficiency of grating written in silica as a function of the pulse energy. Pitch of grating is 10 μm (taken from Ref. [28]).

silica sample. This is essentially two beam interference technique. They used 50 μs pulses with a duration of 500 fs. A photomicrograph of the resulting gratings are shown in Fig. 10.27.

One limitation of the former method is that it will write gratings with a pitch smaller than 1 μm at an arbitrary depth will be a problem. However, this is not a limitation of the interference method.

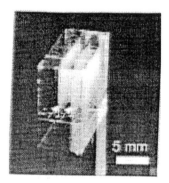
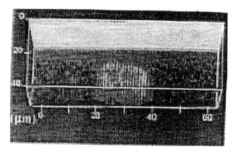

Figure 10.27 Photo of holographic writing of volume-type micro-gratings in silica (taken from Ref. [29]).

10.6.4. Waveguide Amplifier

Another required aspect of a fully functional 3-D optical structure is some way to produce gain, either a laser or an amplifier. Sikorski et al. [30] wrote a 1 cm long waveguide in a Nd-doped glass using the fs-laser writing method. They reported a gain of 1.5 dB/cm at the signal wavelength of 1054 nm using 346 mW of 514 nm pump. The gain was measured as the ratio between the signal power with the pump turned on and with the pump off.

10.6.5. Optical Storage

There have been a few attempts to utilize the ability to produce highly spatially localized optically altered regions as a basis for a 3-D memory. Glazer et al. [31] showed the results of patterned damage in a cube of fused silica using a 0.5 μJ, 100 fs pulse focused through a 0.65 NA microscope objective. The schematic of the idea is shown in Fig. 10.28. The bit is read by use of transmitted light in a microscope with a 0.95 NA objective. The spacing between bits was 2 μm. The spacing between recorder layers was 15 μm. From their analysis of the recorded bit, the lateral extent was the order of 1 μm, although this was limited by the resolution of the

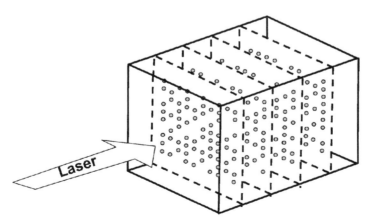

Figure 10.28 Schematic diagram of 3-D optical storage device [31].

optical microscope, and the longitudinal extent was 2.5 μm. Using an SEM technique to look at an exposed recorded spot, the estimate of the size was considerably smaller and led to an estimate of the ability to record 10^{13} bits/cm^3. This technique relies on physical damage, some sort of microvoid in the silica.

The other attempts have proposed to use a photo-thermal material as the recording medium (see Chapter 3, Sec. 3.4.3). By photo-thermal is meant that the initial exposure initiates the effect in the material, but the index change is not manifest until a subsequent thermal treatment is performed. Kondo et al. [32] used a Ce/Ag-based photosensitive glass as their medium. The exposure was with a 630-nm laser with 150 fs pulses, and pulse energy of 80 nJ at 1 kHz for 5 s. The beam was focused into the glass using a 50 mm focal length lens. They then scanned the beam at a rate of 25 μm/s to form lines of exposure. The glass was then given a heat treatment, 540°C for 30 min to develop the NaF phase. The resolution of the developed lines of NaF phase was the order of 10 μm.

Streltsov and Borrelli [33] used essentially the same material and method. Their exposure was at 800 nm with much shorter pulses of 30 fs and correspondingly much lower pulse energy. They also used a shutter so to make distinct exposure spots as would be the case for a memory application. The exposure pattern in a given plane is shown in Fig. 10.29a. The dots correspond to the subsequently thermally formed NaF phase. Figure 10.29b shows the developed pattern looking in from the side. The length of the NaF-containing region comes from the Rayleigh range of the focusing objective. In Fig. 10.29c is shown the diffraction pattern produced by the 3-D pattern.

In another novel variation on the optical memory application, a report by the ERATO group in Japan (REF) describes the ability to change the valence state of Sm^{+3} in glass after exposure to focused fs-pulses at 800-nm at a power level of 200 mW. The photoluminescence of the Sm-ion in the exposed region compared to the unexposed region is shown in Fig. 10.30. The sharp emission lines exhibited in the exposed region is attributed to Sm^{+2}, whereas the weak luminescence in the unexposed region is that of Sm^{+3}. It appears that an

(a)

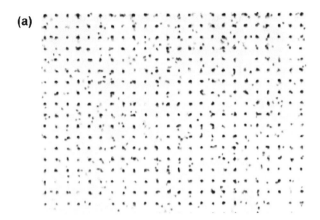

(b)

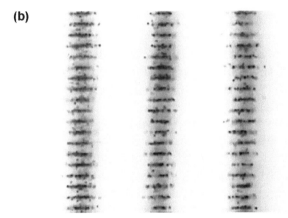

(c)

electron has been excited through a multiphoton process and is trapped by the Sm^{+3} to form Sm^{+2}. They propose that this is a possible scheme to make what amount to a 4-D memory. The spatial location of an altered spot as described above, and in addition the a spot with an different emission wavelength. Of course the read mechanism now would have to correspondingly more complicated.

10.6.6. Miscellaneous Applications

There have been a few rather novel applications involving the use of fs-laser exposure reported in the recent literature. Marcinkevicius et al. [35] describe a novel way to micromachine a silica glass. They essentially created a damage track of a line of small voids. They used 120 fs pulses focused by a 1.3 NA oil immersion objective to make the damage track. They then subjected the sample to a HF etching bath for a period of a few hours. The action of the HF solution was to dissolve the thin glass regions between the voids. They ultimately were able to obtain a 12 μm diameter track through a sample of silica that was 0.12 mm thick.

Streltsov [36] were able produce a much longer channel by using a much lower NA objective. They were able to obtain a 18 μm diameter channel to a distance of 0.8 mm after the HF treatment. They used an exposure protocol of 40 fs pulses at 10 μ9 with a translational speed of 0.1 mm/s. A photomicrograph of their result is shown in Fig. 10.31. They extended the technique to make holes of much larger diameter by using a helical damage track pattern. In this case they were able to produce a 0.12 mm diameter channel through a 2 mm thick sample of silica. A photomicrograph of such a channel is shown in Fig. 10.32. An important application of this new technique of drilling holes in fused silica in shown in Fig. 10.33 where an

----

Figure 10.29 Series of photomicrographs of 3-D structure written in a photosensitive glass, see text. A is an end-on view of photo-produced CaF_2 crystals, B side view showing depth of exposure, and C, a diffraction pattern produced by the 3-D grating [33].

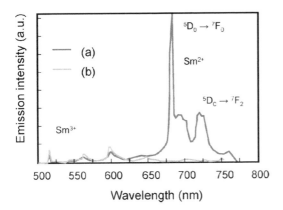

Figure 10.30 Emission spectra of (a) the waveguide formed by the laser irradiation and (b) the non irradiated area in a Sm^{3+} doped fluoride glass. The excitation wavelength is 514 nm. This shows that the photo-reduction of Sm^{3+} to Sm^{2+} has occurred simultaneously with the refractive index change.

Figure 10.31 Photomicrograph of fs-laser hole drilling in silica. See text for explanation of the technique.

Figure 10.32 Photomicrograph picture of single hole shown in the above in Fig. 10.31. Diameter is 125 µm.

Figure 10.33 Standard optical waveguide fiber inserted into hole made by fs-laser drilling.

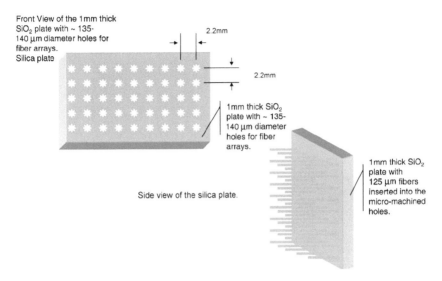

Figure 10.34 Schematic of fiber array holder.

optical waveguide is shown threaded through the laser-made channel. The technique enables one to precisely position holes in a 2-D array in a 2 mm thick piece of fused silica thus providing a way to position a 2-D array of single mode optical fibers as schematically shown in Fig. 10.34.

10.7. INTERACTION WITH CRYSTALS

All that we have discussed above involved the interaction of the fs-laser pulses with glass. The interaction with crystal appears to more difficult to generalize in terms of the ability to alter the refractive index without causing physical damage. The mechanisms that have been proposed for the refractive index change in glass do not readily carry over to crystals. This is primarily because the crystal structure, unlike the glass, has much less ability to accommodate changes. Consider the example of stress. If the effect of the fs-laser interaction were to in some way lead to a volume change, the crystal if it were not to damage would have to find a way to structurally adapt. One possible way would be to create

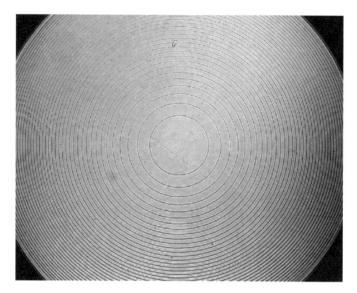

Figure 10.35 Fresnel-like lens made in single crystal CaF$_2$ plate by Fs-laser exposure.

Figure 10.36. Image formed by lens in Fig. [10].

slip-planes, or other oriented changes that would still allow the crystal to maintain the crystal structure. In general, it is safe to say that it is much harder to produce, and control refractive index changes in crystals than it is in glass, and the mechanism may be quite different from that proposed for glasses.

As an example we show the refractive index pattern written in a (111) oriented CaF_2 single crystal with a 800 nm fs-laser exposure in Fig. 10.35. The index pattern essentially forms a Fresnel lens structure. We show the image formed by this lens in Fig. 10.36 with an estimate of the effective focal length of 0.5 mm. The induced refractive index change is estimated to be 0.01 and corresponds to a decrease. Electron microscopic studies indicate that the exposed region does not show a diffraction pattern. This lack of a discernable diffraction pattern could mean that the number of point-like defects produced by the irradiation is so large that the crystalline structure is essentially gone.

In quartz, one also sees a similar decease in the refractive index. In $LiNbO_3$ it is not so clear as to the nature of the refractive index change indicating that the mechanism in this case is even more complex.

10.8. APPENDIX A

10.8.1. Mode-Locked Lasers: Ultra-short Pulses

One of the significant advances in lasers over the last 10 years has been the development of the ultra-short pulse mode-locked laser. Mode locking is an old laser concept, however, the ability to produce an array of stable sub-picosecond pulses is much more recent. Mode locking is based on establishing a temporal relation for modes propagating in the cavity. In general many modes would experience some level of gain. The objective of mode locking is to provide some controlled loss mechanism in the cavity (or external) to create a stable temporal train of pulses The loss in combination with the gain lifetime, will determine the width of the pulse. In simple terms, in a pumped laser cavity, when the gain of any given mode is sufficient to overcome the loss, one gets a laser spike.

The gain is depleted, but the pump is still on so the gain recovers till threshold is reached again and another spike occurs. This would proceed in a chaotic way, but with the appropriate way of controlling the loss, one can stabilize this behavior into a well-behaved train of pulses.

A typical gain cavity is shown in Fig. 10A.1a. The cavity can support many modes, all those who fit according to the wavelength and the physical distance between the mirrors,

$$L = m\lambda/2, \quad m \text{ is an integer} \tag{A10.1}$$

The round trip transit time then is simply

$$T = 2L/c \tag{A10.2}$$

We sketch the gain spectrum of the laser medium, and the modal separation as $\Delta\nu$ in Fig. 10A.1b. In Fig. 10A.1c we show the pulses with period T and pulse width τ. A breakthrough toward the narrowing of the pulse duration to sub-picosecond came with the discovery of what is called passive mode locking. The use of the term passive comes from the comparison to the prior approach that used electo-optic or acouto-optic modulators in the cavity to establish the mode locked state. The modulator was placed in the cavity and driven at an rf frequency that satisfied the following relation:

$$f = 2m/T \quad \text{where } m \text{ is an integer} \tag{A10.3}$$

The initial passive system used an organic dye as the gain medium, rhodamine-6G, and a saturable absorber, DODCI. The mechanism for the pulse formation is indicated in Fig. 10A.2. As one can see when one pumps the system the intensity in the cavity grows. The gain is high because the inversion is high due to the added loss of the absorber. When the intensity is high enough the absorber saturates, indicated by the arrow in the diagram. The inversion now is suddenly reduced leading to the start of the laser pulse. The pulse will persist until the gain drops below threshold. One can see that the shape of the pulse is governed on one edge by the speed at which the absorber saturates and on the other edge by the recovery time of the gain medium.

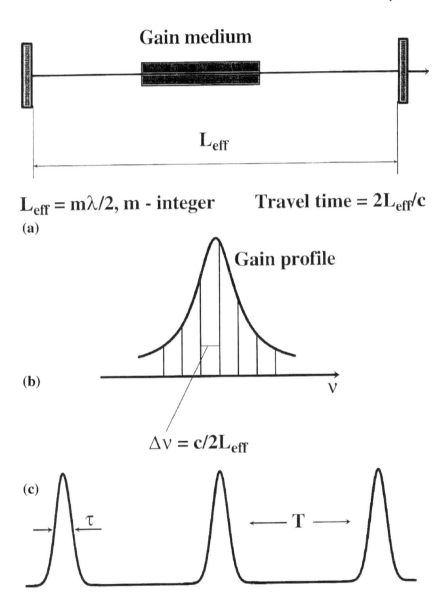

Figure 10A.1 (a) Schematic of gain medium in an optical cavity,
(b) Skeych of a gain curve of tube of the active medium with the
modal separation indicated, (c) Schematic of pulse train generated
showing spacing and width of the pulses.

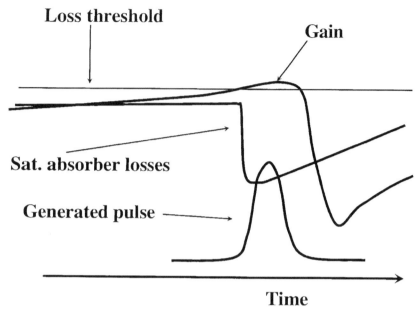

Figure 10A.2 Schematic representation of how pulse is formed when saturable absorber is used. Intensity in cavity grows until sufficient to saturate the dye which initiates the leading edge of the pulse, inversion is then depleted which determines the trailing edge of the pulse.

Kerr lens mode locking

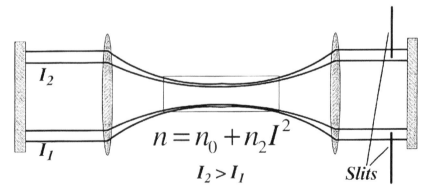

Figure 10A.3 Schematic representation of the Kerr Lens mechanism for mode locking. See text for explanation.

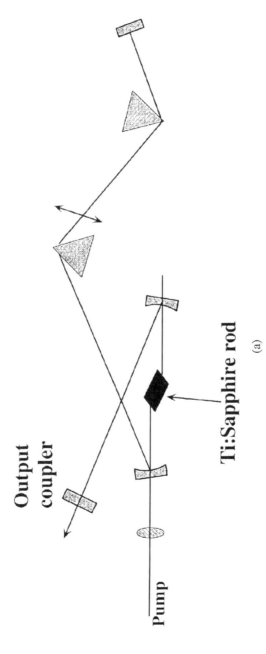

Figure 10A.4 (a) Representation of an actual Kerr Lens optical cavity showing the prisms used to provide dispersion compensation, (b) Typical mode locked pulse, (c) Typical spectral width of a pulse.

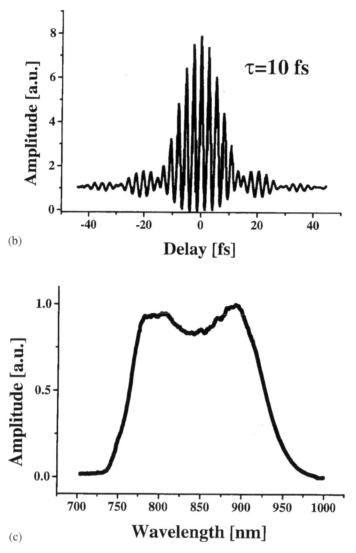

(b)

(c)

Figure 10A.4 Continued

Although the dye-based system could produce sub-picosecond pulses, there were some undesirable features. For a number of reasons, the gain medium of choice was the Ti–sapphire laser. It was an efficient laser material with a much broader gain bandwidth, but it had a longer excited state lifetime and a smaller gain cross-section than the dye. Because the lifetime was too long, the recovery time was too slow for the use of the saturable absorber method. What was found was an entirely new way to produce short pulses which is called Kerr Lens Mode Locking. A simple diagram of how it works is shown in Fig. 10A.3. The lenses focus and re-collimate the light in the cavity so that the focus in inside the laser crystal. The aperture provides the loss. When the intensity gets high enough, then the refractive index of the sapphire crystal is raised as indicated by the high-frequency Kerr effect. So it is a very fast switch since it is a purely electronic phenomenon. The problem that ensues is that there is also considerable dispersion produced by the nonlinearity (self-phase modulation) which broaden the pulses. To compensate for this a number of prisms are added to the path as shown in Fig. 10A.4a. A representative temporal and spectral output is shown in Fig. 10A.4b & c.

REFERENCES

1. Miura, K.; Qui, J.; Inouye, H.; Mitsuyu, T.; Hirao, K. Photowritten optical waveguides in various glasses with ultrashort pulse laser. Appl. Phys. Lett. **1997**, *71* (23), 3329–3331; Hirao, K.; Miura, K. Writing waveguides and gratings in silica and related materials by a femtosecond laser. J. Non-Crystalline Sol., **1998**, *239*, 91–95.

2. Schaffer, C.B.; Brodeur, A.; Mazur, E. Laser-induced breakdown and damage in bulk transparent materials induced by tightly focused femtosecond laser pulses. Measurement Sci. Technol. **2001**, *12*, 1784–1794.

3. Streltsov, A.; Borrelli, N. Fabrication and analysis of a directional coupler written in glass by nanojoule femtosecond laser pulses. Opt. Lett. **2001**, *26* (1), 42–43.

4. Stuart, R.C.; Feit, M.D.; Rubenchik, A.M.; Perry, M.D. Optical ablation by high-power short-pulse lasers. J. Opt. Soc. Am. B. **1996**, *13* (2), 459–468.

5. Kautek, W.; Kruger, J. Laser ablation of dielectrics with pulse duration between 20fs and 3ps. Appl. Phys. Lett. **1996**, *69* (21), 3146–3148.

6. Liu, X.; Mourou, G. Ultrashort laser pulses tackle precision machining. Laser Focus World August **1997**, 101–118.

7. Nolte, S.; Momma, C.; Tunnermann, A. Ablation of metals by ultrshort laser pulses. J. Opt. Soc. Am. B **1997**, *14* (10), 2716–2722.

8. Mourou et al. G.A. US Patent 5, 656, 186, August 1997.

9. Lenzner, M.; Kruger, J.; Sartania, S.; Cheng, Z.; Spielmann, C.; Mourou, G.; Kautek, W.; Kraus, F. Femtosecond optical breakdown in dielectrics. Phys. Rev. Lett. **1998**, *80* (18), 4076–4079.

10. Li, M.; Menon, S.; Nibarger, J.P.; Gibson, G. Ultrafast electron dynamics in femtosecond optical breakdown of dielectrics. Phys. Rev. Lett. **1999**, *82* (11), 2394–2397.

11. Lenzner, M. Femtosecond laser-induced damage of dielectrics. Int. J. Mod. Phys. **1999**, *13* (13), 1559–1578.

12. Liu, X.; Mourou, G. Laser ablation and micromachining with ultrashort laser pulses. IEEE, JQE **1997**, *18* (11).

13. Technical digest. ERATO Hirao Active Glass Project, 1998.

14. Watanabe, W.; Tuma, T.; Yamada, K. Optical seizing and merging of voids in silica glass with fs-laser pulses. Opt. Lett. **2000**, *38* (22).

15. Miura, K.; Qui, J.; Inouye, H.; Mitsuyu, T.; Hirao, K. Photo-written optical waveguides in various glasses with ultrashort pulse laser. Appl. Phys. Lett. **1997**, *71* (23), 3329–3331.

16. Varel, H.; Ashkenasi, D.; Roesnfeld, A.; Herrmann, R.; Noack, F.; Campbell, E. Appl. Phys. A **1998**, *62*, 293–294.

17. Keldysh, L.V. Ionization in the field of a strong em-wave. Sov. Phys. JETP. **1965**, *20*, 1307–1314.

18. Kaiser, A.; Rethfeld, B.; Vicanek, M.; Simon, G. Microscopic processes in dielectrics under irradiation by subpicosecond laser pulses. Phys. Rev. **1998**, *61* (11), 437–450.

19. Davis, K.M.; Miura, K.; Sugimoto, N.; Hirao, K. Writing waveguids in glass with fs-laser. Opt. Lett. **1996**, *21* (21), 1729–1731.

20. Hill, K.O.; Meltz, G. J. Lightwave Techn., Fiber Bragg grating technology fundamentals and overview. **1997**, *15* (8), 1263–1267..

21. Borrelli, N.F.; Smith, C.M.; Allan, D.C. Excimer-laser-induced densification in binary silicate glasses. Opt. Lett. **1999**, *24* (20), 1401–1403.

22. Chan, J.; Huser, T.; Risbud, S.; Krol, D. Structural changes in fused silica after exposure to focused femtosecond pulses. Opt. Lett. **2001**, *26*, 17826–17828.

23. Homoelle, D.; Wielandy, S.; Gaeta, A.L.; Borrelli, N.F.; Smith, C. Infrared photosensitivity in silica glasses exposed to femtosecond laser pulses. Opt. Lett. **1999**, *24* (18), 1311–1313.

24. Florea, C.; Winick, K.A. Characterization of waveguides and gratings written in glass with near-IR fs-laser pulses. J. Light Techn. **2003**, *21*, 246–300.

25. Streltsov, M.; Borrelli, N. Fabrication and analysis of a directional coupler written in glass by nanojoule femtosecond laser pulses. Opt. Lett. **2001**, *26* (1), 42–43.

26. Minoshima, K.; Kowalevicz, A.; Ippen, P.; Fujimoto, J. Fabrication of coupled photonic devices to glass by nonlinear fs-laser material processing. Opt. Express **2002**, *10* (15), 645.

27. Streltsov, A.M.; Borrelli, N.F. Characterization of laser-written waveguides in glasses. OSA Annual meeting, 2001, ThAA4.

28. Florea, C.; Winick, K.A.; Sikorski, Y.; Said, A.A.; Bado, Ph. "Optical waveguide amplifier in Nd-doped glass written with near-IR fs-laser" CLEO 2000, *Tech. Digest Tops 39*.

29. Kawamura, K.; Hirao, M.; Kamiya, T.; Hosono, H. Holographic writing of volume-type micrograting in silica glass by a single chirped laser pulse. Appl. Phys. Lett. **2002**, *81* (6), 1137.

30. Sikorski, Y.; Said, A.A.; Bado, P.; Maynard, R.; Florea, C.; Winick, K. Optical waveguide amplifier in Nd-doped glass written with near-IR fs-laser pulses. Opt. Lett.

31. Glazer, E.N.; Milosavljevic, M.; Huang, L.; Finlay, R.J.; Her, T.-H.; Callan, I.P.; Mazur, E. 3-D optical storage inside transparent materials. Opt. Lett. **1996**, *23*, 2023–2025.

32. Kondo, Y.; Suzuki, T.; Inouye, H.; Mituyu, T.; Hirao, K. 3-D microscopic crystallization in photosensitive glass by fs-laser pulses at nonresonant wavelengths. Jpn. J. Appl. Phys. **1998**, *37*, 94–96.

33. Streltsov, A.; Borrelli, N.F. Private data.

34. Technical Digest, ERATO Hirao Active Glass Project Symposium, Kyoto, Sept. 1999.

35. Marcinkevicius, A.; Jnuodkazis, S.; Watanabe, M.; Miwa, M.; Matsuo, S.; Misawa, H. Femtosecond laser-assisted 3-D microfabrication in silica. Opt. Lett. **2001**, *26* (5), 277–279.

36. Streltsov, A. Private data.

11

Negative Refractive Index Materials

11.1. BACKGROUND

The exposition of the consequences of light propagation in a material that exhibits both a negative dielectric constant and a negative magnetic permeability was first published by Veselago in 1968 [1]. It set the framework for a number of interesting and unusual properties that derive from this hypothetical condition. Since then a number of articles have appeared dealing with the idea [2,3], we will review his mathematical approach below, as well as how this formalism leads to the interesting physical manifestations. This will be followed by a review of the most recent work that is suggestive of ways to find materials with the desired negative quantities and how well the experiments or simulation conform to what was predicted.

It should be realized that there is nothing physically impossible about a material that exhibits a negative dielectric constant, or a negative magnetic permeability. Nonetheless, the odds of finding both at the same time is rare and requires some thought as to how it may be brought about.

One starts with Maxwell's equations for a lossless dielectric

$$\nabla x E = \frac{-1}{c}\frac{\partial B}{\partial t} \tag{11.1}$$

$$\nabla x H = \frac{1}{c}\frac{\partial D}{\partial t} \tag{11.2}$$

One defines the normal linear constitutive relations in the following form:

$$D = \varepsilon E, \qquad B = \mu H \tag{11.3}$$

One can combine Eqs. (1) and (2) in the standard way and substitute Eq. (3) and obtain the following familiar wave equations:

$$\nabla^2 E = \frac{\varepsilon\mu}{c^2}\frac{\partial^2 E}{\partial t^2} \tag{11.4}$$

$$\nabla^2 H = \frac{\varepsilon\mu}{c^2}\frac{\partial^2 H}{\partial t^2} \tag{11.5}$$

Comparing this to the form of the classic wave equation one concludes that $\varepsilon\mu/c^2$ is equal to the reciprocal of the square of the velocity of the wave and thus determines the refractive index as

$$n^2 = \varepsilon\mu \tag{11.6}$$

In the typical case for the common materials in the optical regime, they are not magnetic, here, then μ is taken as unity and we get the commonly seen relation between the refractive index and the dielectric constant.

To examine the consequence of having the negative sign of ε and μ, we go back to Eqs. (1) and (2) and use the solutions

$$\begin{pmatrix} E \\ H \end{pmatrix} = \begin{pmatrix} E(x,y) \\ H(x,y) \end{pmatrix} \exp i(kz - \omega t) \tag{11.7}$$

One obtains the following two expressions relating E and H to the wave vector, k:

$$kx E = \frac{\omega}{c}\mu H \tag{11.8}$$

$$kx H = -\frac{\omega}{c}\varepsilon E \tag{11.9}$$

The vector product has a convention associated with it. If one writes $\mathbf{a} \times \mathbf{b} = \mathbf{c}$, one refers to the triple $(\mathbf{a}, \mathbf{b}, \mathbf{c})$ as a positive triple linked to a right-handed xyz coordinate system, and a negative triple if associated with a left-handed based coordinate system. One can be more quantitative with respect to this classification by the value of the determinant formed by the direction cosines of the vectors in the order of the triple. If the value of the determinant is $+1$, it is a positive triple (right-hand system, and -1 if it is a negative triple left-hand system). An easy way to keep this straight is to point the thumb of the right hand in the direction of the vector $\mathbf{a}$, the forefinger in the direction of $\mathbf{b}$, and then the middle finger will point in the direction of the product, $\mathbf{c}$.

From Eqs. (8) and (9), it is clear that $\mathbf{E}$, $\mathbf{H}$, and $\mathbf{k}$, form a right-handed system when ε and μ are both positive. Following from this fact and the definition of the Poynting vector which gives the direction of energy flux carried by the wave, one can write the following:

$$\mathbf{S} = (c/4\pi)\,\mathbf{E} \times \mathbf{H}, \tag{11.10}$$

The energy flux of the wave for a right-handed system is in the same direction as $\mathbf{k}$, the propagation vector. If we now consider the situation where both ε and μ are negative, 'Eqs. (8) and (9) reverse their signs and one is led to a left-hand system of reckoning of the cross products. As a consequence, the direction of the energy flux carried by the wave is now opposite to the propagation direction defined by $\mathbf{k}$. That is to say the Poynting vector and the propagation vector are antiparallel.

In an anisotropic material for certain propagation directions one can have the Poynting vector and propagation vector making some angle with respect to other, the angle proportional to the magnitude of the anisotropy of the refractive index (see Fig. 8.10 for an example). One must remember that Snell's law refers to the phase index, while it is not relevant to the group index, or Poynting vector. We shall briefly discuss this below for the case of photonic crystal structures.

A number of ordinary phenomena take a novel twist as a consequence of this change in handedness, the most important of which is the concept of a negative refractive index.

11.2. EFFECT OF MAGNETIC PERMEABILITY

One can also see the consequence of including the magnetic permeability of the material into the expression for the Brewster angle. For the general case that includes the magnetic permeability, one has for s-polarized light (light polarized perpendicular to the plane of incidence) the following expression where one is going from vacuum to the material with values of ε, and μ [4]:

$$\tan \theta_B = \sqrt{\frac{\mu^2 - n^2}{n^2 - 1}} \tag{11.11}$$

For the p-polarization direction, one has the following expression:

$$\tan \theta_B = \sqrt{\frac{\varepsilon^2 - n^2}{n^2 - 1}} \tag{11.12}$$

Where we have used Eq. (6) to define the refractive index. Equation (12) is the ordinary definition for the Brewster angle condition for a medium/air interface. In the case where $\mu = 1$, there can be no solution for Eq. (1). However, for the case where $\mu > !$, and in particular for the situation where $\mu^2 > n^2$ one can have finite reflectivity at the Brewster angle for the s-polarization. One can also show that at normal incidence from a vacuum/material interface there will be no reflection if $|\varepsilon| = |\mu|$. This is a well-known effect in the microwave region [5].

11.3. NEGATIVE REFRACTIVE INDEX

We now examine the consequence of having both ε and μ negative. One of the most striking manifestations of this situation is the way light is refracted as it passes from a right-hand medium to a left-hand medium. This result is based on the left-hand way of reckoning the vectors, **E, H, k**. For the refraction problem, the boundary conditions that must be satisfied at the boundary are the following:

$$\mathbf{E}_{t,1} = \mathbf{E}_{t,2}, \quad \mathbf{H}_{t1} = \mathbf{H}_{t,2}, \quad \varepsilon_1 \mathbf{E}_{n,} = \varepsilon_2 \mathbf{E}_{n,2}, \quad \mu_1 \mathbf{H}_{n1} = \mu_2 \mathbf{H}_{n2} \tag{11.13}$$

Here, the subscript "*t*" refers to the transverse component and "*n*" to the normal component, and the numerical subscripts refer to the medium.

This is equivalent to stipulating that the *z*-component (parallel to the interface) of **k** must be conserved across the boundary. But as one passes from a right-hand material to a left-hand material, the direction of the k_z is antiparallel. According to Fig. 11.1, this would require that

$$n_1 \sin \Theta_1 = -n_2 \sin \Theta_2 \tag{11.14}$$

This means that we define a negative refractive index in the left-hand materials and that for such materials we must take the negative value of the square root of Eq. (11.6). Using Eq. (11.14) as the new version of Snell's law, one gets the interesting phenomenon where light bends to the other side of the

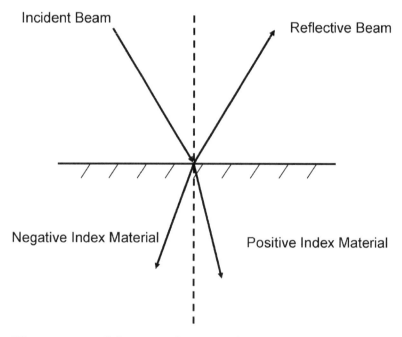

Figure 11.1 Schematic diagram showing direction of refracted beam from a positive refractive index medium to a positive medium (ordinary Snell's law) and into a negative medium where the sign of Snell's law is now negative.

normal as shown in Fig. 11.1 contrasted to the normal case with a right-hand system.

Initially, there was some concern about this phenomenon, but it has been shown experimentally in the microwave regime [6]. We discuss the structure that was used to produce the negative index material in a later section.

11.4. NEGATIVE INDEX MATERIALS

One can make a distinction between the negative index materials as described above by possessing an negative ε and μ, compared to propagation behavior in photonic crystals where one can have modes with negative effective refractive index. In the latter case, certain propagating modes in periodic dielectric structures are made up of materials with positive dielectric properties, the negative index behavior derives from very strong anomalous dispersion near the Brillouin boundaries. So really there are two classes of materials that show unusual refraction phenomena, and ultimately negative refractive index, or left-hand behavior. The first is that of materials with a negative refractive index (negative ε and μ) which we will discuss below, and the second is that of photonic crystals which can exhibit a negative effective index through strong anomalous dispersion produced at the Brillouin boundary.

A convenient graphical way to show the unusual propagation behavior in photonic crystal structures is through the use of what is called the wave-vector diagram [7]. It is a 2-D plot of the wave-vector as a function of angle for a given frequency and state of polarization. Typically, it is shown as a k_x vs. k_y plot which is equivalent to the effective index in the respective directions. For an isotropic material, it is just a circle as shown in Fig. 11.2a. For a birefringent material, it is an ellipse. For photonic crystal structures, as discussed in Chapter 10, the pattern can be quite ornate, reflecting the symmetry of the local structure and strong frequency dependence. For the hexagonal pattern of holes in a dielectric medium, it has at a frequency removed from a band edge, the

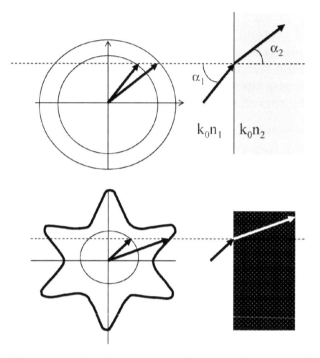

Figure 11.2 Wave-vector diagram (k_x vs. k_y plot) with constant frequency contours, (a) homogeneous material, and (b) photonic crystal.

form of a six-pointed star pattern shown in Fig. 11.2b. As the frequency becomes closer to a band edge (see Chapter 9), the sharper features of the pattern smooth out and eventually becomes a circle again. It is this change in behavior that will make the distinction between anomalous refraction and negative refractive index behavior.

One can use the wave-vector diagram to track the propagation through boundaries between different media. In Fig. 11.2, we have added an inner circle which represents an interface with an isotropic index such as air. The radius of the circle is proportional to the refractive index of the medium it represents. As one can see from the construction, drawing lines from the origin to the points of intersection of the circles with an arbitrary horizontal line (momentum conservation) measures the propagating condition (because of symmetry, the intersection with a dashed line at negative k_y yields the reflecting direction). The same construction is shown for the photonic crystal structure and clearly it results in a much more complicated behavior as a consequence of the periodic structure. Snell's law is no longer determining the relation between the propagation directions unless we define an effective index of the photonic crystal of the medium.

The other interesting aspect of this construction is that the normal to the wave-vector surface at any point is the group velocity

$$v_g = \nabla_k \omega(k) \qquad\qquad (11.15)$$

One can show that this vector $[\partial\omega/\partial k_x, \partial\omega/\partial k_y]$ is every-where normal to the contour formed by $\omega(k_x, k_y) = \omega_0$.

In Fig. 11.3, we show how the propagation changes as a function of incident direction. This is not to be confused with a negative refractive index effect. In the negative index medium, the group velocity and the momentum of propagation vector are traveling in opposite directions as the left-hand system would require. In this case, the two are an angle with respect to each other, in the same half-plane. An actual image of this behavior is shown in Fig. 11.4.

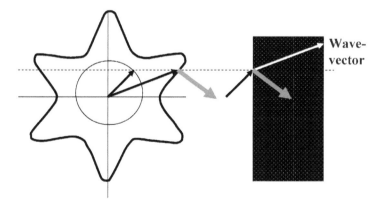

Figure 11.3 Construction with wave-vector diagram showing the determination of the group velocity. Inner circle represents air.

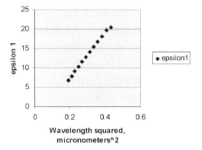

Figure 11.4 Real part of the dielectric constant vs. the square of the wavelength for metallic silver.

As Notomi [10] has pointed out, the difference between these two behaviors (anomalous refraction and true negative refraction) is a matter of the strength of the dispersion produced by a given structure or, for that matter, any strong dispersion producing phenomenon. In the case of photonic crystals, this strength is determined by the proximity of the frequency to the forbidden band gaps. In a very real way, strong anomalous dispersion is the source of all negative refractive index phenomena in all cases. When it gets strong enough to make the refractive index negative at a frequency close to a resonance, then we have the left-hand system where the energy flow is in the opposite direction to the momentum of the wave.

11.5. METHODS OF PRODUCING NEGATIVE REFRACTIVE INDEX BEHAVIOR

As we have mentioned above, there are two distinct ways to produce the phenomenon associated with a negative refractive index. The first way is to find, or design a material, or structure that exhibits the property of negative ε and negative μ. As we will see there are physical phenomena that can be exploited to produce negative ε, and more exotic ways to produce a negative magnetic permeability greater than unity at optical frequencies [8,9]. The second way, as we have alluded to in the previous section, is to use the periodic dielectric structures discussed in Chapter 9 that are made up locally of positive refractive index with sufficiently high contrast to permit the propagation of a mode with a negative effective phase velocity [10,11]. In a given periodic structure not all propagating modes have this property and even those that do will do so only in a limited frequency range. We first deal with the methods that have been suggested to produce both negative ε and μ, sometimes called "metamaterials".

11.5.1. Use of Resonant Structures

The behavior of electron plasma can easily exhibit a wavelength region of negative ε. The classical expression for the frequency of an electron plasma oscillation is given by

$$\omega_p^2 = \frac{N_e{}^2}{m} \tag{11.16}$$

In metals where the electron density is high and one can approximate the effect of the positive nuclei by using an effective mass for the electron, one describes these collective electron oscillations as plasmons. The dielectric function then can be expressed as the following:

$$e(\omega) = \varepsilon_1 + i\varepsilon_2 = \varepsilon_i - \frac{\omega_p^2}{\omega(\omega + i\gamma)} \tag{11.17}$$

Here ε_1 and ε_2 represent the real and imaginary parts of the dielectric and ω_p is the plasma frequency. In the free electron case ε_I is equal to unity, but in this case, it is used to include the interband contributions to the dielectric function, as well as the so-called free electron contribution. One can see that if the frequency is sufficiently below the plasma frequency, $\omega \ll \omega_\pi$, then ε can become negative. To demonstrate this we show in Fig. 11.4 the measured value of the real part of the dielectric constant as a function of wavelength for silver colloids in glass. One can see the characteristic wavelength squared dependence described by Eq. (11.20) over a spectral region >500 nm. The surface plasmon frequency for silver in a glass matrix is found to correspond to a wavelength of 400 nm.

Generating negative magnetic permeability at optical frequencies is a much greater challenge. For magnetic materials, the magnetic permeability diminishes at high frequencies. A suggested method to produce magnetic behavior at optical frequencies is using nonmagnetic materials in a nanostructured array.

Smith et al. [12] were the first to critically look at a structure that yields a negative refractive index. They proposed a split ring structure (SRR) as shown in Fig. 11.5. A time varying magnetic field applied normal to the plane of the ring induces a current. These currents produce a magnetic field that either opposes or adds to the incident field. This is the source of the desired magnetic permeability. The specific design of the structure is explained in the reference mentioned above, in other words what purpose the split serves, and why two rings with their openings opposed. This

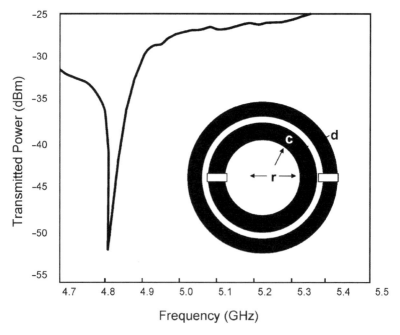

Figure 11.5 Resonance curve of copper split ring, $c = 0.8$ mm, $d = 0.2$ mm, $r = 1.5$ mm (taken from Ref. [12]).

has to do with lowering the resonant frequency and overall optimizing the interaction with the field. The dimensions of the SRR must be small relative to the free space wavelength which in this case is in the microwave regime. Here, the structure is designed to have a resonance at 4.8 GHz.

Smith et al. [12] chose to show the results of this analysis of the properties of the SRR in the form of a frequency vs. propagation constant diagram as reproduced here in Fig. 11.6a–d. This is an interesting way to look at the phenomenon in that if either μ or ε go negative no propagation is possible, thus the appearance of a gap is conformation of

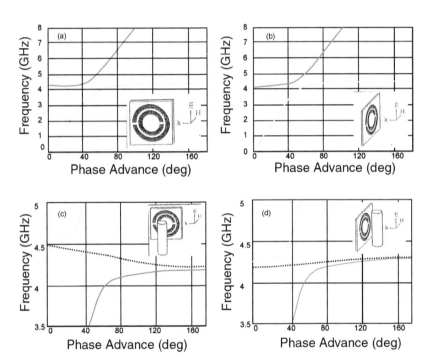

Figure 11.6 Dispersion curves of the split ring of Fig. 11.5, for the two orientations of the magnetic field relative to the axis of the ring, (a) parallel and (b) perpendicular. The gaps are attributed to the region where either ε or μ is negative. The figures (c) and (d) are for the case where a wire structure has been added to the split ring structure in the orientations as shown. The wire structure introduces an allowed state in the gap of (c) (taken from Ref. [12]).

a negative value of either μ or ε. The upper two figures, a and b, are for the orientation of the magnetic field as indicated. In both cases, a narrow frequency gap is predicted. They interpret the gap in (a) as due to a negative value of μ and that of (b) to a negative value of ε. This distinction is made by adding thin wire structure to the split rings as shown in (c) and (d). It is known from the analysis of the thin wire structure alone that one can reduce the plasma resonance into the desired microwave regime. This means that one can obtain a negative value of ε through Eq. (11.20) in the range of interest. By doing so they show that a defect band is produced in the gap corresponding to the configuration of (c) with the characteristic negative dispersion. This would confirm that in this case one has both negative ε as well as negative μ. For the other orientation shown in (d), the defect state lies outside the gap.

Some other examples of proposed structures are shown in Fig. 11.7 [13,14]. The structures are designed to work at or near resonance, such that large magnetic moments are possible. The preferred mechanism for optical frequency excitation of the metal structure is through surface plasmon resonance. The example of the approach that we show here is that proposed by Panina et al. [4]. This involves the use of either an open ring or the parallel wire structure as shown in Fig. 11.8. The reader can go to the original papers to follow the exact mathematical derivation of the magnetic permeability. It is sufficient here to give qualitative description. There are restrictions on the size of the structures, lengths, thickness, etc. to allow the analysis to be carried out. Suffice to say that the dimensions are all to be smaller than the wavelength of light. (In the example given by Panina, the length of the loop is 100 nm and the cross-section 10 nm). The circulating currents when large enough, then become the source of the magnetic field interaction and hence the desired permeability.

Utilizing a 2-D array of such parallel wire sets has an effect depending on how one adds the contributions We reproduce one result here as Fig. 11.9, where the electric field is oriented along the wire axis. They [4] calculate the real and

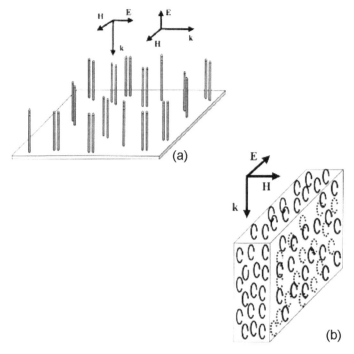

Figure 11.7 Other suggested composite medium structures which could lead to negative ε and μ at optical frequency (taken from Ref. [4]).

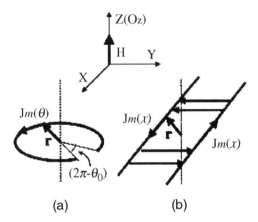

Figure 11.8 Detailed drawings of the structures indicating the directions and quantities used in the analysis, (a) split ring, (b) wire pair (taken from Ref. [4]).

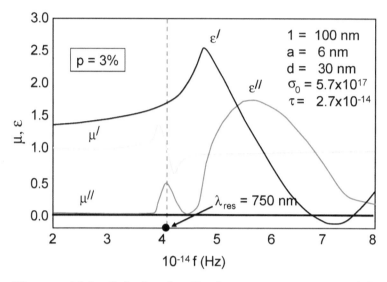

Figure 11.9 Calculated effective magnetic permeability and dielectric constant vs. frequency of composites containing wire pairs after Ref. [4].

imaginary part of the dielectric and magnetic polarizability, for the single pair and then estimate the result for an array of these two-wire elements. The separation distance is labeled as 30 nm in the figure. They show the result using an effective medium approach which broadens the resonance and produces no negative value of μ'. One must keep in mind, that the goal is to produce a negative value of μ' and ε'' while maintaining a reasonably low loss, that is, a small value of ε''. These data for this structure do not indicate a region of negative ε and μ.

11.5.2. Photonic Crystals

Certain modes in a photonic crystal propagate with a negative refractive index. Foteinopoulou et al. [14] and others [10,11] have done a simulation of the modes that propagate in a 2-D photonic crystal structure made up of locally positive materials. They used a hexagonal structure with the hole radius to pitch ratio of 0.35. They let the structure be air holes in a medium with a value of $\varepsilon = 12.96$. For a particular unspecified mode they calculated the effective refractive index (k_x/k_0) as a function of normalized frequency (pitch/wavelength). This is shown in Fig. 11.10. One can see that the index goes abruptly negative at a normalized frequency of approximately 0.61 and continues to grow more negative as the frequency decreases.

The suggested requirement for such a mode or modes to exist is that they be relatively isotropic with respect to the wavevector direction. In other words, such modes cannot have negative k in one direction in reciprocal space and positive in another. A map where we show the normalized frequency as a function of the position in the Brillouin for the first seven bands (modes) is shown in Fig. 11.11 We show in Fig. 11.12

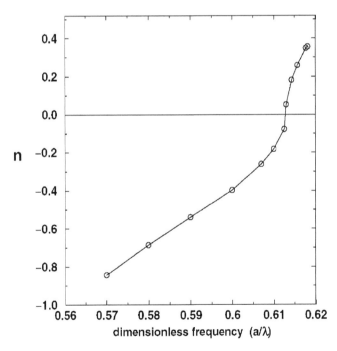

Figure 11.10 Calculated effective refractive index vs. normalized frequency for a 2-D triangular photonic crystal structure. Specific mode is not specified (taken from Ref. [16]).

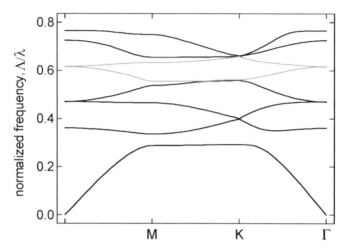

Figure 11.11 Band diagram for the first eight modes for triangular lattice photonic crystal structure; fifth mode counting from the bottom exhibits negative refractive index.

a map of the frequency surfaces as a function of k_x and k_y for the sixth band which exhibits a negative effective index [16]. Here, one clearly sees the required smooth isotropic nature of the surface.

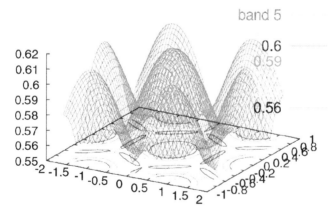

Figure 11.12 Full equi-frequency wave-vector diagram for the fifth mode shown in Fig. 11.11. Dome-like shape shows required isotropic nature in all direction of k-space.

Foteinopoulou et al. [14] also did an interesting simulation of the incidence of a monochromatic wave onto a PC structure which exhibits a negative effective index. Their simulation is reproduced here as Fig. 11.13. Here, one sees a somewhat unusual development of the field as the incident light strikes the surface. There is a point in time where the field seems undecided how to propagate, then at a somewhat longer time the field distribution clearly emerges in the negative direction. The concern about any violation of causality is clearly taken care of by the prediction of a transition time where the refracted wave is trapped wherein it has time to reorganize before heading in the negative direction.

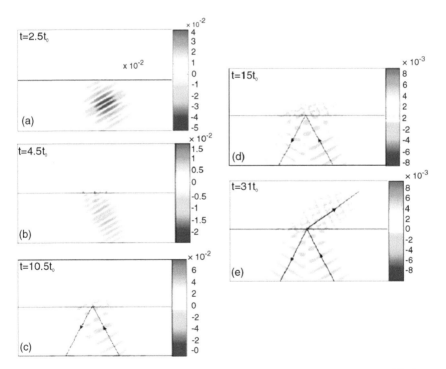

Figure 11.13 Time sequence simulation of the refraction of light as it passes from a positive medium into a negative medium. Important aspect is the indication of a transition time where phases reorganize (taken from Ref. [14]).

11.6. APPLICATIONS

It is premature to talk about applications in particular in the optical frequency regime, but nonetheless it is interesting to discuss some of the more fascinating ideas even if they never do come to reality.

11.6.1. Behavior of Negative Refractive Index Lens

Velesago [1], and more recently Pendry [15] have pointed out the unusual phenomenon of the imaging capability of a flat piece of material that possesses a negative refractive index. Actually, if one refers back to Eq. (2.4) in Chapter 2 one can see how this comes about from the transfer matrix formalism. We reproduce the transfer matrix here for the convenience of the reader

$$M = \begin{pmatrix} 1 - \frac{T}{nf_1} & \frac{T}{n} \\ \frac{T}{nf_1f_2} - \frac{1}{f_1} - \frac{1}{f_2} & 1 - \frac{T}{nf_2} \end{pmatrix} \tag{11.18}$$

The f's refer to the radius of curvatures of the first and second surfaces, $R_c = (n-1)f$. The convention used here is that if the curvature is convex, f is positive, so for a double convex surface both f_1 and f_2 are taken as positive. From this definition of f, it changes sign under a curvature change, say convex to concave and also from positive to negative index change. Consequently, a simultaneous change in both leaves f unchanged. We can use these simple rules to create the transfer matrix for a medium with a negative refractive index.

Bi-convex

$$\begin{pmatrix} 1 + \frac{T}{nf} & -\frac{T}{n} \\ \frac{-T}{nf^2} - \frac{2}{f} & 1 + \frac{T}{nf} \end{pmatrix} \tag{11.19a}$$

Bi-concave

$$\begin{pmatrix} 1 - \frac{T}{nf} & -\frac{T}{n} \\ \frac{-T}{nf^2} + \frac{2}{f} & 1 - \frac{T}{nf} \end{pmatrix} \tag{11.19b}$$

We show the schematic ray trace for a number of simple situations in Fig. 11.14. The values of n in all the matrices are to be taken as the absolute value. A general rule is that the imaging properties of a negative lens with a negative index is like a positive lens with a positive index.

The most interesting manifestation of the behavior of a negative refractive index medium is that a flat plate can also image light. One can see this quite easily by letting the value of f in either Eq. (11.16a), or (11.16b) go to infinity, and then trace the ray $(1, m_1)$ to obtain the picture shown in Fig. 11.15

$$\begin{pmatrix} y_2 \\ m_2 \end{pmatrix} = \begin{pmatrix} 1 & -T/n \\ 0 & 1 \end{pmatrix} \begin{pmatrix} 1 \\ m_1 \end{pmatrix} \tag{11.20}$$

From this one gets the relationship between the working distances, as indicated in the figure as the following:

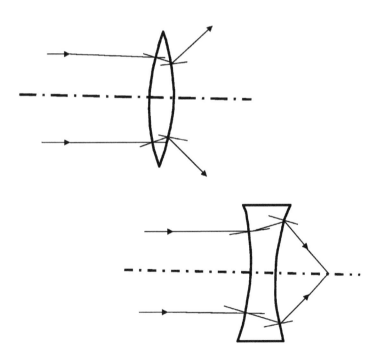

Figure 11.14 Ray trace of light through a bi-convex lens made of a negative material (a) and through a bi-concave lens of a negative material.

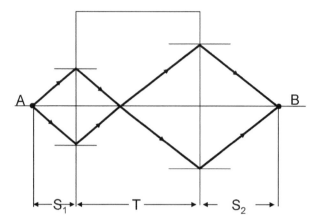

Figure 11.15 Rays traced through a parallel-faced slab of negative material showing imaging.

$$s_2 = \frac{T}{n} - s_1 \qquad (11.21)$$

One can see that as long as $T/n > s_1$, a real image will appear at a distance s_2. Note that one cannot collimate light with such a lens.

However, Pendry et al. [15] did show that a slab of a negative refractive index material has the capability to focus all Fourier components of a two-dimensional image, even the evanescent wave. The latter fact is the surprising result in that he shows that as a consequence of the negative index medium, the normally decaying evanescent is *amplified* as it travels through the negative medium. The reader is referred to the paper for the derivation of this result.

11.6.2. Optical Waveguide

Another interesting realization of a negative index medium is to consider the existence of a waveguide with an air core and a cladding made up of a material with a negative refractive index. Shedrivov et al. [16] were the first to investigate a very similar situation where the core of a slab waveguide was a negative material and the cladding was a positive material. We shall see that this is really no different in result, so we will follow this approach with the slight difference of the core being the positive medium.

We investigate the conditions for bound modes in a slab waveguide where the cladding exhibits a negative refractive index, or more precisely, where the cladding medium has $\varepsilon < 0$ and $\mu < 0$. We start with the following picture and definitions.

ε_2, μ_2

ε_1, μ_1 $\qquad\qquad x\uparrow \to z$ $\qquad\qquad \begin{array}{c}\uparrow\\2L\\\downarrow\end{array}$

The y-direction is out of the paper. The appropriate differential equation for the TE modes is

$$\frac{\partial^2 E_y}{x^2} + \frac{\partial^2 E_y}{\partial z^2} = \frac{\varepsilon_m \mu_m}{c^2}\frac{\partial^2 E_y}{\partial t^2}, \quad m = 1, 2 \tag{11.22}$$

We assume E_y takes the form

$$E_y = E(x)\exp[i(\omega t - \beta z)] \tag{11.23}$$

Substituting Eq. (23) into Eq. (22) gives the following differential equation in each region:

$$\frac{d^2 E}{dx^2} + (k_0^2 \varepsilon_m \mu_m - \beta^2)E = 0 \tag{11.24}$$

where $k_0 = \omega/c$. The solutions of Eq. (3) in each region can be written as

$$E_m = C_m \exp\left[i(k_0^2 \varepsilon_m \mu_m - \beta^2)^{1/2}x\right]$$
$$+ D_m \exp\left[-i(k_0^2 \varepsilon_m \mu_m - \beta^2)^{1/2}x\right] \tag{11.25}$$

We first consider the case where $k_0^2 \varepsilon_1 \mu_1 > \beta^2$ or $\beta^2/k_0^2 \equiv n_\beta^2 < \varepsilon_1 \mu_1$. For the simple layer structure, we have symmetric and antisymmetric solutions for the guiding layer

$$E_y(x) = A\cos(hx) \tag{11.26a}$$

and

$$E_y(x) = A\sin(hx)$$

where

$$h = (k_0^2 \varepsilon_1 \mu_1 - \beta^2)^{1/2} = k_0^2(\varepsilon_1 \mu_1 - n_\beta^2)^{1/2} \tag{11.26b}$$

For the cladding $|x| > L$, we ignore the positive exponential, since this corresponds to an unbound mode

$$E_y = B\exp(-p|x|), \tag{11.27a}$$

where

$$p = (\beta^2 - k_0^2 \varepsilon_2 \mu_2)^{1/2} = k_0^2(n_\beta^2 - \varepsilon_2 \mu_2)^{1/2} \tag{11.27b}$$

We initially are looking for solutions for β such that

$$k_0^2 \varepsilon_2 \mu_2 < \beta^2 < k_0^2 \varepsilon_1 \mu_1 \tag{11.28}$$

or

$$\varepsilon_2 \mu_2 < n_\beta^2 < \varepsilon_1 \mu_1$$

This is the normal condition for waves that are guided by total internal reflection. One has two boundary conditions

$$E_{1y}(L) = E_{2y}(L) \tag{11.29a}$$

and

$$H_{1z}(L) = \frac{i}{\omega \mu_1} \frac{\partial E_{1y}}{\partial x} = \frac{i}{\omega \mu_2} \frac{\partial E_{2y}}{\partial x} = H_{2z}(L) \tag{11.29b}$$

If one applies the boundary conditions, one gets the following expressions for the allowable values of β. The first is for the symmetric modes and the second is for the antisymmetric modes

$$h \frac{\mu_2}{\mu_1} \tan(h) = p \tag{11.30}$$

and

$$-h \frac{\mu_2}{\mu_1} \frac{1}{\tan(h)} = p,$$

where p and h are defined in Eq. (26b) and (27b). It is also useful to define the quantity V, which yields a relation between p and h

$$V^2 = h^2 + p^2 = k_0^2 (\varepsilon_1 \mu_1 - \varepsilon_2 \mu_2) \tag{11.31}$$

which implies $p^2 = V^2 - h^2 = k_0^2 (\varepsilon_1 \mu_1 - \varepsilon_2 \mu_2) - h^2$.

In a similar manner for the TM modes, one can write H_y as we did for E_y. Similarly, the boundary conditions require the continuity of H_y and D_z at $x = L$. The latter is expressed in terms of E_z by the following expression:

$$\frac{-i}{\omega \varepsilon_1} \frac{\partial H_{1y}}{\partial x} = \frac{-i}{\omega \varepsilon_2} \frac{\partial H_{2y}}{\partial y} \tag{11.32}$$

We now proceed to calculate the allowable modes from Eq. (11.30). We can look at both cases, the case where both ε, μ are both positive (normal case) and where they are negative. The only distinction we need make is in the signs we use in (11.30). We plot the left-hand side of the both equations, for the case where μ_2/μ_1 is positive and equal to 0.8 and also for the case where the ratio is negative. This is shown in Fig. 11.16a where h is plotted vs. p. To obtain the allowable values of β, we also plot Eq. (11.32) for various values of V. The intersection of these segments of circles with the tangent functions yields the allowable values of β.

These modes derived from the negative index case are not unusual in that they look like modes of the normal positive material. It should be noted that at this point it makes

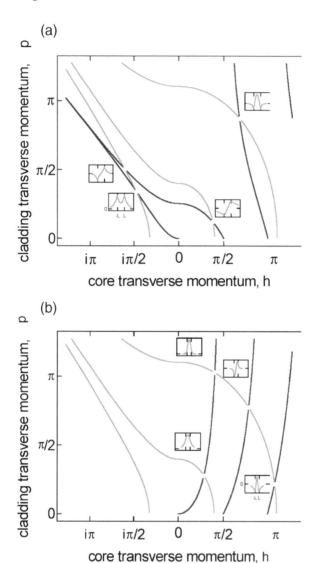

Figure 11.16   (a) Transverse wave vector for cladding, "p" plotted
against that in core, "h" according to Eq. (11.30). The intercepts with
the lines representing Eq. (11.32) are the allowable values of β Value
of $\mu2/\mu1$ was taken as +0.8. (b) Same as in "a" except $\mu2/\mu1$ is taken
to be negative and thus the plot is of p vs. the real part of "h" corresponding to Eq. (11.34). The intercept with the lines corresponding
to (11.35) are the allowed values of β for this case.

no difference whether we make the core or the cladding the negative medium. What we did assume was that both h and p were real, that is,

$$\varepsilon_2\mu_2 < n_\beta^2 < \varepsilon_1\mu_2 \tag{11.33}$$

This is the normal assumption that is made in waveguides because it is the only region that real values of β occur. This is easily seen from where the points of intersection occur in Fig. 11.16a and b.

However, the unique aspect afforded by one of the layers having a negative refractive index is that there is another region of h–p where a real value of β exists. To examine this region we recognize that to be a bound solution the field must decay into the cladding layer. In the normal positive index case the field in the core is sinusoidal, and the appropriate boundary conditions can be met at the boundaries, in particular the continuity of B_z for the TE case and D_z for the TM case where the value is positive on both sides of the boundary. However, what is different in the case where one of the layers has a negative index is that the slope of the field at the boundary must have slopes with opposite sign which permits a new situation, that of a mode that decays from the boundary both into the core and the cladding layer. Mathematically, one can see this situation by letting h ($k_\perp$ in the core field) be an imaginary number. This can occur by allowing $\beta^2 > k_0^2 n_1^2$ in Eq. (11.30). This makes $h = i\eta$ imaginary and thus putting this into Eq. (11.30)

$$i\eta \frac{\mu_2}{\mu_1} \tanh(\eta L) = p \tag{11.34}$$

and

$$-i\eta \frac{\mu_2}{\mu_1} \frac{1}{\tanh(\eta L)} = p$$

Using the same graphical technique as before although now plotting η vs. p and recognizing that Eq. (11.31) is now written as

$$V^2 = p^2 - \eta^2 \tag{11.35}$$

we obtain the results shown in Fig. 11.16b.

One can see that a solution for β does exist while having $p > 0$. The solution in this case yields a sine hyperbolic form in the core, thus fulfilling the feature of decay into the core. The full solutions for the y-component of the electric field and the z-component of the magnetic field for this antisymmetric case, for the inside and outside layers are the following all multiplied by $\exp[i(\omega t - \beta z)]$

$$E_y(x) = \exp(-pL)\sinh(\eta x)$$

$$H_z(x) = \frac{i}{\omega\mu_1}(i\eta)\exp(-pL)\cosh(\eta x), \quad -L < x < L$$

$$H_x(x) = \frac{-i}{\omega\mu_1}\frac{\partial E_y}{\partial z} = \frac{-i}{\omega\mu_1}(-i\beta)\exp(-pL)\sinh(\eta x) \quad (11.36)$$

$$E_y(x) = \sinh(\eta L)\exp(-p|x|)$$

$$H_z(x) = \frac{-i}{\omega\mu_2}p\cosh(\eta L)\exp(-p|x|), \quad |x| > L$$

$$H_x(x) = \frac{-i}{\omega\mu_2}(-i\beta)\sinh(\eta L)\exp(-p|x|) \quad (11.37)$$

For the symmetric case, the equations would look similar with cosh replacing sinh. It would appear that for the TM modes the solutions would be the same except to replace the μ with ε.

To examine the energy flow, one can use the Poynting vector for each region. This was done by Shedrivov et al. [16] and is reproduced here in Fig. 11.17.

The actual realization of this type of waveguide that is having the core or the cladding mode exhibit a negative effective index, is not obvious. One could imagine a hollow core waveguide as discussed in Chapter 9, Sec. 9.4.5, where at some frequency, there were extended modes (cladding modes) with a negative effective refractive index of the cladding.

Photonic crystal structures can effectively act as a negative index medium. Parimi et al. [17] have actually demonstrated imaging in the microwave regime (9 GHz) by a 2-D flat lens

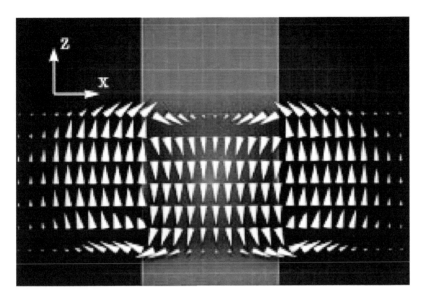

Figure 11.17 Poynting vector field for propagation in a slab of negative index medium surrounded by a positive index medium as computed in Ref. [16].

(see Eq. (11.26) made up of a periodic structure of aluminum rods on a square lattice, that is a photonic crystal. This was a clear demonstration. The same group also have shown negative index behavior in a prism of the same structure [18].

11.6.3. Negative Refraction Devices

As we mentioned above in Sec. 11.4, the phenomenon known as negative refraction also has some useful manifestations. We discussed above how the photonic crystal structures significantly enhance the negative refraction effect, which will lead to large group velocity dispersion. This can be seen from the behavior shown in Fig. 11.3 where the sharp changes in curvature with the propagation direction produce large dispersion of the group velocity. The direction of the group velocity can deviate significantly from that direction. One can then use this large dispersion of the group velocity to produce what has been called a "superprism" [19].

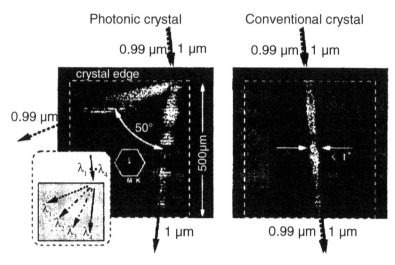

Figure 11.18 Simulation of the negative refraction in a photonic crystal structure showing large difference in group velocity direction for different wavelength; superprism concept. Taken from Ref. [19].

Kosaka et al. have shown and is reproduced here in Fig. 11.18 [19]. The simulation shows the result for a photonic crystal structure, which produces a 50 angle between two wavelengths that are only 0.01 µm apart. The inset shows how one could use thus to spatially separate a large number of wavelengths.

Another practical property that can be derived from the negative refraction behavior is a way to spatially separate polarization states. In Fig. 11.19 we show a simulation of this behavior. Here, an unpolarized input beam is incident on a triangular lattice structure at a particular angle such that the TM and TE modes travel significantly different paths. This is no difference in concept to what we described in Sec. 8.4.2 except the 2-D periodic structure enhances the effect (for example see Fig. 8.14).

The larger size of the effect in the 2-D structure allows one to easily make a polarizer that is a structure with sufficient path difference to allow one polarization state to be rejected. This is shown in Fig. 11.20.

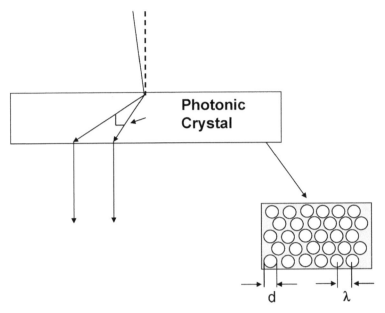

Figure 11.19 Schematic of polarization separator device using negative refraction. Arbitrary input polarization splits into two paths, one for TE and one for TM. Optimum design of photonic crystal structure produces largest path difference.

APPENDIX A. NEGATIVE EFFECTIVE REFRACTIVE INDEX

The analogy of negative phase velocity in a periodic structure for some or all modes at some point in the BZ to that of negative effective mass of an electron as a consequence of curvature of $E(\mathbf{k})$, makes it tantalizing to try to figure out some simple mathematical correlation. One approach is to recall the Taylor series expansion of the dispersion relation of the propagation constant in a fiber, that is $k_z \equiv \beta$, which is related to the phase index, $n_{\text{eff}} = \beta c / \omega$.

$$\beta = n_{\text{eff}} \frac{\omega}{c} = \beta_0 + \frac{\partial \beta}{\partial \omega}(\omega_0 - \omega) + \frac{\partial^2 \beta}{\partial \omega^2}(\omega_0 - \omega)^2 + \cdots$$

$$(A11.1)$$

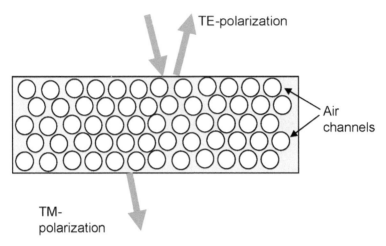

Figure 11.20 Schematic of negative refraction in a photonic crystal structure acting as a plarizer. Similar concept to the polarization separator, but here TE mode is reflected while TM mode is transmitted.

It shows that the effective refractive index could go negative if $dk/d\omega$, (reciprocal group velocity), and or $d^2k/d\omega^2$ were negative enough. There is also a more general way to write the group velocity for which we are not restricting the propagation to just the z-direction.

$$v_g = \nabla_k \omega \tag{A11.2}$$

Of course in one dimension $1/v_g = d\omega/dk$ as we have in Eq. (1). What this suggests is to write Eq. (1) in a vector form, that is, just replace k with $\mathbf{k}$. Thus, for each of the three Cartesian directions one would have the following expression:

$$k_i = k_i^0 + \frac{\partial k_i}{\partial \omega}(\omega_0 - \omega) + \frac{\partial^2 k_i}{\partial \omega^2}(\omega_0 - \omega)^2 \quad \text{for } i = x, y, z \tag{A11.3}$$

There is another way to write this, so that one can more easily use the calculated equi-frequency surfaces, $\omega(\kappa)$

$$k_i = k_0 + \frac{1}{\partial\omega/\partial k_i}(\omega_0 - \omega) + \frac{\partial}{\partial\omega}\left(\frac{1}{\partial\omega/\partial k_i}\right)(\omega_0 - \omega)^2$$

$$(A11.4)$$

The evaluation of the sign and magnitude of the coefficients of the frequency term in Eq. (A11.4) for each direction would indicate whether k could be negative. One can recognize the group velocity in the linear term, and the dispersion of the group velocity in the squared term. This shows that it is these terms, their magnitude and sign that can give rise to the negative effective index effects discussed above.

It is not hard to imagine the similarity between band theory as it deal with electrons in a crystal compared with photons in a periodic dielectric structure. Both deal with a motion of a particle influenced by a periodic disturbance. The dynamics of electrons in a crystal are dominated by the nature of the energy surface $E(k)$. One can draw many parallels both in terminology, as well as physical phenomenon [20,21].

REFERENCES

1. Vesalago, V.G. The electrodynamics of substances with simultaneous negative values of ε and μ. Sov. Phys. (Uspekhi) **1968**, *10* (4), 509–514.

2. Shelby, R.A.; Smith, D.R.; Schultz, S. Experimental verification of a negative refractive order. Science **2001**, *292*, 77.

3. Pendry, J. B. Optics: Positively Negative. Nature p. 22, *423*, May **2003**.

4. Panina, L.V.; Grigorenko, A.N.; Makhnovsky, D.P. Optomagnetic composite medium with conducting nanoelements. PR B **2002**, *66*, 155411–1 to 155411–17.

5. Wangness, R.K. In *Electromagnetic Field*. Wiley: New york, 1979.

6. Houck, A.A.; Brock, J.B.; Chuang, I.I. Experimental observation of a left-handed material that obeys Snell's law. PRL **2003**, *90* (13), 137401–137411.

7. Russell, P.StJ.; Birks, T.A.; Loyd-Lucas, D. Photonic bloch waves and photonic band gaps. In *Confined Electrons and Photons*; Burstein, E., Weisbuch, C., Eds.; Plenum Press: London, 1995.

8. Pendry, J.B.; Holden, A.J.; Stewart, W.J.; Youngs, L. Extremely low frequency plasmons in metallic mesostructures. PRL **1996**, *76* (25), 4773–4776.

9. Podolsky, V.A.; Sarychev, A.K.; Shalaev, V.M. Plasmon modes and negative refraction in metal nanowires composites. Opt. Express **2003**, *7* (11), 735–745.

10. Notomi, M. Negative refraction in photonic crysrtals. Opt. Quant. Electron. **2002**, *34*, 133–143.

11. Schvets, G. Photonic approach to making a material with a negative index of refraction. PR B **2003**, *67*, 035109–1 to 035109–11.

12. Smith, D.R.; Padilla, W.J.; Wier, D.C.; Nemat-Nasser, S.C.; Schultz, S. PRL **2000**, *84*, 4184–4187.

13. Sievenpiper, D.F.; Sickmiller, M.E.; Yablonovitch, E. 3D wire mesh photonic crystals. PRL **1996**, *76* (14), 2480.

14. Foteinopoulou, S.; Economou, E.N.; Soukoulus, M. Refraction in media, with a negative refractive index. l PRL **2003**, *90* (10), 107402–107411.

15. Pendry, J.B. Negative refraction makes a perfect lens. PRL **2000**, *85* (18), 3966.

16. Shedrivov, I.V.; Sukhorukov, A.A.; Kivshar, Y.S. Guided modes in negative-refractive index waveguides. PR E **2003**, *47*, 057602–1 to 057602–11.

17. Parimi, P.V.; Lu, W.T.; Vodo, P.; Strindar, S. Imaging by flat lens using negative refraction. Nature **2003**, *426*, 404.

18. Parimi, P.V.; Lu, W.T.; Vodo, P.; Sokoloff, J.; Derov, J.S.; Sridhar, S. Negative refraction and left-handed electromagnitism in microwave photonic crystals. Phys. Rev. Lett. **2004**, *92* (12), 127401–127411.

19. Kosaka, H.; Kawashima, T.; Tomita, A.; Notomi, M.; Tamamura, T.; Sato, T.; Kawakami, S. Photonic crystals for micro lightwave circuits using wavelength-dependent angular beam steering. Appl. Phys Lett. **1999**, *74* (10), 1370–1372.

20. Russell, P.St.J.; Birks, T.A.; Lloyd-Lucas, F.D. Photonic bloch waves and photonic band-gaps. In *Confined Electrons and Photons*; Burstein, E., Weisbuch, C., Eds.; Plenum Press: London, 1995.

21. Ziman, J. *Principles of the Theory of Solids*. Cambridge Press: Cambridge, 1964, 155 pp.

Index

T - #0091 - 111024 - C552 - 229/152/26 - PB - 9780367393533 - Gloss Lamination